Molecular Mechanisms of Leaf Morphogenesis

Molecular Mechanisms of Leaf Morphogenesis

Editor

Tomotsugu Koyama

MDPI • Basel • Beijing • Wuhan • Barcelona • Belgrade • Manchester • Tokyo • Cluj • Tianjin

Editor
Tomotsugu Koyama
Bioorganic Research Institute
Suntory Foundation for Life
Sciences
Kyoto
Japan

Editorial Office
MDPI
St. Alban-Anlage 66
4052 Basel, Switzerland

This is a reprint of articles from the Special Issue published online in the open access journal *International Journal of Molecular Sciences* (ISSN 1422-0067) (available at: www.mdpi.com/journal/ijms/special_issues/leafmorphogenesis).

For citation purposes, cite each article independently as indicated on the article page online and as indicated below:

LastName, A.A.; LastName, B.B.; LastName, C.C. Article Title. *Journal Name* **Year**, *Volume Number*, Page Range.

ISBN 978-3-0365-3163-2 (Hbk)
ISBN 978-3-0365-3162-5 (PDF)

Contents

Preface to "Molecular Mechanisms of Leaf Morphogenesis"

Leaf morphology is obviously determined in a plant. By contrast, its morphology is often changeable when the plant copes with various environmental changes. We can speculate that leaf morphogenesis is based on the regulatory mechanisms with remarkable robustness and flexibility. Recent research has increasingly investigated the regulatory network of leaf morphogenesis and revealed some important regulators functioning in the leaf development but not obtained its full view of the molecular mechanisms of leaf morphogenesis.

To update our understanding of the leaf morphogenesis, this book contains nine academic papers that focus on the regulation of genes, proteins, hormones, and other metabolites for leaf morphogenesis in various plant species. It further provides important insights in biochemical, developmental, evolutional, and physiological events operating during morphogenesis. Emphasis is also placed on the perspective views of how these molecular mechanisms contribute to the survival of plants and are applicable to improve plant traits. I hope readers will be fascinated and interested to find beautiful mechanisms in leaf morphogenesis.

Tomotsugu Koyama
Editor

International Journal of
Molecular Sciences

MDPI

Article

Establishment of the Embryonic Shoot Meristem Involves Activation of Two Classes of Genes with Opposing Functions for Meristem Activities

Mitsuhiro Aida [1,*], Yuka Tsubakimoto [2], Satoko Shimizu [2], Hiroyuki Ogisu [2], Masako Kamiya [2], Ryosuke Iwamoto [2], Seiji Takeda [2,3], Md Rezaul Karim [2,4], Masaharu Mizutani [5], Michael Lenhard [6] and Masao Tasaka [2]

1 International Research Organization for Advanced Science and Technology (IROAST), Kumamoto University, 2-39-1 Kurokami, Chuo-ku, Kumamoto 860-8555, Japan

2 Graduate School of Biological Sciences, Nara Institute of Science and Technology, 8916-5 Takayama, Nara 630-0192, Japan; moritsubaki.panda@gmail.com (Y.T.); s-s.combination@docomo.ne.jp (S.S.); h.ogisu@icloud.com (H.O.); masadecosmos@gmail.com (M.K.); r.ftvm22@ymobile.ne.jp (R.I.); seijitakeda@kpu.ac.jp (S.T.); mrkarim1996@yahoo.com (M.R.K.); m-tasaka@bs.naist.jp (M.T.)

3 Graduate School of Life and Environmental Sciences, Kyoto Prefectural University, Hangi-cho 1-5, Shimogamo, Sakyo-ku, Kyoto 606-8522, Japan

4 Department of Horticulture, Bangladesh Agricultural University, Mymensingh 2202, Bangladesh

5 Graduate School of Agricultural Science, Kobe University, 1-1 Rokkodai, Nada-ku, Kobe 657-8501, Japan; mizutani@gold.kobe-u.ac.jp

6 University of Potsdam, Institute for Biochemistry and Biology, Karl-Liebknecht-Str. 25-26, 14476 Potsdam-Golm, Germany; michael.lenhard@uni-potsdam.de

* Correspondence: m-aida@kumamoto-u.ac.jp

Received: 31 July 2020; Accepted: 12 August 2020; Published: 15 August 2020

check for updates

Abstract: The shoot meristem, a stem-cell-containing tissue initiated during plant embryogenesis, is responsible for continuous shoot organ production in postembryonic development. Although key regulatory factors including *KNOX* genes are responsible for stem cell maintenance in the shoot meristem, how the onset of such factors is regulated during embryogenesis is elusive. Here, we present evidence that the two *KNOX* genes *STM* and *KNAT6* together with the two other regulatory genes *BLR* and *LAS* are functionally important downstream genes of *CUC1* and *CUC2*, which are a redundant pair of genes that specify the embryonic shoot organ boundary. Combined expression of *STM* with any of *KNAT6*, *BLR*, and *LAS* can efficiently rescue the defects of shoot meristem formation and/or separation of cotyledons in *cuc1 cuc2* double mutants. In addition, *CUC1* and *CUC2* are also required for the activation of *KLU*, a cytochrome P450-encoding gene known to restrict organ production, and *KLU* counteracts *STM* in the promotion of meristem activity, providing a possible balancing mechanism for shoot meristem maintenance. Together, these results establish the roles for *CUC1* and *CUC2* in coordinating the activation of two classes of genes with opposite effects on shoot meristem activity.

Keywords: shoot meristem; embryogenesis; stem cell; boundary; transcription factor; cytochrome P450; *CUC*; *STM*; *LAS*; *BLR*; *KNAT6*; *KLU*; *CYP78A5*

1. Introduction

Primary growth in plant shoots depends on the activity of stem-cell-containing tissue called the shoot meristem, which is located at the tip of the stem [1]. The shoot meristem is initially formed during embryogenesis and is activated upon germination to produce shoot organs such as leaves, stems, and floral organs, while it maintains an undifferentiated stem cell population at its center. Once activated, the shoot meristem keeps the balance between cell proliferation and differentiation to maintain an appropriate size of the stem cell population within it; factors essential for this process have been identified [2,3]. Although activation of these maintenance factors is associated with shoot meristem initiation during embryogenesis, how the process is coordinated is unknown.

Several key regulators for shoot meristem initiation have been reported [4–11]. Among them, the NAM/CUC3 type of NAC-domain transcription factors represents a class of regulators that are required for specification of shoot organ boundaries, which are sites for shoot meristem formation in embryonic and postembryonic development [12–16]. In *Arabidopsis thaliana*, the three NAM/CUC3 genes *CUC1*, *CUC2*, and *CUC3* are expressed in cells along the boundary between two cotyledon primordia and promote shoot meristem formation and the separation of cotyledons [5,6,17]. As shoot meristem formation proceeds, expression of these genes is downregulated from the meristem center and becomes restricted to the adaxial and lateral boundaries of cotyledons. In postembryonic development, the three *CUC* genes are expressed at the adaxial and lateral boundaries of leaf primordia and are required for the formation of axillary shoot meristem as well as for the separation of leaves [6,18].

Several genes whose expression is dependent on *CUC* gene activities have been identified. Expression of the two *KNOTTED1-like homeobox* (*KNOX*) genes *SHOOT MERISTEMLESS* (*STM*) and *KNOTTED1-like from Arabidopsis thaliana 6* (*KNAT6*), which are required for shoot meristem maintenance, is absent from the *cuc1 cuc2* double mutant [17,19] and ectopic expression of the *CUC* genes induces *STM* expression [5,20,21]. The *LIGHT-DEPENDENT SHORT HYPOCOTYLS* (*LSH*) genes *LSH3* and *LSH4*, which encode nuclear proteins of the *Arabidopsis* LSH1 and *Oryza* G1 (ALOG) family, have been identified as direct transcriptional targets of the CUC1 protein and their overexpression induces ectopic shoot meristem formation [22]. Genome-wide mapping of protein–DNA interactions among boundary-enriched genes has identified the *GRAS* family gene *LATERAL SUPPRESSOR* (*LAS*) and the microRNA gene *MIR164C* as direct transcriptional targets of CUC2 [23]. *LAS* encodes a putative transcriptional regulator and is required for axillary shoot meristem formation [24]. Together, these analyses indicate that *CUC* genes regulate multiple genes involved in shoot meristem activity or boundary specification, or both. However, the functional relationship between these downstream genes and *CUC* gene activity remains elusive.

Here, we selected a set of genes whose expression is dependent on *CUC1* and *CUC2* during embryogenesis and demonstrated that the combined activities of *STM*, *KNAT6*, *BLR*, and *LAS* are important for promoting shoot meristem formation and cotyledon separation downstream of *CUC1* and *CUC2*. Moreover, *CUC1* and *CUC2* are also required for the expression of *KLUH* (*KLU*)/*CYP78A5*, a cytochrome P450-encoding gene involved in the rate of shoot organ production and organ size [25,26]. Genetic analysis indicates that *KLU* restricts shoot meristem activity and counteracts *STM* function. Our results thus indicate that the activation of two classes of genes with opposing functions, one positively and the other negatively affecting meristem activity, is an important step for shoot meristem formation.

2. Results

2.1. Selection of Candidate CUC1 and CUC2 Downstream Genes

It has been reported that the two *KNOX* genes *STM* and *KNAT6*, the GRAS gene *LAS*, and the two *ALOG* genes *LSH3* and *LSH4* show overlapping expression patterns to those of *CUC1* and *CUC2* in the boundary region of cotyledons, and their expression is absent in the corresponding region of *cuc1 cuc2* double-mutant embryos [5,6,17,19,22]. We identified six additional candidate downstream genes positively regulated by *CUC1* and/or *CUC2* from microarray-based screening combined with quantitative real-time polymerase chain reaction (qRT-PCR) and in situ hybridization experiments (Supplementary Text S1; Supplementary Tables S1–S3). These genes were expressed in the boundary region that overlapped with the *CUC* gene expression domain [5,17] and were downregulated specifically in the corresponding region of *cuc1 cuc2* embryos (Figure 1A–D; Supplementary Figure S1), indicating the dependence of their expression on *CUC1* and *CUC2* activities during embryogenesis.

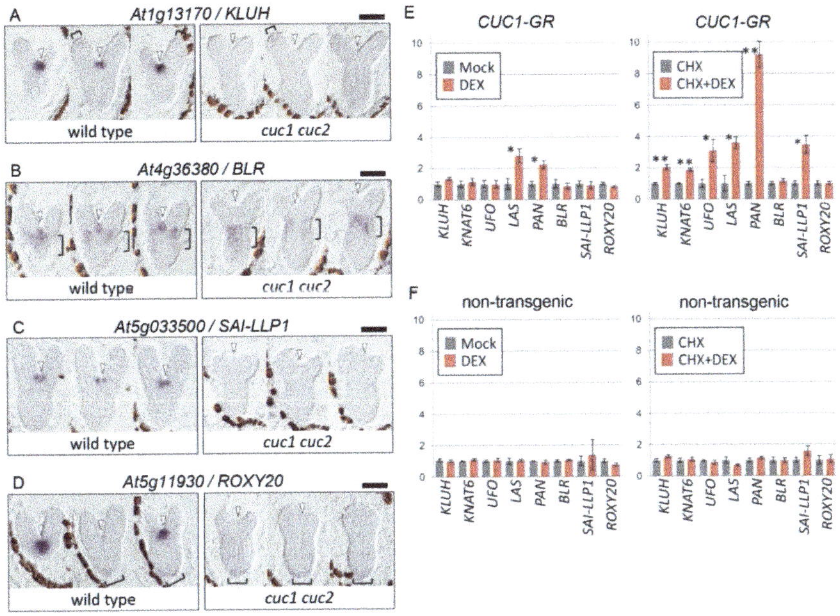

Figure 1. Regulation of candidate downstream genes by *CUC1* and *CUC2*. (**A–D**) In situ hybridization of newly identified candidate downstream genes. Four of the six candidates are shown. Three serial longitudinal sections of wild-type L*er* (**left**) and *cuc1-1 cuc2-1* double-mutant (**right**) embryos at the late heart stage. Arrowheads indicate the position of the cotyledon boundary region. Brackets in (**A**), (**B**), and (**D**) indicate the position of expression outside the boundary region. Bars = 50 μm. (**E**,**F**) Transcriptional responses of candidate downstream genes upon dexamethasone (DEX) treatment in *CUC1-GR* (**E**) and non-transgenic (**F**) plants in the absence (**left**) or presence (**right**) of the protein synthesis inhibitor cycloheximide (CHX). Three biological replicates of 7-day-old seedlings. Single and double asterisks indicate $p < 0.05$ and $p < 0.01$, respectively, in comparisons between samples with and without DEX (Welch's *t*-test).

To gain insight into how CUC1 regulates expression of these candidate genes, we used the glucocorticoid receptor (GR) system, in which the activity of CUC1 is induced by the exogenous application of dexamethasone (DEX) [27]. Using this system, we previously found evidence for the direct activation of the *LSH3*, *LSH4*, and *STM* genes by CUC1 [22,28]. Among the remaining eight genes, we found that only *LAS* and *PAN* were significantly upregulated upon DEX treatment alone in *CUC1-GR* plants (Figure 1E, left panel). By contrast, treating with both DEX and the protein synthesis inhibitor cycloheximide (CHX), which blocks secondary transcriptional responses caused by genes directly activated by CUC1, significantly upregulated not only *LAS* and *PAN*, but also four additional genes among the eight tested. Treatment with CHX alone did not alter their expression levels (Figure 1E, right panel). Together, the results suggest that the six genes are under direct transcriptional regulation by the CUC1-GR protein, but that the action of CUC1 is counteracted by a CHX-sensitive negative factor with respect to activation of four of the genes (*KLUH*, *KNAT6*, *UFO*, and *SAI-LLP1*). Another possibility is that DEX treatment alone indirectly promotes expression of genes that negatively affect CUC1-dependant activation of the four genes. In control non-transgenic plants, which do not express CUC1-GR, none of the genes were upregulated by DEX treatment in the presence or absence of

CHX (Figure 1F), indicating that the induction of the downstream genes was not a secondary effect of DEX or CHX.

2.2. Combined Expression of STM with LAS, BLR, and KNAT6 is Sufficient to Rescue the Embryonic Shoot Phenotypes of cuc1 cuc2

To examine the functional significance of the candidate genes in processes downstream of *CUC1* and *CUC2*, we expressed each gene in the *cuc1 cuc2* double-mutant background under the control of the *CUC2* promoter (*ProCUC2*), which drives expression in the boundary region in both wild-type and double-mutant embryos (Figure 2A,B). In wild type, seedlings develop a shoot that continuously produces leaves immediately after germination and have two completely separated cotyledons (Figure 2C). By contrast, the *cuc1 cuc2* double mutant has two cotyledons fused along their margins and fails to form a shoot, and this phenotype is fully penetrant (Figure 2D) [4]. When the coding sequence of *CUC2* was used as a positive control (*ProCUC2:CUC2*), 41.7% of the T1 seedlings showed a strongly rescued phenotype with no or slight delay in shoot formation and with completely separated cotyledons, 50.0% showed a mildly rescued phenotype with delayed or no shoot formation and half-separated cotyledons, and the remaining 8.3% showed non-rescued phenotype identical to that of *cuc1 cuc2* (Table 1, Supplementary Figure S2A,B).

Among the 10 downstream genes, the *ProCUC2:STM* and *ProCUC2:LAS* transgenes were able to mildly rescue the *cuc1 cuc2* phenotype, resulting in the occasional formation of a functional shoot that can produce leaves as well as in the partial separation of cotyledons (Table 1, Figure 2E, and Supplementary Figure S2C–E). In cleared seedlings, wild type has the dome-shaped shoot meristem with a few leaf primordia, whereas *cuc1 cuc2* plants lack either structure (Figure 2F,G) [4]. On the other hand, plants partially rescued by *ProCUC2:STM* showed variable phenotypes: some lacked a shoot meristem and leaf primordia, some developed small undifferentiated tissue, and the other produced the shoot meristem and leaf primordia (Figure 2H). These results indicate that *STM* and *LAS* play prominent roles in shoot meristem formation and cotyledon separation and that their individual activities can partially bypass the requirements for *CUC1* and *CUC2* for embryonic shoot meristem formation and cotyledon separation.

The rescue of the *cuc1 cuc2* mutant phenotype by the *STM* or *LAS* transgene alone was only mild and partial, thus we next tested their combined activities. In the F_2 generation of the cross between the lines with the *STM* and *LAS* transgenes, only plants carrying both showed a rescued phenotype (Table 2). These rescued plants showed either partial (Figure 2J) or complete separation of cotyledons (Figure 2K), with the latter forming leaves with only a slight delay compared with the timing in the wild type, indicating that combined expression of *STM* and *LAS* is sufficient to compensate for the loss of *CUC1* and *CUC2* activities.

We next selected five other genes encoding transcription factors or transcriptional co-regulators, and tested their ability to rescue the *cuc1 cuc2* phenotype in combination with the *STM* transgene (Table 3). *BLR* and *KNAT6* were able to achieve rescue when combined with *STM* (Figure 2K,L), while the rest failed to do so. Plants expressing both *STM* and *BLR* produced nearly normal shoots with completely separated cotyledons (Figure 2K), indicating that, similarly to *LAS*, *BLR* can efficiently support the ability of *STM* to promote shoot meristem formation and cotyledon separation in the absence of *CUC1* and *CUC2*. By contrast, plants expressing both *STM* and *KNAT6* only rescued the cotyledon phenotype, but not that of shoot formation (Figure 2L), indicating that *KNAT6* can support the *STM* activity only in the limited developmental pathway downstream of the *CUC* genes.

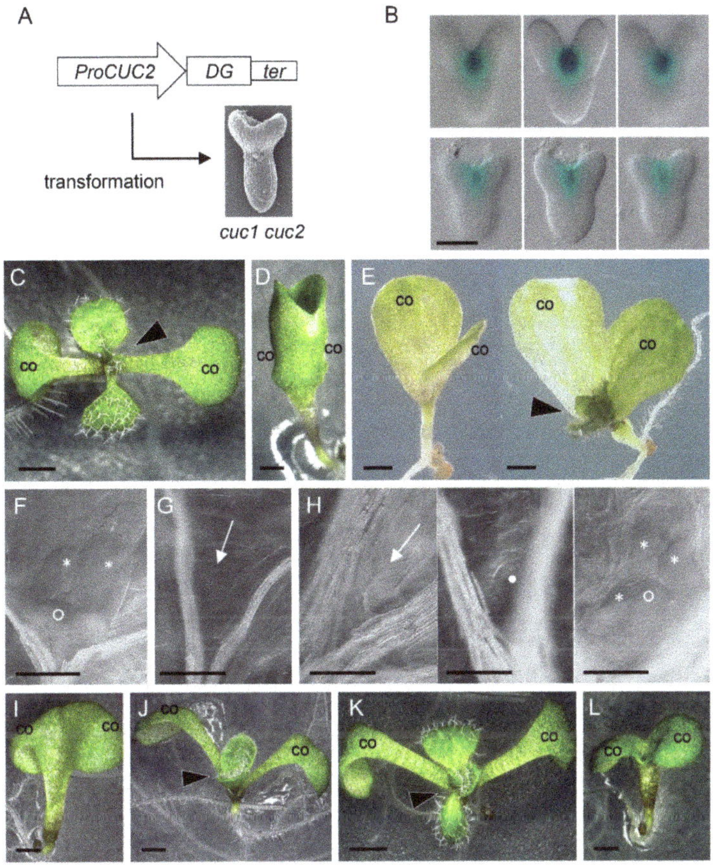

Figure 2. Rescue of *cuc1 cuc2* phenotype by candidate downstream genes. (**A**) Schematic diagram of rescue experiments. *CUC2* promoter (*ProCUC2*), cDNA of downstream gene (*DG*), and nos terminator (*ter*). (**B**) *CUC2* promoter activity detected by β-glucuronidase (GUS). Longitudinal views of GUS-stained wild-type Col (**top**) and *cuc1-5 cuc2-3* (**bottom**) embryos in three different optical sections, showing expression in the boundary region. (**C**,**D**) Seedlings of wild-type Col (**C**) and *cuc1-5 cuc2-3* (**D**), 7 days after germination (dag). Wild type has two separated cotyledons (co) and develops a shoot between them (arrowhead), whereas *cuc1-5 cuc2-3* has cotyledons (co) fused along their margins. (**E**) A *cuc1-5 cuc2-3* seedling mildly rescued by the *STM* transgene at 9 dag (**left**) and 16 dag (**right**). Cotyledons (co) are partially fused on one side. Note that the shoot is only visible at 16 dag (arrowhead). (**F**,**G**) Shoot apices in cleared seedlings (11 dag) of wild type (**F**) and *cuc1-5 cuc2-3* (**G**). Wild type develops the shoot meristem (open circle) and leaf primordia (asterisks), whereas *cuc1-5 cuc2-3* lacks these structures at the corresponding postion (arrow). (**H**) Shoot apices in cleared seedlings (11 dag) of *cuc1- 5 cuc2-3* rescued by *STM*, showing variable phenotypes: no visible shoot meristem and leaf primordia (**left**, arrow), small undifferentiated tissue (**middle**, closed circle), and shoot meristem with leaf primordia (**right**, open circle, and asterisks). (**I–L**) *cuc1-5 cuc2-3* seedlings (9 dag) rescued by combined transgenes: *STM* and *LAS* (**I**,**J**), *STM* and *BLR* (**K**), and *STM* and *KNAT6* (**L**). Arrowheads represent emerging shoot. co, cotyledon. Bars = 50 μm (**B**,**F–H**); 1 mm (**C–E**,**I–L**).

Table 1. Rescue of *cuc1 cuc2* seedling phenotype by downstream gene expression under the control of *CUC2* regulatory sequence.

Transgene	No Rescue [a] (%)	Mild Rescue [b] (%)	Strong Rescue [c] (%)	Total Number of T1 Seedlings
KLU	100	0	0	25
KNAT6	100	0	0	15
UFO	100	0	0	20 [d]
LAS	86.5	13.5	0	37
STM	54.5	45.5	0	11
PAN	100	0	0	12
LSH4	100	0	0	31
BLR	100	0	0	15
SAI-LLP1	100	0	0	12
ROXY20	100	0	0	12
CUC2	8.3	50.0	41.7	12
GUS	100	0	0	4

[a] Cotyledons were fused along both sides with only a small split at their tips and no shoot was formed. [b] Cotyledons were fused along one or both sides with more than half of their margins split. No shoot was formed or a shoot became visible only after 9 dag. [c] Cotyledons were completely separated and a shoot was visible by 9 dag. [d] One seedling had flat and round green tissue on top of the hypocotyl instead of a cup-shaped cotyledon. This is classified as "No rescue", as neither shoot formation nor cotyledon separation occurred.

Table 2. Effect of combined expression of *STM* and *LAS* on the phenotype of *cuc1 cuc2*.

Transgene A	Transgene B	No Rescue [a] (%)	Mild Rescue [b] (%)	Strong Rescue [c] (%)	Total Number of F2 Seedlings	Group [d]
STM	-	100	0	0	11	a
-	*LAS*	100	0	0	26	a
STM	*LAS*	0	71.1	28.9	38	b

[a] Cotyledons were fused along both sides with only a small split at their tips and no shoot was formed. [b] Cotyledons were fused along one or both sides with more than half of their margins split. No shoot was formed or a shoot became visible only after 9 dag. [c] Cotyledons were completely separated and a shoot was visible by 9 dag. [d] Different letters indicate statistically significant differences ($p < 0.01$, Fisher's exact test with Holm multiple testing correction).

Table 3. Effect of combined expression of *STM* and other downstream genes on the phenotype of *cuc1 cuc2*.

Transgene A	Transgene B	No Rescue [a] (%)	Mild Rescue [b] (%)	Strong Rescue [c] (%)	Total Number of F1 Seedlings
STM	–	92.3	7.7	0	13
–	*KNAT6*	100	0	0	6
STM	*KNAT6*	42.9	57.1	0	14*
STM	–	100	0	0	5
-	*UFO*	100	0	0	14
STM	*UFO*	100	0	0	12
STM	–	100	0	0	4
–	*PAN*	100	0	0	5
STM	*PAN*	100	0	0	9
STM	–	93.8	6.3	0	32
–	*LSH4*	100	0	0	19
STM	*LSH4*	92.9	7.1	0	14
STM	–	100	0	0	7
–	*BLR*	100	0	0	3
STM	*BLR*	16.7	0	83.3	6 **
STM	–	100	0	0	10
–	*GUS*	100	0	0	9
STM	*GUS*	100	0	0	10

[a] Cotyledons were fused along both sides with only a small split at their tips and no shoot was formed. [b] Cotyledons were fused along one or both sides with more than half of their margins split. No shoot was formed or a shoot became visible only after 9 dag. [c] Cotyledons were completely separated and a shoot became visible by 9 dag. Asterisks indicate significant differences in the ratio of the phenotypes of plants carrying the transgene B alone and those carrying both transgenes A and B (* $p < 0.05$, ** $p < 0.01$; Fisher's exact test).

2.3. Combined Loss of Function of STM, LAS, BLR, and KNAT6 Severely Impairs Shoot Meristem Formation and Cotyledon Separation

We then tested the combined effect of the loss-of-function mutations in the four genes (*STM, LAS, BLR,* and *KNAT6*) in young seedlings. Among the single mutants of these genes, only *stm* shows defects in shoot development with complete penetrance. In the case of the strong allele *stm-1C* [29], the mutant typically stops leaf formation after producing the first two leaves (Figure 3A–D). It has been reported that mutations in *BLR* or *KNAT6* alone do not cause visible phenotypes in seedlings, but enhance the defects of *stm* mutants [19,30]. Consistent with this, we found that the *stm-1C blr* double mutant produced fewer leaves than *stm-1C* (Figure 3A,C,D) and showed delayed first-leaf growth (Figure 3C,E). In addition, *stm-1C blr* showed a higher frequency of cotyledon fusion than *stm-1C* (Table 4). The *stm-1C knat6* double mutant showed strong cotyledon fusion and the lack of leaf production and both phenotypes were observed with complete penetrance (Figure 3F). These results confirmed the previous results that *BLR* and *KNAT6* are required for shoot meristem formation and cotyledon separation.

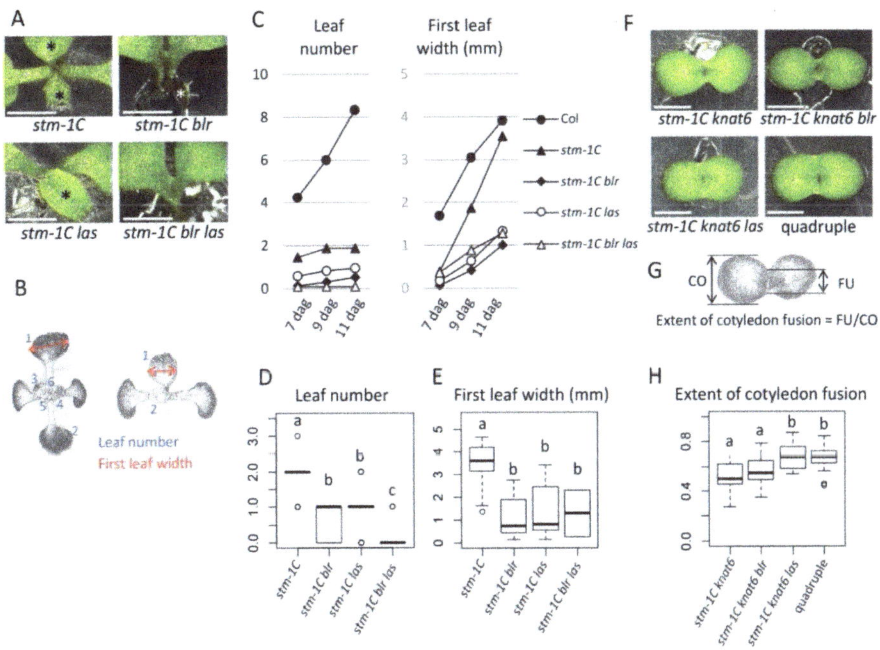

Figure 3. Genetic interactions of *stm, las, blr,* and *knat6* mutants. (**A**) The *las* mutation enhances the shoot production phenotype of *stm-1C* and *stm-1C blr* mutants. Shoot apex at 9 dag. Asterisks indicate developing leaves. (**B**) Examples of shoot phenotype measurements in the wild-type Col (**left**) and *stm-1C* (**right**). (**C**) Change in leaf number (**left**) and first leaf width (**right**). (**D**,**E**) Box plots of leaf number (**D**) and first leaf width (**E**) at 11 dag. (**F**) *las* enhances the cotyledon fusion phenotype of *stm-1C knat6* and *stm-1C knat6 blr* mutants. Seedlings at 7 dag. (**G**) Quantification of cotyledon fusion. (**H**) Box plot showing extent of cotyledon fusion at 7 dag in each genotype. Different letters in box plots indicate statistically significant differences ($p < 0.01$, Steel–Dwass method for (**D**); $p < 0.05$, Tukey–Kramer method for (**E**) and (**H**)). Sample size is 24, 23, 17, and 23 for (**C**) (**left**) and (**D**); 24, 12, 13, and 2 for (**C**) (**right**) and (**E**); and 11, 23, 12, and 19 for (**H**). Bars in (**A**) and (**E**), 2 mm.

Table 4. Frequency of cotyledon fusion in *stm* mutant combined with *blr* and *las* mutants.

Genotype	Frequency of Fusion (%)	Total	Group *
Col	0	55	a
stm	12.5	24	b
blr stm	91.3	23	c
las stm	35.3	17	b
blr las stm	95.7	23	c

* Different letters indicate statistically significant differences ($p < 0.01$, Fisher's exact test with Holm multiple testing correction).

Similar to *blr* and *knat6* single mutants, young seedlings of the *las* single mutant reportedly show a normal appearance, except for a very small proportion of plants with fused cotyledons (0.26%) [6]. However, we found that the *las* mutation enhanced defects in shoot meristem activity when combined with the *stm-1C* single or *stm-1C blr* double mutant (Figure 3A–E). In addition, the *las* mutation enhanced the cotyledon fusion phenotype of both *stm-1C knat6* and *stm-1C blr knat6* mutants (Figure 3F–H). These results show that the *LAS* gene contributes to embryonic shoot meristem formation and cotyledon separation independently of *STM*, *KNAT6*, and *BLR*. The *blr knat6 las* triple mutant was phenotypically normal. Together, our results demonstrate that the four transcription factor-encoding genes *STM*, *KNAT6*, *BLR*, and *LAS* play key roles for embryonic shoot meristem formation and cotyledon separation downstream of CUC1 and CUC2.

2.4. *The KLUH Gene Restricts the Embryonic Shoot Meristem and Counteracts STM*

Among the downstream target genes, *KLU*/*CYP78A5* encoding a cytochrome P450 enzyme of the *CYP78A* family plays postembryonic roles in shoot organ size and organ production rate [25,26], raising the possibility that this gene also affects shoot meristem activity. Moreover, the mutation in rice *PLA1*, a member of the same family, causes an enlarged shoot meristem phenotype [31,32]. Indeed, we found that the two independent *klu* insertion alleles (*klu-019348* and *klu-4*) were associated with precocious leaf initiation in young seedlings (Figure 4A–F), suggesting enhanced shoot meristem activity. In addition, the width of the shoot meristem was significantly greater in *klu* than in the wild type (Figure 4G). Furthermore, embryos of the *klu* mutants displayed an enlarged cotyledon boundary region (Figure 4H–J) and this phenotype was associated with an enlarged expression domain of the shoot stem cell marker *CLV3::GUS* [33] (Figure 4K,L). These results indicate that the *KLU* gene is required for restricting shoot meristem size and activity during embryogenesis and postembryonic development.

The identification of *KLU* as a downstream gene of *CUC1* and *CUC2* was unexpected because *KLU* negatively affects shoot meristem activity, whereas the *CUC* genes are positive regulators of shoot meristem formation. To further investigate the relationship between *KLU* and the *CUC* genes, we crossed the *klu* mutant with the *cuc1 cuc2* double mutant and examined their genetic interactions. In single mutants of *klu*, *cuc1*, and *cuc2*, all seedlings produced a functional shoot as the wild type, and except for a small fraction of *cuc1* mutants with partially fused cotyledons, their cotyledons were completely separated (Table 5; Figure 5A). Similarly, seedlings of *klu cuc1* and *klu cuc2* double mutants all produced a fully functional shoot and most of them developed completely separated cotyledons, whereas small fractions had partially fused cotyledons (Table 5; Figure 5B). Importantly, the *klu cuc1 cuc2* triple mutant was indistinguishable from *cuc1 cuc2* double mutants in that they formed strongly fused cup-shaped cotyledons and lacked a shoot meristem (Table 5; Figure 2D,G and Figure 5C–E). These results show that the *cuc1 cuc2* double mutations are epistatic to *klu* and indicate that the *KLU* gene can affect shoot meristem activity only when the functional *CUC1* and *CUC2* genes are present.

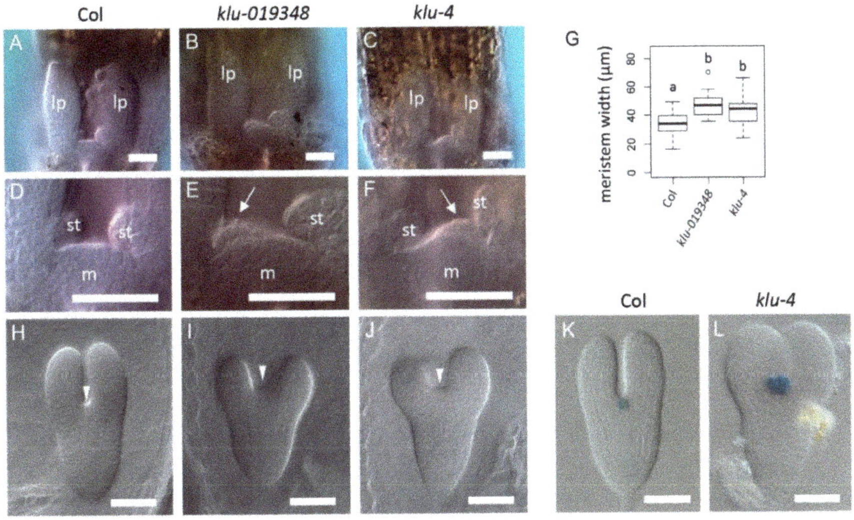

Figure 4. Mutations in *KLU* cause shoot meristem enlargement and precocious leaf formation. (**A–F**) Shoot apex of wild type (**A**,**D**), *klu-019348* (**B**,**E**), and *klu-4* (**C**,**F**) in cleared seedlings at 3 dag. (**D–F**) are close-up views of (**A–C**), respectively. Optical sections are photographed at a position slightly off-center to reveal the precociously formed third-leaf primordia in the *klu* mutants (arrows). lp, the first two leaf primordia; m, shoot meristem; st, stipule. (**G**) Shoot meristem width of wild type and two *klu* mutant alleles at 3 dag. Different letters in box plots indicate statistically significant differences ($p < 0.01$, Tukey–Kramer method). Sample size is 34, 28, and 42 for **G**. (**H–J**) Torpedo-stage embryos of wild type (**H**), *klu-019348* (**I**), and *klu-4* (**J**). Arrowheads indicate the position of the cotyledon boundary region. (**K**,**L**) *CLV3::GUS* expression in wild-type (**K**) and *klu-4* (**L**) torpedo-stage embryos. Scale bar = 50 µm.

Table 5. Genetic interactions among *klu-4*, *cuc1-5*, and *cuc2-3*.

Genotype	Phenotype			Total Number of Seedlings
	Normal [a] (%)	Weak [b] (%)	Strong [c] (%)	
Col	100	0	0	76
klu	100 *	0	0	289
cuc1	95.6	4.4	0	135
cuc2	100	0	0	118
klu cuc1	95.9	4.1	0	172
klu cuc2	97.2	2.8	0	217
cuc1 cuc2	0	0	100	50
klu cuc1 cuc2	0	0	100	54

[a] Cotyledons were completely separated and a shoot was formed immediately after germination. [b] Cotyledons were partially fused and a shoot was formed immediately after germination. [c] Cotyledons were strongly fused, forming a cup shape, and no shoot was formed after two weeks of observation. * A very small fraction (1.7%) had three cotyledons instead of two.

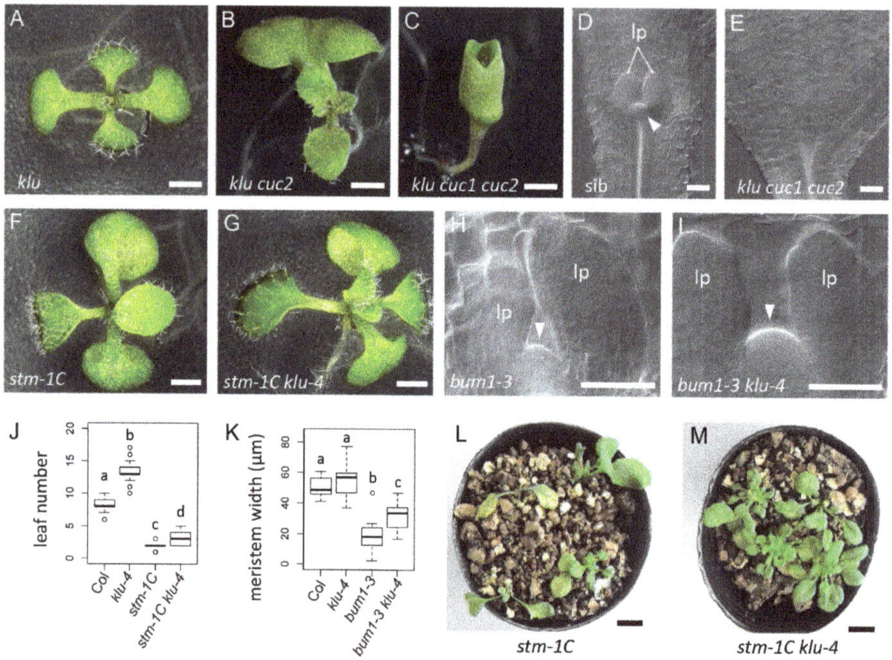

Figure 5. *KLU* acts downstream of *CUC1* and *CUC2*, and counteracts *STM* in the regulation shoot meristem activity. (**A–C**) Seven-day-old seedlings of *klu-4* (**A**), *klu-4 cuc2-3* (**B**), and *klu-4 cuc1-5 cuc2-3* (**C**). (**D**,**E**) Shoot apices of the progeny from *klu-4 cuc1-5 cuc2-3*/+ parent plants. A sib seedling with the shoot meristem (**D**) and a *klu cuc1 cuc2* seedling without it (**E**). (**F**,**G**) Nine-day-old seedlings of the strong *stm* allele *stm-1C* (**F**) and *stm-1C klu-4* (**G**). (**H**,**I**) Shoot apex of the weak *stm* allele *bum1-3* (**H**) and *bum1-3 klu-4* (**I**) at 4 dag. (**J**) Leaf number at 11 dag. (**K**) Shoot meristem width at 4 dag. (**L**,**M**) Twenty-nine-day-old plants of *stm-1C* (**L**) and *stm-1C klu-4* (**M**). Four plants are grown in each pot. Different letters in box plots indicate statistically significant differences ($p < 0.01$, Steel–Dwass method for (**J**); $p < 0.05$, Tukey–Kramer method for (**K**)). Sample size is 55, 30, 24, and 14 for (**J**); 14, 18, 15, and 9 for (**K**). Arrowheads indicate the shoot meristem. lp, leaf primordia. lp, leaf primordia. Scale bar, 1 mm for (**A**) to (**C**,**F**,**G**); 50 µm for (**D**,**E**,**H**,**I**); 10 mm for (**L**) and (**M**).

Next, we examined the genetic interaction of *KLU* with *STM*, which represents a functionally important class of *CUC* downstream genes that positively affects shoot meristem activity. Young seedlings of the *stm klu* double mutant produced more leaves and had an enlarged shoot meristem compared with *stm* single mutants (Figure 5F–K). Moreover, whereas the strong *stm-1C* mutant allele typically arrested shoot growth after producing a few leaves (Figure 5L), the *stm-1C klu-4* double mutant showed prolonged vegetative shoot growth (Figure 5M). Taken together, these results indicate that the *KLU* gene activity counteracts that of *STM* in postembryonic shoot development.

To clarify the mechanism by which *CUC1* and *CUC2* regulate *KLU* gene expression, we examined the expression of reporter genes containing *cis*-regulatory sequences of *KLU*. A reporter construct that carried regions 2 kb upstream and 0.6 kb downstream (*Pro2kb*) showed activities in the cotyledon boundary region, cotyledon margins, and root pole (Figure 6A,B). This expression pattern resembled that of *KLU* mRNA detected by in situ hybridization, except that the reporter activity in cotyledons was broader and that in the root pole was detected for a prolonged time (compare Figure 6B with Figure 1A and Figure S1A), indicating that the *Pro2kb* reporter contained a set of *cis*-regulatory elements sufficient for driving the native expression pattern of the gene at least in the cotyledon boundary region. When this construct was introduced into the *cuc1 cuc2* double-mutant background, its expression disappeared specifically in the cotyledon boundary region (Figure 6C), which corresponds to the region where the *CUC* genes act in normal development. Moreover, deletion of the region between 1 and 2 kb upstream of the gene resulted in the loss of expression in the cotyledon boundary region as well as in cotyledon margins (Figure 6A,D; *Pro1kb*). These results indicate that this 1 kb region contains *cis*-regulatory elements required for CUC1- and CUC2-dependent transcription in the shoot apex.

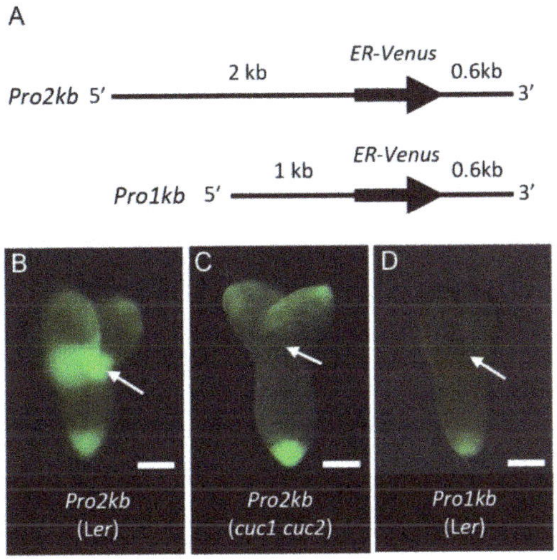

Figure 6. *KLU* expression is regulated by *CUC1* and *CUC2* via a specific promoter region. (**A**) Schematic diagram of the *KLU* reporter genes. (**B–D**) Expression of *KLU* reporter genes in embryos. *Pro2kb* in wild type L*er* (**B**), *Pro2kb* in *cuc1-1 cuc2-1* (**C**), and *Pro1kb* in wild type L*er* (**D**). Arrows indicate the position of the cotyledon boundary region of wild type (**B**,**D**) and the corresponding region of *cuc1-1 cuc2-1* (**C**). Scale bar, 50 μm.

3. Discussion

In this work, through functional analyses of genes acting downstream of *CUC1* and *CUC2*, we found that the combined activities of *STM* with *LAS*, *BLR*, and *KNAT6* were important for shoot meristem formation and cotyledon separation. The abilities of these genes to rescue the *cuc1 cuc2* mutant phenotypes when expressed under the boundary-specific promoter, together with the strong shoot meristem and cotyledon phenotypes observed in the quadruple mutant, support a model in which the *CUC* genes promote shoot meristem formation and cotyledon separation mainly through the activation of these four genes. It was previously shown that *STM* is a direct transcriptional target of *CUC1* [28]. The experiments using the DEX-inducible CUC1-GR plants and CHX treatments in our current work indicate that *LAS* and *KNAT6* are additional direct targets of the CUC1 protein, whereas the regulation of *BLR* by CUC1 may be indirect.

Our analysis highlights the importance of the activation of *STM* expression by the *CUC* genes in shoot meristem formation and cotyledon separation. *STM* encodes a KNOX transcription factor [34] whose activity is continuously required for shoot meristem maintenance in postembryonic development through regulating various aspects of shoot meristem properties including pluripotency, self-maintenance, promotion of cell cycle, repression of differentiation, and hormone metabolism [28,35–37]. How *STM* promotes cotyledon separation is currently unknown, but its ability to repress growth and promote leaf dissection, when ectopically expressed, may be involved in the repression of growth at the cotyledon boundary [38].

The possible molecular mechanisms by which *STM* acts in concert with the rest of the four genes may vary. *KNAT6* encodes a KNOX protein closely related to STM, so its ability to enhance the rescuing activity of *STM* can simply be explained by functional redundancy [19,39]. By contrast, *BLR* encodes a BEL-class homeodomain protein, which physically interacts with STM [30], and nuclear localization of STM requires its interaction with BLR [40–42]. These results indicate that the coexpression of *BLR* with *STM* provides a sufficient amount of BLR–STM complex to the nucleus, thereby efficiently promoting shoot meristem formation and cotyledon separation. In axillary shoot meristem formation, another BEL protein, ATH1, which is functionally redundant with BLR, forms a heterodimer with STM and directly promotes the transcription of *STM*, thus forming a self-activation loop [43]. It is also possible that the formation of BLR–STM heterodimer during embryogenesis is critical for initiating the *STM* self-activation loop, allowing self-maintenance of the shoot meristem.

Our functional analyses also demonstrate that the *LAS* gene contributes to embryonic shoot meristem formation and cotyledon separation downstream of CUC1 and CUC2. *LAS* encodes a putative transcriptional regulator of the GRAS family and acts as a central hub in the gene regulatory network for axillary meristem formation downstream of CUC2 [23,24]. The precise mechanisms by which *LAS* regulates these processes remain elusive; however, the mutation in the *LAS* ortholog in tomato affects the levels of hormones involved in shoot meristem activity, such as auxin, gibberellin (GA), and cytokinin [44,45]. Recently, it has been shown that the LAS protein binds to the promoter of *GA2ox4*, which encodes a GA deactivation enzyme, and promotes its expression [46]. These results raise the possibility that *LAS* reduces the levels of GA in the boundary region to promote shoot meristem activity, thereby contributing independently of *STM*, *KNAT6*, and *BLR* to the process downstream of the *CUC* genes.

In contrast to the above four genes, which are positive regulators of shoot meristem activity, *KLU/CYP78A5* represents the functionally opposite class of downstream genes regulated by *CUC1* and *CUC2*, as shown by the enhanced shoot meristem size and activity in the *klu* mutant as well as its genetic interaction with the *stm* mutant. Our data thus indicate that, by activating the two classes of genes with opposing functions during embryogenesis, the *CUC* genes create an optimal microenvironment for the shoot meristem to maintain its appropriate size and to produce organs at an appropriate rate. It has been well established that the balance between the self-renewal of stem cells and differentiation of their progeny is critical for postembryonic shoot meristem and that this balance is maintained by the WUS–CLV3 feedback system, which is supported by multiple transcription factors and plant hormones [3,47]. Our results provide an additional level of regulation for the balancing mechanism of shoot stem cell maintenance. Detailed functional analysis of the *KLU* gene as well as its relationship to previously known stem cell regulators will further improve our understanding of shoot meristem regulation.

4. Materials and Methods

4.1. Plant Materials

Arabidopsis thaliana accessions Columbia (Col) and Landsberg *erecta* (L*er*) were used as the wild type. *CUC1-GR* was established in the L*er* background using the same construct as described previously [22]. *CLV3::GUS* was reported previously [33]. Expression analyses of genes downstream of *CUC* were carried out in *cuc1-1 cuc2-1* [4,5]. For rescue experiments, *cuc1-5 cuc2-3* was used [6]. Loss-of-function mutants of the downstream genes were as follows: *klu-019348* (Sallk_019348) and *klu-4* for *KLU* [25]; *knat6-2* for *KNAT6* [19]; *las-101* for *LAS* [6]; *stm-1C* for *STM* [29]; and *pny-40126* for *BLR* [48]. Details of the mutants are described in Supplementary Table S4.

4.2. Constructs

For *ProCUC2:LAS*, we amplified the *LAS* coding sequence derived from Col with the PCR primers BamHI-LAS-F (5′-TATCTGGATCCATGCTTACTTCCTTCAAATC-3′) and EcoRI-LAS-R (5′-TTCTCGAATTCTCATTTCCACGACGAAACGG-3′) and placed it upstream of the *35S* terminator in a modified UAS cassette vector [49] using the EcoRI and BamHI sites. The BamHI-NotI fragment containing *LAS* and the terminator was then placed downstream of the *CUC2* promoter of *pBS-gC2*, which contains a 5.9 kb fragment of the *CUC2* genomic sequence [50], yielding *ProCUC2:LAS BS*. The SalI-NotI fragment of *ProCUC2:LAS BS* was then inserted into pBIN50, a modified binary vector carrying a kanamycin resistance gene [29]. To obtain *ProCUC2:BLR*, cDNA derived from L*er* was amplified using the primers BLRcDNAfull-F (5′-TTTCCCATGGCTGATGCATA-3′) and BLRcDNAfull-R (5′-TCAACCTACAAAATCATGTA-3′), cloned into pCR™-Blunt II-TOPO (Invitrogen, [Waltham, MA, USA]), and the resulting EcoRI fragment was then placed upstream of the 35S terminator of the modified UAS cassette. Fusion with the *CUC2* promoter and transfer to a binary vector was carried out in the same manner as that for *ProCUC2:LAS*, except that pBIN60, a modified binary vector with a sulfadiazine resistance gene, was used. To obtain the other chimeric constructs of the CUC2 promoter and downstream genes, the 3.1 kb SalI-BglII fragment of the CUC2 promoter was blunt-ended and cloned into the blunt-ended HindIII site of the gateway destination vector pGWB1 carrying kanamycin and hygromycin resistance genes [51], resulting in *ProCUC2 pGWB1*. To generate entry clones, cDNAs were first amplified with gene-specific primers and then with the *attB* adaptor primers listed in Supplementary Table S5 and cloned into pDONR221 by BP reaction. The inserts were then transferred to *ProCUC2 pGWB1* by LR reaction. For *ProCUC2:GUS*, the *GUS* gene fragment was transferred from the entry clone pENTR-gus (Invitrogen [Waltham, MA, USA]) to *ProCUC2 pGWB1* via LR reaction. Plant transformation was carried out by the floral dip method [52]. The *Pro2kb* and *Pro1kb* reporter constructs of the *KLU* gene contain −2066 to +16 and −1061 to +16 sequences from the start codon, respectively, at their 5′ end of *vYFPer*, an ER-localized version of Venus, and −85 to +338 sequence from the stop codon at their 3′ end and are cloned in the binary vector pBarMAP [53]. Both constructs were transformed to L*er*.

4.3. Rescue Experiments

Double-homozygous *cuc1-5 cuc2-3* plants were seedling lethal and did not produce flowers, thus each construct for the rescue experiments was transformed to *cuc1-5 cuc2-3/+* plants via the floral dip method. T_1 seeds were selected for drug resistance on Murashige–Skoog plates [54] and the phenotypes of the resistant seedlings were scored. Subsequently, double-homozygous plants were identified by PCR-based genotyping. For analyses of the combined effect of *STM* with *LAS*, one transgenic line containing the *ProCUC2:STM* transgene was maintained and T_2 plants with the *cuc1-5 cuc2-3/+* genotype were crossed with two independent lines of *ProCUC2:LAS* with the *cuc1-5 cuc2-3/+* genotype. The seedling phenotype was first scored in the F_2 generation and the genotype of the *CUC2* locus as well as the presence of each transgene was subsequently analyzed. The same *ProCUC2:STM* line was used for analysis of the combined effect of *STM* with the rest of the genes and was crossed with two independent transgenic lines for each gene. Scoring of the seedling phenotype and subsequent genotyping were carried out in the F_1 generation. To classify the phenotypes into the mild and strong categories, we examined the presence or absence of visible shoot under a binocular. The absence of shoot production was confirmed at 14 dag or later.

4.4. Histological Analysis

In situ hybridization was carried out as described previously [17]. As a template for the *UFO* probe, we used pDW221.1 [55]. For generating templates for other probes, gene-specific fragments were PCR-amplified using the primers listed in Supplementary Table S6 and cloned into pCR™-Blunt II-TOPO. For GUS detection, embryos were dissected and immediately stained in staining solution with 5 mM ferricyanide and ferrocyanide [56]. Embryos and seedling apices were visualized after clearing as described previously [17]..

4.5. DEX Induction and qRT-PCR

Induction of CUC1-GR with DEX and expression analysis by qRT-PCR were carried out as described previously [22]. Primers used for qRT-PCR are listed in Supplementary Table S7.

Supplementary Materials: Supplementary Materials can be found at http://www.mdpi.com/1422-0067/21/16/5864/s1. Table S1: List of 52 Genes Downregulated in cuc1 cuc2 Embryos in Microarray Experiments. Table S2: Quantitative RT-PCR of 51 Selected Genes in Heart Stage Embryos. Table S3: Quantitative RT-PCR of 51 Selected Genes in Bending Cotyledon Stage Embryos. Table S4: Mutants used in this analysis. Table S5: Gene-Specific Primers for Rescue Constructs. Table S6: Primers used for cloning probe templates for in situ hybridization. Table S7: Primers used for qRT-PCR experiments. Figure S1: Expression of candidate downstream genes in wild-type Ler and cuc1-1 cuc2-1 embryos. Figure S2: Examples of rescued and non-rescued plants. Text S1: Screening of additional genes regulated by CUC1 and CUC2.

Author Contributions: Conceptualization, M.A. and M.T.; Transcriptome analysis and qRT-PCR verification, Y.T.; In situ hybridization, M.A.; Expression analysis of *CUC1-GR*, S.S.; Rescue experiments and loss-of-function analysis of *CUC* downstream genes, H.O., M.K., M.A., S.T., M.R.K. and M.M.; Genetic analysis of *KLU* and *STM*, M.A.; Reporter analysis of *KLU*, R.I., M.A. and M.L.; Original draft preparation, M.A. All authors have read and agreed to the published version of the manuscript.

Funding: This work was supported by the Ministry of Education, Culture, Sports, Science, and Technology of Japan (grant numbers 23012031, 2411400901, 18H04842, 20H04889 to M.A.); the Japan Society for the Promotion of Science (Grant No. 23370023, 16K07401 to M.A.); and Takeda Science Foundation (to M.A.).

Acknowledgments: We thank Detlef Weigel for pDW221.1, Chiyoko Machida for the *knat6-2* allele, Ayako Yamaguchi and *Arabidopsis* Biological Resource Center (ABRC) for the *klu* alleles, Rüdiger Simon for *CLV3::GUS*, and RIKEN Biological Resource Center (BRC) for RAFL cDNAs (pda12509 and pda11315). We also thank Mizuki Yamada for critical reading and Edanz (www.edanzediting.co.jp) for editing the English text of a draft of this manuscript. Finally, we thank Etsuko Habe, Seiko Ishihara, Naoko Fujihara, Eriko Tanaka, Maki Niidome, Mie Matsubara, and Kazuko Onga for technical assistance.

Conflicts of Interest: The authors declare no conflict of interest. The funders had no role in the design of the study; in the collection, analyses, or interpretation of data; in the writing of the manuscript, or in the decision to publish the results.

References

1. Weigel, D.; Jürgens, G. Stem cells that make stems. *Nature* **2002**, *415*, 751–754. [CrossRef]
2. Barton, M.K. Twenty years on: The inner workings of the shoot apical meristem, a developmental dynamo. *Dev. Biol.* **2010**, *341*, 95–113. [CrossRef]
3. Gaillochet, C.; Lohmann, J.U. The never-ending story: From pluripotency to plant developmental plasticity. *Development* **2015**, *142*, 2237–2249. [CrossRef]
4. Aida, M.; Ishida, T.; Fukaki, H.; Fujisawa, H.; Tasaka, M. Genes involved in organ separation in Arabidopsis: An analysis of the *cup-shaped cotyledon* mutant. *Plant Cell* **1997**, *9*, 841–857. [CrossRef]
5. Takada, S.; Hibara, K.; Ishida, T.; Tasaka, M. The *CUP-SHAPED COTYLEDON1* gene of *Arabidopsis* regulates shoot apical meristem formation. *Development* **2001**, *128*, 1127–1135.
6. Hibara, K.; Karim, M.R.; Takada, S.; Taoka, K.; Furutani, M.; Aida, M.; Tasaka, M. *Arabidopsis CUP-SHAPED COTYLEDON3* regulates postembryonic shoot meristem and organ boundary formation. *Plant Cell* **2006**, *18*, 2946–2957. [CrossRef]
7. McConnell, J.R.; Emery, J.; Eshed, Y.; Bao, N.; Bowman, J.; Barton, M.K. Role of *Phabulosa* and *Phavoluta* in determining radial patterning in shoots. *Nature* **2001**, *411*, 709–713. [CrossRef] [PubMed]
8. Emery, J.F.; Floyd, S.K.; Alvarez, J.; Eshed, Y.; Hawker, N.P.; Izhaki, A.; Baum, S.F.; Bowman, J.L. Radial patterning of *Arabidopsis* shoots by class III HD-ZIP and KANADI genes. *Curr. Biol.* **2003**, *13*, 1768–1774. [CrossRef]
9. Prigge, M.J.; Otsuga, D.; Alonso, J.M.; Ecker, J.R.; Drews, G.N.; Clark, S.E. Class III homeodomain-leucine zipper gene family members have overlapping, antagonistic, and distinct roles in *Arabidopsis* development. *Plant Cell* **2005**, *17*, 61–76. [CrossRef] [PubMed]
10. Zhang, Z.; Tucker, E.; Hermann, M.; Laux, T. A molecular framework for the embryonic initiation of shoot meristem stem cells. *Dev. Cell* **2017**, *40*, 264–277.e264. [CrossRef] [PubMed]
11. Knauer, S.; Holt, A.L.; Rubio-Somoza, I.; Tucker, E.J.; Hinze, A.; Pisch, M.; Javelle, M.; Timmermans, M.C.; Tucker, M.R.; Laux, T. A protodermal miR394 signal defines a region of stem cell competence in the *Arabidopsis* shoot meristem. *Dev. Cell* **2013**, *24*, 125–132. [CrossRef] [PubMed]

12. Aida, M.; Tasaka, M. Genetic control of shoot organ boundaries. *Curr. Opin. Plant Biol.* **2006**, *9*, 72–77. [CrossRef] [PubMed]
13. Aida, M.; Tasaka, M. Morphogenesis and patterning at the organ boundaries in the higher plant shoot apex. *Plant Mol. Biol.* **2006**, *60*, 915–928. [CrossRef] [PubMed]
14. Takeda, S.; Aida, M. Establishment of the embryonic shoot apical meristem in *Arabidopsis thaliana*. *J. Plant Res.* **2011**, *124*, 211–219. [CrossRef]
15. Rast, M.I.; Simon, R. The meristem-to-organ boundary: More than an extremity of anything. *Curr. Opin. Genet. Dev.* **2008**, *18*, 287–294. [CrossRef]
16. Wang, Q.; Hasson, A.; Rossmann, S.; Theres, K. *Divide et impera*: Boundaries shape the plant body and initiate new meristems. *New Phytol.* **2016**, *209*, 485–498. [CrossRef]
17. Aida, M.; Ishida, T.; Tasaka, M. Shoot apical meristem and cotyledon formation during *Arabidopsis* embryogenesis: Interaction among the *CUP-SHAPED COTYLEDON* and *SHOOT MERISTEMLESS* genes. *Development* **1999**, *126*, 1563–1570.
18. Raman, S.; Greb, T.; Peaucelle, A.; Blein, T.; Laufs, P.; Theres, K. Interplay of miR164, *CUP-SHAPED COTYLEDON* genes and *LATERAL SUPPRESSOR* controls axillary meristem formation in *Arabidopsis thaliana*. *Plant J.* **2008**, *55*, 65–76. [CrossRef]
19. Belles-Boix, E.; Hamant, O.; Witiak, S.M.; Morin, H.; Traas, J.; Pautot, V. *KNAT6*: An *Arabidopsis* homeobox gene involved in meristem activity and organ separation. *Plant Cell* **2006**, *18*, 1900–1907. [CrossRef]
20. Hibara, K.; Takada, S.; Tasaka, M. *CUC1* gene activates the expression of SAM-related genes to induce adventitious shoot formation. *Plant J.* **2003**, *36*, 687–696. [CrossRef]
21. Daimon, Y.; Takabe, K.; Tasaka, M. The *CUP-SHAPED COTYLEDON* genes promote adventitious shoot formation on calli. *Plant Cell Physiol.* **2003**, *44*, 113–121. [CrossRef] [PubMed]
22. Takeda, S.; Hanano, K.; Kariya, A.; Shimizu, S.; Zhao, L.; Matsui, M.; Tasaka, M.; Aida, M. CUP-SHAPED COTYLEDON1 transcription factor activates the expression of *LSH4* and *LSH3*, two members of the ALOG gene family, in shoot organ boundary cells. *Plant. J.* **2011**, *66*, 1066–1077. [CrossRef] [PubMed]
23. Tian, C.H.; Zhang, X.N.; He, J.; Yu, H.P.; Wang, Y.; Shi, B.H.; Han, Y.Y.; Wang, G.X.; Feng, X.M.; Zhang, C.; et al. An organ boundary-enriched gene regulatory network uncovers regulatory hierarchies underlying axillary meristem initiation. *Mol. Syst. Biol.* **2014**, *10*. [CrossRef] [PubMed]
24. Greb, T.; Clarenz, O.; Schäfer, E.; Müller, D.; Herrero, R.; Schmitz, G.; Theres, K. Molecular analysis of the *LATERAL SUPPRESSOR* gene in *Arabidopsis* reveals a conserved control mechanism for axillary meristem formation. *Genes Dev.* **2003**, *17*, 1175–1187. [CrossRef] [PubMed]
25. Anastasiou, E.; Kenz, S.; Gerstung, M.; MacLean, D.; Timmer, J.; Fleck, C.; Lenhard, M. Control of plant organ size by *KLUH/CYP78A5*-dependent intercellular signaling. *Dev. Cell* **2007**, *13*, 843–856. [CrossRef] [PubMed]
26. Wang, J.W.; Schwab, R.; Czech, B.; Mica, E.; Weigel, D. Dual effects of miR156-targeted *SPL* genes and *CYP78A5/KLUH* on plastochron length and organ size in *Arabidopsis thaliana*. *Plant Cell* **2008**, *20*, 1231–1243. [CrossRef]
27. Aoyama, T.; Chua, N.H. A glucocorticoid-mediated transcriptional induction system in transgenic plants. *Plant J.* **1997**, *11*, 605–612. [CrossRef]
28. Scofield, S.; Murison, A.; Jones, A.; Fozard, J.; Aida, M.; Band, L.R.; Bennett, M.; Murray, J.A.H. Coordination of meristem and boundary functions by transcription factors in the SHOOT MERISTEMLESS regulatory network. *Development* **2018**, *145*. [CrossRef]
29. Takano, S.; Niihama, M.; Smith, H.M.; Tasaka, M.; Aida, M. *gorgon*, a novel missense mutation in the *SHOOT MERISTEMLESS* gene, impairs shoot meristem homeostasis in *Arabidopsis*. *Plant Cell Physiol.* **2010**, *51*, 621–634. [CrossRef]
30. Byrne, M.E.; Groover, A.T.; Fontana, J.R.; Martienssen, R.A. Phyllotactic pattern and stem cell fate are determined by the *Arabidopsis* homeobox gene *BELLRINGER*. *Development* **2003**, *130*, 3941–3950. [CrossRef]
31. Itoh, J.I.; Hasegawa, A.; Kitano, H.; Nagato, Y. A recessive heterochronic mutation, *plastochron1*, shortens the plastochron and elongates the vegetative phase in rice. *Plant Cell* **1998**, *10*, 1511–1522. [CrossRef] [PubMed]
32. Miyoshi, K.; Ahn, B.O.; Kawakatsu, T.; Ito, Y.; Itoh, J.I.; Nagato, Y.; Kurata, N. *PLASTOCHRON1*, a timekeeper of leaf initiation in rice, encodes cytochrome P450. *Proc. Natl. Acad. Sci. USA* **2004**, *101*, 875–880. [CrossRef] [PubMed]

33. Brand, U.; Grünewald, M.; Hobe, M.; Simon, R. Regulation of *CLV3* expression by two homeobox genes in Arabidopsis. *Plant Physiol.* **2002**, *129*, 565–575. [CrossRef] [PubMed]
34. Long, J.A.; Moan, E.I.; Medford, J.I.; Barton, M.K. A member of the KNOTTED class of homeodomain proteins encoded by the *STM* gene of *Arabidopsis*. *Nature* **1996**, *379*, 66–69. [CrossRef]
35. Jasinski, S.; Piazza, P.; Craft, J.; Hay, A.; Woolley, L.; Rieu, I.; Phillips, A.; Hedden, P.; Tsiantis, M. KNOX action in *Arabidopsis* is mediated by coordinate regulation of cytokinin and gibberellin activities. *Curr. Biol.* **2005**, *15*, 1560–1565. [CrossRef]
36. Yanai, O.; Shani, E.; Dolezal, K.; Tarkowski, P.; Sablowski, R.; Sandberg, G.; Samach, A.; Ori, N. *Arabidopsis* KNOXI proteins activate cytokinin biosynthesis. *Curr. Biol.* **2005**, *15*, 1566–1571. [CrossRef]
37. Scofield, S.; Dewitte, W.; Nieuwland, J.; Murray, J.A.H. The Arabidopsis homeobox gene *SHOOT MERISTEMLESS* has cellular and meristem-organisational roles with differential requirements for cytokinin and CYCD3 activity. *Plant J.* **2013**, *75*, 53–66. [CrossRef]
38. Kierzkowski, D.; Runions, A.; Vuolo, F.; Strauss, S.; Lymbouridou, R.; Routier-Kierzkowska, A.L.; Wilson-Sánchez, D.; Jenke, H.; Galinha, C.; Mosca, G.; et al. A growth-based framework for leaf shape development and diversity. *Cell* **2019**, *177*, 1405–1418.e1417. [CrossRef]
39. Hake, S.; Smith, H.M.; Holtan, H.; Magnani, E.; Mele, G.; Ramirez, J. The role of *knox* genes in plant development. *Annu. Rev. Cell Dev. Biol.* **2004**, *20*, 125–151. [CrossRef]
40. Bhatt, A.M.; Etchells, J.P.; Canales, C.; Lagodienko, A.; Dickinson, H. VAAMANA—A BEL1-like homeodomain protein, interacts with KNOX proteins BP and STM and regulates inflorescence stem growth in *Arabidopsis*. *Gene* **2004**, *328*, 103–111. [CrossRef]
41. Cole, M.; Nolte, C.; Werr, W. Nuclear import of the transcription factor SHOOT MERISTEMLESS depends on heterodimerization with BLH proteins expressed in discrete sub-domains of the shoot apical meristem of *Arabidopsis thaliana*. *Nucleic Acids Res.* **2006**, *34*, 1281–1292. [CrossRef] [PubMed]
42. Rutjens, B.; Bao, D.; van Eck-Stouten, E.; Brand, M.; Smeekens, S.; Proveniers, M. Shoot apical meristem function in Arabidopsis requires the combined activities of three BEL1-like homeodomain proteins. *Plant J.* **2009**, *58*, 641–654. [CrossRef] [PubMed]
43. Cao, X.; Wang, J.; Xiong, Y.; Yang, H.; Yang, M.; Ye, P.; Bencivenga, S.; Sablowski, R.; Jiao, Y. A self-activation loop maintains meristematic cell fate for branching. *Curr. Biol.* **2020**. [CrossRef] [PubMed]
44. Tucker, D.J. Endogenous growth regulators in relation to side shoot development in the tomato. *New Phytol.* **1976**, *77*, 561–568. [CrossRef]
45. Schumacher, K.; Schmitt, T.; Rossberg, M.; Schmitz, G.; Theres, K. The *Lateral suppressor* (*Ls*) gene of tomato encodes a new member of the VHIID protein family. *Proc. Natl. Acad. Sci. USA* **1999**, *96*, 290–295. [CrossRef]
46. Zhang, Q.Q.; Wang, J.G.; Wang, L.Y.; Wang, J.F.; Wang, Q.; Yu, P.; Bai, M.Y.; Fan, M. Gibberellin repression of axillary bud formation in *Arabidopsis* by modulation of DELLA-SPL9 complex activity. *J. Integr. Plant Biol.* **2020**, *62*, 421–432. [CrossRef]
47. Somssich, M.; Je, B.I.; Simon, R.; Jackson, D. CLAVATA-WUSCHEL signaling in the shoot meristem. *Development* **2016**, *143*, 3238–3248. [CrossRef]
48. Smith, H.M.; Hake, S. The interaction of two homeobox genes, *BREVIPEDICELLUS* and *PENNYWISE*, regulates internode patterning in the Arabidopsis inflorescence. *Plant Cell* **2003**, *15*, 1717–1727. [CrossRef]
49. Sabatini, S.; Heidstra, R.; Wildwater, M.; Scheres, B. SCARECROW is involved in positioning the stem cell niche in the *Arabidopsis* root meristem. *Genes Dev.* **2003**, *17*, 354–358. [CrossRef]
50. Taoka, K.; Yanagimoto, Y.; Daimon, Y.; Hibara, K.; Aida, M.; Tasaka, M. The NAC domain mediates functional specificity of *CUP-SHAPED COTYLEDON* proteins. *Plant J.* **2004**, *40*, 462–473. [CrossRef]
51. Nakagawa, T.; Kurose, T.; Hino, T.; Tanaka, K.; Kawamukai, M.; Niwa, Y.; Toyooka, K.; Matsuoka, K.; Jinbo, T.; Kimura, T. Development of series of gateway binary vectors, pGWBs, for realizing efficient construction of fusion genes for plant transformation. *J. Biosci. Bioeng.* **2007**, *104*, 34–41. [CrossRef] [PubMed]
52. Clough, S.J.; Bent, A.F. Floral dip: A simplified method for *Agrobacterium*-mediated transformation of *Arabidopsis thaliana*. *Plant J.* **1998**, *16*, 735–743. [CrossRef] [PubMed]
53. Adamski, N.M.; Anastasiou, E.; Eriksson, S.; O'Neill, C.M.; Lenhard, M. Local maternal control of seed size by *KLUH/CYP78A5*-dependent growth signaling. *Proc. Natl. Acad. Sci. USA* **2009**, *106*, 20115–20120. [CrossRef] [PubMed]

54. Fukaki, H.; Fujisawa, H.; Tasaka, M. *SGR1, SGR2, SGR3*: Novel genetic loci involved in shoot gravitropism in *Arabidopsis thaliana. Plant. Physiol.* **1996**, *110*, 945–955. [CrossRef] [PubMed]
55. Lee, I.; Wolfe, D.S.; Nilsson, O.; Weigel, D. A *LEAFY* co-regulator encoded by *UNUSUAL FLORAL ORGANS*. *Curr. Biol.* **1997**, *7*, 95–104. [CrossRef]
56. Sessions, A.; Weigel, D.; Yanofsky, M.F. The *Arabidopsis thaliana MERISTEM LAYER 1* promoter specifies epidermal expression in meristems and young primordia. *Plant J.* **1999**, *20*, 259–263. [CrossRef]

International Journal of
Molecular Sciences

MDPI

Review

Synergistic Interaction of Phytohormones in Determining Leaf Angle in Crops

Xi Li [1,2], Pingfan Wu [1,2], Ying Lu [1,2], Shaoying Guo [1,2], Zhuojun Zhong [2,3], Rongxin Shen [2,3,*] and Qingjun Xie [1,2,*]

1 State Key Laboratory for Conservation and Utilization of Subtropical Agro-Bioresources, Guangdong Provincial Key Laboratory of Plant Molecular Breeding, South China Agricultural University, Guangzhou 510642, China; 20183137023@stu.scau.edu.cn (X.L.); 201713070123@stu.scau.edu.cn (P.W.); luying@stu.scau.edu.cn (Y.L.); syguo@scau.edu.cn (S.G.)
2 Guangdong Laboratory of Lingnan Modern Agriculture, Guangzhou 510642, China; zhongzhuojun@stu.scau.edu.cn
3 State Key Laboratory for Conservation and Utilization of Subtropical Agro-Bioresources, College of Life Sciences, South China Agricultural University, Guangzhou 510642, China
* Correspondence: shenrongxin@scau.edu.cn (R.S.); qjxie@scau.edu.cn (Q.X.)

Received: 17 June 2020; Accepted: 15 July 2020; Published: 17 July 2020

check for updates

Abstract: Leaf angle (LA), defined as the angle between the plant stem and leaf adaxial side of the blade, generally shapes the plant architecture into a loosen or dense structure, and thus influences the light interception and competition between neighboring plants in natural settings, ultimately contributing to the crop yield and productivity. It has been elucidated that brassinosteroid (BR) plays a dominant role in determining LA, and other phytohormones also positively or negatively participate in regulating LA. Accumulating evidences have revealed that these phytohormones interact with each other in modulating various biological processes. However, the comprehensive discussion of how the phytohormones and their interaction involved in shaping LA is relatively lack. Here, we intend to summarize the advances in the LA regulation mediated by the phytohormones and their crosstalk in different plant species, mainly in rice and maize, hopefully providing further insights into the genetic manipulation of LA trait in crop breeding and improvement in regarding to overcoming the challenge from the continuous demands for food under limited arable land area.

Keywords: Leaf angle; Phytohormones; crop yield; BR; Crosstalk

1. Introduction

To overcome the challenge of ever-increasing global demands for food, feedstock, and bioenergy products, breeders have been forced to select and breed cultivars with a key feature that can be planted at higher densities in order to increase grain yield with the limited availability of arable land area [1,2]. To this end, genetic improvement of crops with ideal plant architecture is considered as one of the most powerful strategies for addressing this issue. The key components of ideal plant architecture in crop generally include plant height, grain architecture, and leaf angle (LA) [3]. Since the 1960s, genetic engineering of decreasing plant height in crops has dramatically boosted the crop yield and definitely benefited millions of people worldwide, which is remarked as the "Green Revolution". The great achievement of Green Revolution is attributed to the utilization of two "Green Revolution" genes, mutant allelic *Semi-dwarf1* (*Sd1*) in rice, which encodes a key enzyme in the gibberellic pathway GA20ox2, and *Reduced height-1* (*Rht-1*) in wheat that encodes a key repressor in gibberellin signaling pathway called DELLA, are responsible for gibberellin metabolism in rice and wheat [4,5], respectively. However, excessive application of nitrogen fertilizer for ensuring the yield and productivity of the semidwarf varieties has brought severe contamination on environment. Recently, a novel gibberellin-GIBBERELLIN INSENSITIVE DWARF1 (GID1)-NITROGEN-MEDIATED TILLER GROWTH RESPONSE 5 (NGR5) signaling pathway has been stated [6], which can be effectively used to boost yield of semi-dwarf crop by simultaneously enhancing the tiller number and nitrogen use efficiency (NUE) in next-generation Green Revolution. Besides plant height, the LA trait has also been substantially selected for high-yield varieties breeding, particularly in maize and rice, due to its vital role for optimal light interception and competition between neighboring plants in natural settings.

A typical grass leaf is consisted of distal blade, proximal sheath, and a boundary called ligular region (also called lamina joint) that separates blade and sheath into distinct parts, all of which contain epidermal, ground, and vascular tissues that are continuous with each other but distinct in cell types and patterns [7,8]. LA, defined as the angle between the stem and adaxial side of the blade (Figure 1), is one of the most important architecture traits selected for crop yield improvement [9]. Crops with architecture of smaller LA and more upright leaves can facilitate higher plant density and enhance the photosynthetic efficiency, thus elevating yield [10–12]. The ligular region, an annular structure outside the joint of the leaf blade and leaf sheath, is the pivotal structure for determining LA in grasses. Accumulating evidences have been implicated that the formation of LA is mediated by the shape of the lamina joint, differences in cell numbers/size at adaxial/abaxial region and distinctive mechanical tissue strength [9,13–15], and thus any influence on them could alter the LA. For example, failure of longitudinal elongation of the adaxial cells in the lamina joint may result in erect leaves [16], whereas excessive expansion of the adaxial cells would increase the LA [17,18].

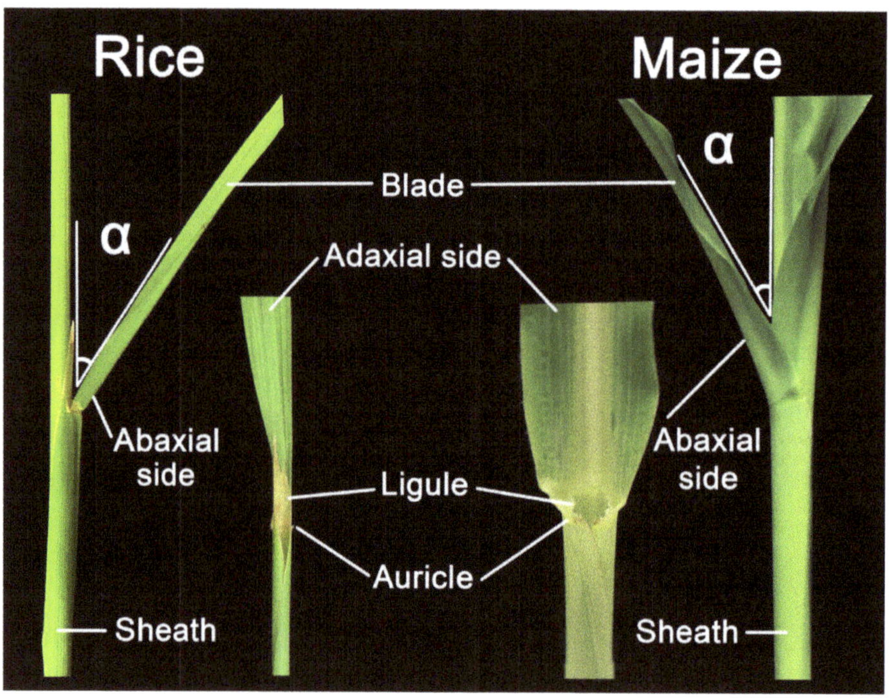

Figure 1. Structural composition of leaf angle in rice and maize. Leaf angle is defined by the angle between the plant stem and leaf adaxial side of the blade, indicating as the α in the figure. This leaf morphology is affected by the development of ligule and auricle as well.

Given the rapid developments of plant functional genomics, a number of genes controlling LA have been cloned and the relevant regulatory network underlying the development of LA in grasses has been well characterized. These studies revealed that phytohormones, such as brassinosteroids (BRs), auxin, and gibberellins (GAs), comprehensively participate in regulating LA by orchestrating the homeostais of their biosynthesis and the expression of signaling transcription factors (TFs). Though each phytohormone and TF has been found to contribute quite similar traits/phenotypes in term of LA, how they interplay with each other is still far beyond understood. This review is an attempt to highlight the regulation mechanism of LA and the genetic interactions among phytohormones in regulating LA formation with particular emphasis on maize and rice.

2. Regulation of Lamina Joint Bending by Brassinosteroid (BR)

BRs are a group of steroid phytohormones involved in many important biological processes, and thus play a vital role in regulating several important agronomic traits, such as leaf angle, plant height and stress resistance. The regulatory pathways of BR biosynthesis, metabolism, and signal transduction have been well established in rice [19–21]. Since LA is a grass-species-specific trait, the role of BR in regulating LA has only been characterized in maize and rice rather than Arabidopsis, indicating that BR is a positive regulator of LA (Figure 2).

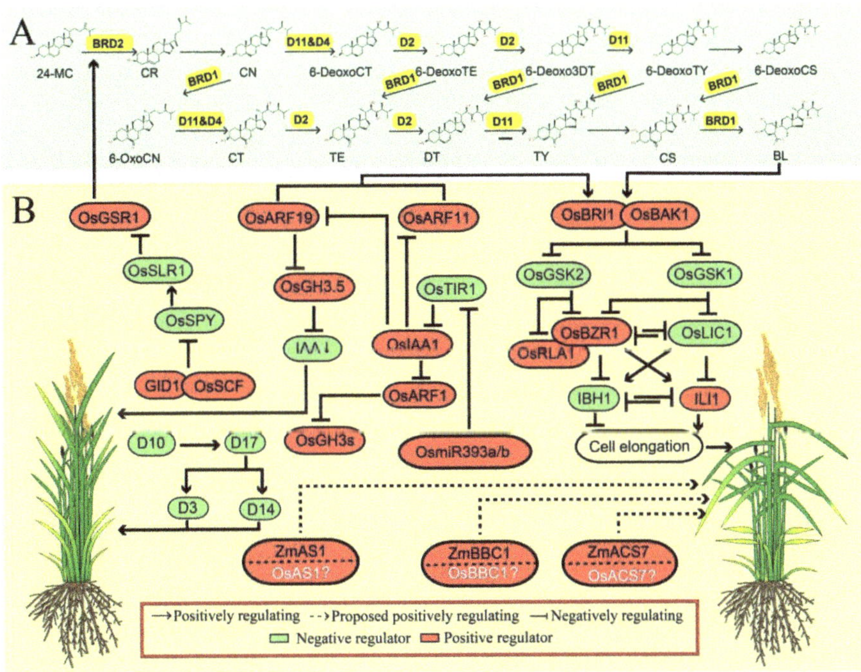

Figure 2. The synergistic regulation mechanisms of leaf angle by phytohormones in rice. (**A**) Brassinosteroid biosynthesis pathway. The corresponding enzyme that catalyze each reaction in yellow color. The secondary structure of the chemical is obtained from ChemSpider (http://www.chemspider.com/Default.aspx). (**B**) Positive and negative regulation of leaf angle by brassinosteroid (BR) signaling pathway and its crosstalk with other regulators involved in other plant hormones pathway, and other phytohormones regulators positively participate in regulation of leaf angle. The proteins in white color and with a question mark represent that they are homologs of those reported to be involved in the regulation of leaf angle. The green arrow represents the downregulation of indoleacetic acid (IAA). The protein in red color box represents as the positive regulator while the one in green color box represents the negative regulator of leaf angle. The dashed line and solid line represent the indirect or direct evidence supporting the responsible regulation, respectively.

Up-regulation of the BR content within lamina joint region or enhanced BR signaling pathway by boosting the expression of BR related regulators resulted in increasing LA (Table 1). For instance, many researches in rice have elaborated that leaf inclination is closely associated with biosynthesis or signaling of BR [19,22]. Loss-of-function of BR biosynthetic genes, such as *Dwarf 2 (D2)*, perturbed the endogenous level of BR, eventually resulting in erect leaf phenotype [11,23–26]. Besides, other BR biosynthetic regulators, such as the Cytochrome P450 family proteins, have also been documented to function in the development of LA. For example, *BR-deficient Dwarf1* (*OsBRD1*) is cytochrome P450 protein and encodes a key enzyme (BR C-6 oxidase) catalyzing BR biosynthesis. Disruption of *OsBRD1* causes pleiotropic effects, including severe dwarf phenotype, tortuous leaves, short panicles, small seeds, etc. [27,28]. Another two Cytochrome P450 proteins, OsDWARF4 and OsDWARF11, also catalyze the rate-limiting reaction (c-22 hydroxylation) of BR biosynthesis. *OsDWARF4* is highly expressed in leaf blade and root, which is inhibited by BR but increased in BR-insensitive or -deficient mutants, indicating there is a feedback regulation on *OsDWARF4* expression. Depletion of *OsDWARF4* caused mild phenotype with erect leaves but not any detrimental effect on the development of leaf, inflorescence, and seed relative to wild-type plant [11,24]. Distinct from *Osdwarf4* mutant, *Osdwarf11* mutant showed a much more severe phenotype, including dwarfism in plants, erection of leaves, pollen abortion, and small grains. Further investigation revealed that *OsDWARF4* and *OsDWARF11* function redundantly in BRs biosynthesis, but *OsDWARF11* performs a major role in BR biosynthesis pathway while *OsDWARF4* plays the complementary role [25], which explained the different effects of them in term of plant growth and development, as well as the LA. It is worthy to mention that the *Osdwarf4* mutant with erect leaf simultaneously promoted biomass production and higher yields than wild type at different planting densities, even without additional fertilization, indicating that this gene/allele is a potential candidate for sustainability increasing crop yield in limited land area [11]. Taken together, these studies have clearly illustrated that BR metabolism is responsible for the LA formation in crop.

In the BR signaling transduction pathway, BR is perceived by extracellular domain of BRASSINOSTEROID INSENSITIVE 1 (BRI1), a single transmembrane leucine-rich repeat receptor-like protein kinase (LRR-RLK), and its co-receptor BRI1-ASSOCIATED KINASE 1 (BAK1). In the absence of BR, BRI1's kinase domain is deactivated by the negative factor BRI1 KINASE INHIBITOR 1 (BKI1); while BR is present, the BR compound binds to the extracellular domain of the BRI1 and BAK1. Subsequently, BRI1 phosphorylates BKI1, which results in the release of BKI1 from the plasma membrane, and then induce the phosphorylation of the kinase domain of BRI1 and BAK1, finally activating the initiation of BRI1-mediated signaling transduction [29]. The activation of BRI1 in turn phosphorylates downstream BR-SIGNALING KINASE1 (BSK1), CONSTITUTIVE DIFFERENTIAL GROWTH 1 (CDG1), and some of their homologs. Both BSK1 and CDG1/CDL1 phosphorylates BRI1-SUPPRESSOR 1 (BSU1) and subsequently activates BSU1. Activated BSU1 dephosphorylates and inactivates BRASSINOSTEROID INSENSITIVE 2 (BIN2) to release the suppression of BRASSINAZOLE-RESISTANT 1 (BZR1) and BRI1-EMS-SUPPRESSOR 1 (BES1), which are two key downstream transcription factors positively mediating BR responses. BZR1 and BES1 are rapidly dephosphorylated by Protein Phosphatase 2A (PP2A) family, leading to their nuclear accumulation and their regulation of thousands of BR-responsive genes expression [30]. Besides, in the absence of BR, the interaction of phosphorylated BZR1 and BES1 with 14-3-3 proteins, a group of conserved phosphopeptide-binding proteins, can lead to their cytoplasm retention and degradation, while in the presence of BR, dissociated BKI1 in cytoplasm can competitively bind to 14-3-3 proteins to inhibit this process [31]. Otherwise, the PPA2 dephosphorylates BRI1, BZR1 and BES1, relieving the binding of BZR1 and BES1 to 14-3-3 proteins, which regulates downstream genes in response to BR and induce diverse BR responses [32]. In addition to the BR biosynthetic genes mentioned above, BR signaling genes also play an essential role in regulating the development of LA in rice. For example, OsBRI1 is an ortholog of Arabidopsis BRI protein in rice. *OsBRI1* participates in regulating various aspects of growth and development processes in rice, including the bending of lamina joint, the intercalary meristem formation, and the longitudinal elongation of internode cells. Therefore, depletion of *OsBRI1* significantly perturbed the internode elongation and bending of the lamina joint [33]. *OsBAK1* (a homologous gene to *Arabidopsis BAK1*) encodes SERK family receptor kinase protein and acts as co-receptor kinase for OsBRI1 to mediate BR signal transduction. Down-regulating the expression level of *OsBAK1* produced a rice variety with erect leaf and normal reproduction so that considered as a promising target for improving rice grain yield [34]. Nevertheless, suppression of *OsBZR1* expression by RNA interference (RNAi) in rice greatly altered the expression of BR-responsive genes and resulted in plant dwarfism and erect leaf, suggesting that *OsBZR1* plays an central downstream role in BR signaling [35]. The rice 14-3-3 proteins have been proven to interact with and retain OsBZR1 in cytoplasm instead of nucleus, ultimately leading to inhibiting the function of *OsBZR1*. However, BR treatment can dissociate their interactions and activate *OsBZR1*, in turn causing the erect leaf phenotype [36]. Similarly, a rice zinc finger transcription factor, *LEAF* and *TILLER ANGLE INCREASED CONTROLLER* (*OsLIC*), also antagonized *OsBZR1* to repress BR signaling in rice. Overexpression of *OsLIC* resulted in erect leaves by eliminating BR response, indicating that *OsLIC1* negatively modulates leaf inclination in rice. *OsLIC* directly regulated the *INCREASED LEAF INCLINATION 1* (*ILI1*), a positive regulator in lamina inclination, to oppose the action of BZR1 [36]. In addition, recent studies have identified an APETALA2 (AP2)/ERF (ethylene-responsive element binding factor) family transcription factor, *Reduced Leaf Angle 1* (*RLA1*) that is identical to the *SMALL ORGAN SIZE 1* (*SMOS1*), as a positive regulator of BR signaling, which physically interacted with OsBZR1 to enhance its transcriptional activity to enlarge LA in rice [37,38]. *BU1* encodes a helix-loop-helix protein and acts as a novel BR positive regulator. Overexpressing *BU1* in rice led to enhanced bending of the lamina joint, increased grain size, and resistance to BR, while repression of *BU1* and its homologs in rice displays erect leaves [39]. Furthermore, a pair of antagonizing HLH/bHLH factors, *OsILI1* that is the homology of *Arabidopsis thaliana Paclobutrazol Resistance 1* (*PRE1*) and *ILI1 binding bHLH 1* (*OsIBH1*), functioned as downstream factors of *OsBZR1* to regulate leaf angle. *OsILI1* positively regulated BR-mediated cell elongation, whereas OsIBH1 directly interacted with OsILI1 and performed an opposite role [40]. In addition, another bHLH transcription factor, *BRASSINOSTEROID-RESPONSIVE LEAF ANGLE REGULATOR 1* (*OsBLR1*), has also been implicated to participate in LA regulation though the BR pathway in rice. Over-expressing *OsBLR1* simultaneously increased leaf angle, grain length and sensitivity to BR, whereas mutation of *OsBLR1* resulted in erect leaf and shorter grain [41]. Similarly, gain-of-function of *OsbHLH079* also caused wide LA, longer grain and hypersensitive to BR in rice, while *OsbHLH079*-RNAi lines showed opposite phenotype [42]. Collectively, these findings suggested that *bHLH* family transcription factors may be broadly involved in BR-mediated LA regulation. Previously, a plant-specific gene family transcription factor, *OsGRAS19* (*GA INSENSITIVE (GAI), REPRESSOR OF GAI (RGA),* and *SCARECROW (SCR) 19*) was suggested to be involved in regulating BR signaling, because the knockdown lines of *OsGRAS19* displayed less sensitivity to the 24-epi-brassinolid (BL) treatment as compared to WT in rice. Higher expression of *OsGRAS19* caused larger LA, narrow leaf and thin culm and panicle in rice [43]. Recently, a novel mutant of *OsGRAS19*, *D26*, has been identified in rice, which displayed typical BR-mediated phenotypes, including semidwarf, wider, and shorter leaf in addition to the erect leaf, as well as longer grain [44]. Taken together, these results further elaborated that the BR signaling is extensively involved in regulating LA in rice.

Until now, the molecular regulation of BR underlying LA in maize has been rarely investigated. Previous reports illustrated that suppression of *ZmBRI1* or knock-out of *Dwarf* and *Irregular Leaf 1* (*ZmDIL1*) displayed dwarf plant and erect leaves [45,46], suggesting a conserved function of BR for plant architecture in monocots. Recently, a SQUAMOSA-PROMOTER BINDING PROTEIN-LIKE (SPL) family protein LIGULELESS 1 (LG1), which is recognized as a conserved key factor in the formation of ligule and auricle in both rice and maize [47,48], confers an important role in controlling the leaf angle through regulating BR and auxin signaling pathway in maize and wheat. In maize, *ZmLG1* activated a B3-domain containing transcription factor *ZmRAVL1*, which sequentially activated *ZmBRD1* that was also designated as *Upright Plant Architecture 1* (*UPA1*), eventually resulting in the alternation of leaf angle. *UPA2* functioned as a distant *cis* element to regulate *ZmRAVL1* expression, which was controlled by the *Drooping Leaf 1* (*DRL1*). This *UPA2-ZmRAV1-UPA1* module fine-tuned the BR pathway and finally determined the plant architecture and LA in maize [49]. In wheat, the *LG1* ortholog, *TaSPL8*, directly bound to the promoter of an auxin response factor *TaARF6* and the BR biosynthesis gene *CYP90D2* (*TaD2*), and subsequently activated their expressions, leading to enhancing lamina joint development [50]. Recently, *ZmILI1*, an ortholog of *OsILI1* that plays an important role in LA in rice, was found to directly bind to the *ZmLG1* and *CYP90D1* promoters to affect the BR biosynthesis and signal, eventually changing the LA [51]. Collectively, these studies supported a notion that components of BR signaling pathway play a dominant and conserved role in the formation of leaf angle in multiple crops and would be the potential targets for genetic improvement of crop architecture and yield.

3. Regulation of Lamina Joint Bending by Indoleacetic Acid (IAA)

Indoleacetic acid (IAA) is another crucial hormone, regulating leaf inclination mainly through patterning the adaxial/abaxial cell growth of leaves. In contrast to the positive role of BR in LA regulation, IAA functioned as a negative regulator since eliminating IAA content resulted in increased leaf inclination while increasing IAA content caused reduction of leaf inclination and upright leaves [52].

Auxin signal transduction is a sophisticated pathway, which is consisted of several key components, including the F-box TRANSPORT INHIBITOR RESPONSE 1/AUXIN SIGNALING F-BOX PROTEIN (TIR1/AFB) auxin co-receptors, the Auxin/INDOLE-3-ACETIC ACID (Aux/IAA) transcriptional repressors, and the AUXIN RESPONSE FACTOR (ARF) transcription factors and a ubiquitin-dependent protein degradation system. When IAA is deficient, Aux/IAA proteins form heterodimers with ARFs and block the function of ARFs to inhibit the auxin signal. Presence of auxin promotes the interaction between TIR1/AFB and Aux/IAA proteins, and then triggers a proteasome-mediated degradation of Aux/IAA by 26S Proteasome, eventually resulting in the release of ARF transcriptional activity. Subsequently, activation of ARF induces the changes of auxin-mediated gene expression pattern and growth responses [53–56]. This auxin signaling transduction model in rice has been implicated to associate with the regulation of lamina joint bending (Figure 1).

For instance, *FISH BONE* (*FIB*) encodes an orthologous of TAA protein, which plays a negative effect on leaf inclination. Loss-function of *FIB* caused a reduction of IAA level and altered auxin polar transport activity, thus producing small leaves with enlarged lamina joint angle [57]. *LEAF INCLINATION1* (*LC1*) encoding an IAA amino synthetase, termed as OsGH3.1 in rice, maintained auxin homeostasis by catalyzing excess IAA binding to various amino acids. A gain-of-function mutant *lc1-D* in rice displayed a reducing content of free IAA and increasing leaf angle due to the promotion of cell elongation in the adaxial surface of lamina joint [58]. In addition, two *OsMIR393a/b* targeted auxin receptors, *OsTIR1* and *AUXIN SIGNALING f-box 2* (*OsAFB2*), have also been proven to be involved in the regulation of leaf angle. Overexpression of *OsmiR393a/b* repressed the expression of *OsTIR1* and *OsAFB2*, which led to greater leaf angle in rice [59]. Further study demonstrated that BR promoted *OsTIR1* and *OsAFB2* to trigger the degradation of OsIAA1, which resulted in de-suppression of OsARF11 and OsARF19 proteins and consequently caused enlarge LA in rice [60]. Furthermore, a SPOC domain-containing transcription suppressor Leaf inclination 3 (LC3) regulated leaf inclination through interacting with a HIT zinc finger domain-containing protein, LC3-interacting protein 1 (LIP1). Meanwhile, LC3 could also directly bind to the promoter regions of *OsIAA12* and *OsGH3.2* to regulate auxin signal transduction and auxin homeostasis, finally influencing the formation of leaf inclination [61].

The maize auxin efflux carrier P-glycoprotein (ZmPGP1) was an adenosine triphosphate (ATP) binding cassette (ABC) transporter, which was involved in the polar transport of auxin and associated with the LA in maize [62]. Notably, multiple haplotypes of *ZmPGP1* were present in various landraces, teosintes, and inbred accessions [63], suggesting a domesticated selection and improvement due to the demand of higher density planting.

4. Regulation of Lamina Joint Bending by Gibberellins (GA)

GA signaling pathway has also been implicated to participate in the regulation of the LA through the BR signaling dependent manner. Previously, transcriptomic profiling identified the co-expression pattern between GA and BR associated genes, however, it was less understood whether these genes were coordinated to regulate the LA. Recently, several researches have verified that certain GA genes interact with BR genes to modulate the LA. For instance, knockdown of the rice *SPINDLY* (*OsSPY*) caused the eliminated expression of BR biosynthesis genes *D11*, *D2*, and *OsDWARF/BRD1*, but increased the *OsDWARF4* expression that functions downstream of the GA gene, *SLENDER RICE 1* (*SLR1*), ultimately resulting in the enhanced leaf angle [64]. A recent study found that the Arabidopsis O-fucosyltransferase SPY could mono-O-fucosylate the DELLA protein, leading to higher affinity of interaction between DELLA and BZR1 [65]. From this point of view, it is supposed that OsSPY also could activate the OsSLR1 encoding a DELLA protein to interact with OsBZR1, thereby affecting LA in rice. The rice GA-stimulated transcript gene (*OsGSR1*) induced by GA is another downstream gene of *SLR1*, and it directly binds to BRD2 to enhance the BR biosynthesis, which subsequently altered the leaf angle [64,66]. In addition to these GA genes, other regulators involved in GA pathway have also been identified to participate in LA regulation. *OsDCL3a* encodes a Dicer-like endoribonuclease involved in generating siRNA. Down-regulation of the *OsDCL3a* resulted in increased leaf angle by modulating the expression of GA and BR associated genes, including *OsGSR1* and *BRD1* [67]. Recently, *OsmiR396d* has been reported to regulate the LA in rice, which was promoted by OsBZR1 and then regulated the expression of BR responsive genes by targeting the *GROWTH REGULATING FACTOR 4* (*OsGRF4*) [68]. Alternatively, as another target of *OsmiR396d*, the *OsGRF6* participated in GA biosynthesis and signal transduction but was not directly involved in BR signaling only modulated the plant height rather than LA [68]. Taken together, it is supposed that GA likely regulates LA dependent on the BR pathway, and thus identification of much more GA components involved in response to BR would extend our knowledge to this issue.

5. Regulation of Lamina Joint Bending by Crosstalk among Various Phytohormones

Recent studies have shown that crosstalk between IAA and BR cooperatively regulates the development of leaf angle. OsIAA1 is a key negative regulator for auxin signal transduction, which is induced by auxin and BR. Over-expression of *OsIAA1* in rice resulted in dwarfism and increased leaf angle with decreased sensitivity to auxin treatment but increased sensitivity to BR treatment. *OsARF1* is an auxin signal positive regulator which is inhibited by *OsIAA1* in the absence of auxin. Further analysis showed that mutation of *OsARF1* reduced sensitivity to BR treatment, resembling the phenotype of *OsIAA1*-overexpression plants, which indicated that BR may interact with auxin through the OsIAA1-OsARF1 module to regulate LA in rice [69,70]. *OsARF19* acted as another coordinator of auxin and BR by positively regulating the expression of *OsGH3.5* to reduce the content of free IAA on the one hand and activating *OsBRI1* to stimulate BR signal cascades on the other hand, thus resulting in increased leaf angle [71]. *RLA1/SMOS1* functioned downstream of the auxin signaling pathway, and enhanced the transcriptional activity of *OsBZR1* by interacting with OsBZR1, suggesting that *RLA1/SMOS1* integrated BR and IAA signal pathway to regulate the development of leaf angle [37,38]. In summary, the above studies showed that the auxin antagonized BR by interfering both BR metabolism and signaling to negatively regulate the leaf angle in rice.

Additionally, the rice *d1* mutant with null function of an α subunit of G-protein (Gα), Rho GTPase activating protein 1 (RGA1), exhibited a dwarfism phenotype with erect leaves, and reduced sensitivity to GA and BR, indicating that *RGA1* was involved in both GA and BR responses [29,72]. Further studies showed that D1/RGA1 interacted with TUD1 to induce *BU1* expression, resulting in the increased leaf inclination [73,74]. BR can also act upstream of GA by modulating GA metabolism to regulate cell elongation. BR activated *OsBZR1* and induced the expression of *D18/GA3ox-2*, one of the GA biosynthetic genes, leading to increased bioactive GA levels in rice seedlings. In contrast, GA extensively inhibited BR biosynthesis and the BR response with a feedback mechanism, so that GA treatment decreased the enlarged leaf angles in plants by attenuated BR biosynthesis or signaling [75]. These results showed that BR and GA were intertwined to regulate the leaf angle in rice.

6. Other Phytohormones Involved in Regulation of Lamina Joint

Phytohormones such as ethylene, strigolactones (SLs), jasmonic acid (JA), and abscisic acid (ABA) were also involved in regulating leaf inclination of rice (Figure 3). An early report demonstrated the interaction of ethylene and BR to regulate leaf inclination of rice, but the underlying mechanism remaining to be elusive [18]. Currently, a study has validated that altering the C terminus of 1-Aminocyclopropane-1-carboxylate (ACC) synthase 7 (ZmACS7) responsible for ethylene biosynthesis in maize led to the stability of this protein and the accumulation of ACC and ethylene contents, as well as the up-regulation of ethylene responsive genes, which finally reduced plant height and increased leaf angle [76]. Similar to *ZmACS7*, overexpression of its closest paralog *ZmACS2* also resulted in flatter leaves [76], further suggesting that ethylene positively regulates LA.

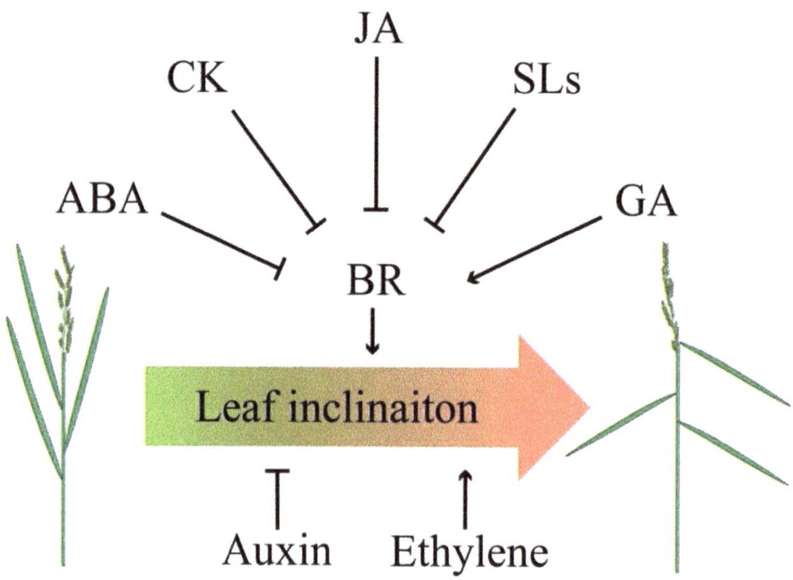

Figure 3. Crosstalk of phytohormones in determining leaf angle. Abscisic acid (ABA), CK, jasmonic acid (JA), and strigolactones (SLs) negatively regulate BR, thereby inhibiting leaf angle, whereas gibberellins (GA) positively coordinates BR to increase leaf angle. Application of auxin results in erect leaf while ethylene leads to flat leaf.

SLs negatively regulate leaf inclination at seedling stage [77]. Interestingly, recent report showed that SLs also mediated leaf inclination in response to nutrient deficiencies in rice [78]. Similar to SLs, JA also showed a negative role in leaf inclination. JA treatment decreased lamina joint inclination by repressing the expression of BR biosynthesis-related genes which thus decreased endogenous BRs levels. Besides, inactivation of a negative regulator of BR signaling, GSK3-like kinase, partly rescued the inhibited effect of JA on lamina joint inclination, indicating that JA may disturb both BR biosynthesis and BR signaling pathway to limit lamina joint inclination [79].

Previous research has revealed that BR antagonized with ABA in regulating seed germination and hypocotyl elongation in Arabidopsis. The study showed that the ABSCISIC ACID INSENSITIVE5 (ABI5) directly interacts with BRASSINOSTEROID INSENSITIVE2 (BIN2), and then was phosphorylated and stabilized by BIN2 upon ABA treatment [80]. Additionally, the BES1 also physically interacted with ABI5 to hinder the expression of ABI5-targeted *EARLY METHIONINE-LABELED 1* (*EM1*) and *EM6* [81], eventually facilitating the seed germination in Arabidopsis. Interestingly, the ABI1 and ABI2 can interact with and dephosphorylate BIN2 to attenuate the BR signaling in Arabidopsis [82], indicating a complicated crosstalk between ABA and BR. Recently, there is research which further revealed that ABA also antagonized BR to regulate the lamina joint inclination in rice by targeting the BR biosynthesis gene *D11* and BR signaling genes *GSK2* and *DLT* [83], and it therefore raises an issue that whether the rice homologs of Arabidopsis *ABI1* and *ABI2* may also interfere the BR-mediated LA. Investigation of the LA phenotype of the *Osabi1* and *Osabi2*, as well as overexpression lines of these two genes, may provide the answer for this hypothesis. Another research preprinted in bioRxiv demonstrated that a transcriptional repressor *ZmCLA4* (the ortholog of *LAZY1* in rice and Arabidopsis) responsible for multiple phytohormone mediated pathways negatively regulated LA by altering mRNA accumulation. Further analysis showed that ZmCLA4 could directly bind to two key components of BR signaling *ZmBZR3* and *14-3-3*, and two important responsive transcription factors of ABA *ZmWRKY4* and *ZmWRKY72* respectively, thus mediating the crosstalk between BR and ABA in LA regulation [84]. In maize, a bHLH transcription factor *ZmIBH1-1* is a negative regulator of LA in maize. Transcriptome analysis suggested that the *ZmIBH1-1*-mediated LA in the leaf ligular region was highly correlated with cytokinin (CK), JA, and ethylene synthesis and signal transduction pathways associated genes, in particular the two CK responsive genes, GRMZM2G145280 (*BBC1*) and GRMZM2G149952 (*ZmAS1*) that were tightly correlated with cell division, implying that CK may modulate LA through the control of cell profile [85]. Notably, it has been demonstrated that CK indirectly interacted with BR through auxin pathway. For example, BR regulated the development of root primordia through increasing the *PIN* genes expression while the CK inhibited the root primordia by repressing *PIN* genes [86]. However, it was also found that CK was accumulated in the young seedling of wheat with BR treatment [87], whereas overexpression of the rice *Isopentyl Transferase* (*IPT*) driven by the promoter of stress- and maturation-inducible gene, *Senescence-associated Receptor Kinase* (*SARK*), resulted in up-regulation of BR genes, including *DWF4*, *BRI1*, and *BZR1* etc., ultimately leading to higher rice yield during water-stress [88]. These studies further suggest that CK interplays with BR in regulating plant growth and stress response. However, it is still unclear whether and how they cooperated in the LA regulation. As mentioned above, the *bHLH* family generally integrated BR signaling to regulate the LA, such as the *OsILI1* and *OsBLR1*, and thus we proposed that the CK may interplay with BR to regulate LA through the *bHLH-BBC1/AS1* module mediated BR signaling.

In summary, almost all of the regulators modulated leaf angle through the phytohormones-dependent manner (Table 1), however, whether there are unknown components independent on phytohormones, the pathway still remains elusive.

Table 1. Cloned genes associated with leaf angle in rice and maize.

Pathway	Arabidopsis	Maize/Rice	Functions in Leaf Angle [1]	Refs
BR	*AtCYP90D1* (AT3G13730)	*OsCYP90D2* (LOC_Os01g10040)	Positively regulating leaf angle and related to the BR biosynthesis in rice.	[11,24–26]
	AtDWARF4 (AT3G50660)	*OsDWARF4* (LOC_Os03g12660)	Positively regulating leaf angle and catalyzing C-22 hydroxylation in BR biosynthesis pathway.	[11,24]
	AtCYP724A1 (AT5G14400)	*OsDWARF11* (LOC_Os04g39430)	Positively regulating leaf angle and catalyzing C-22 hydroxylation in BR biosynthesis pathway.	[23,25]
	AtBR6OX2 (AT3G30180)	*OsBRD1* (LOC_Os03g40540)	Positively regulating leaf angle and catalyzing BR biosynthesis.	[27,28]
	AtBR6OX2 (AT3G30180)	*ZmBRD1* (GRMZM2G103773)	Positively regulating leaf angle and catalyzing C-6 oxidation in BR biosynthesis.	[49]
	AtDWARF1 (AT3G19820)	*OsBRD2* (LOC_Os10g25780)	Positively regulating leaf angle and participating in the complementary pathway of BR synthesis.	[64,66]

Table 1. *Cont.*

Pathway	Arabidopsis	Maize/Rice	Functions in Leaf Angle [1]	Refs
BR	*AtSERK2* (AT1G34210)	*OsBAK1* (LOC_Os08g07760)	Positively regulating leaf angle and mediating BR signal transduction.	[34]
	AtBRI1 (AT4G39400)	*OsBRI1/OsDWARF61* (LOC_Os01g52050)	Positively regulating leaf angle and stimulating BR signal cascade to regulate organ development by controlling cell division and elongation, but is not necessary for organ initiation.	[71]
	AtBRI1 (AT4G39400)	*ZmBRI1a* (GRMZM2G048294)	Positively regulating leaf angle.	[46]
	AtBRI1 (AT4G39400)	*ZmBRI1b* (GRMZM2G449830)		
	AtBZR2 (AT1G19350)	*OsBZR1* (LOC_Os07g39220)	Positively regulating leaf angle and acting downstream of BR signaling.	[35]
	AT1G75340 (AT1G75340)	*OsLIC1* (LOC_Os06g49080)	A direct target of OsBZR1 and negatively modulating leaf inclination.	[36]
	AT2G41710 (AT2G41710)	*OsRLA1/OsSMOS1* (LOC_Os05g32270)	Positively regulating leaf angle and direct downstream of GSK2.	[37,38]
	AtIBH1 (AT2G43060)	*OsIBH1* (LOC_Os04g56500)	Interacting with OsILI1 and negatively regulating leaf angle.	[40]
	AtIBH1 (AT2G43060)	*ZmIBH1-1* (GRMZM2G388823)	Negative regulator of LA by modulating cell wall lignification and cell elongation in the ligular region.	[85]
	NA NA	*ZmDIL1* (NA)	Positively regulating leaf angle.	[45]
	AtPRE5 (AT3G28857)	*OsILI1* (LOC_Os04g54900)	Positive control of leaf angle.	[40]
	AtBS1 (AT1G74500)	*ZmILI1* (GRMZM2G072820)	Positively regulating leaf angle.	[51]
	AtSPL8 (AT1G02065)	*ZmLG1* (GRMZM2G036297)	Positively regulating leaf angle and directly activating *ZmBRD1* expression, leading to increased BR and leaf angle.	[49]
	AtNGA1 (AT2G46870)	*ZmRAVL1* (GRMZM2G102059)	Positively regulating leaf angle by regulating *ZmBRD1*.	[49]
	AtPGP1 (AT2G36910)	*ZmPGP1/ZmBR2* (GRMZM2G315375)	Positively regulating leaf angle and being involved in the polar transport of auxin.	[62]
	AtGPA1 (AT2G26300)	*OsD1/OsRGA1* (LOC_Os05g26890)	Positively regulating leaf angle by interacting with OsTUD1 to induce *OsBU1*, leading to increasing leaf inclination.	[73,74]
	AtPUB30 (AT3G49810)	*OsTUD1* (LOC_Os03g13010)	Positively regulating leaf angle.	[73,74]
	AtKDR (AT1G26945)	*OsBU1* (LOC_Os06g12210)	Positively regulating leaf angle.	[73,74]
	AtBIN2 (AT4G18710)	*OsGSK1* (LOC_Os01g10840)	Negatively regulating leaf angle and BR.	[36]
	AtGSK1 (AT1G06390)	*OsGSK2* (LOC_Os05g11730)	Negatively regulating leaf angle and the expression of downstream BR response genes.	[83]
	AT1G63100 (AT1G63100)	*OsSMOS2/OsGS6* (LOC_Os06g03710)	Positively regulating leaf angle and BR-mediated signaling pathway.	[83]
	AtDCL3 (AT3G43920)	*OsDCL3a* (LOC_Os01g68120)	Negatively regulating leaf angle.	[67]

Table 1. *Cont.*

Pathway	Arabidopsis	Maize/Rice	Functions in Leaf Angle [1]	Refs
IAA	*AtGH3.6* (AT5G54510)	*OsGH3.1/OsLC1* (LOC_Os01g57610)	Positively regulating leaf angle and maintaining auxin homeostasis by catalyzing excess IAA binding to various amino acids.	[58]
	AtGH3.2 (AT4G37390)	*OsGH3.2* (LOC_Os01g55940)	Positively regulating leaf angle and auxin signal transduction and auxin homeostasis.	[61]
	OsmiR393	-	Positively regulating leaf angle but negatively regulating *OsTIR1* and *OsAFB2*.	[59]
	AtTIR1 (AT3G62980)	*OsTIR1* (LOC_Os05g05800)	Negatively regulating leaf angle and being the target of *OsmiR393*.	[59]
	AtAFB2 (AT3G26810)	*OsAFB2* (LOC_Os04g32460)	Negatively regulating leaf angle and being the OsmiR393 target.	[59]
	AtIAA17 (AT1G04250)	*OsIAA1* (LOC_Os01g08320)	Positively regulating leaf angle and inhibiting *OsARF11* and *OsARF19*	[60]
	AtIAA3 (AT1G04240)	*OsIAA12* (LOC_Os03g43410)	Positively regulating leaf angle and auxin signal transduction and homeostasis.	[61]
	AT5G11430 (AT5G11430)	*OsLC3* (LOC_Os06g39480)	Negatively regulating leaf angle and inhibiting expressions of *OsIAA12* and *OsGH3.2*.	[61]
	AT5G63830 (AT5G63830)	*OsLIP1* (LOC_Os10g37640)	Negatively regulating leaf angle and interacting with LC3 to inhibit *OsIAA12* and *OsGH3.2*.	[61]
	AtARF2 (AT5G62000)	*OsARF1* (LOC_Os11g32110)	Negatively regulating leaf angle and inhibited by OsIAA1 in the absence of auxin.	[69,70]
	AtARF5 (AT1G19850)	*OsARF11* (LOC_Os04g56850)	Positively regulating leaf angle and suppressed by OsIAA1.	[60]
	AtARF19 (AT1G19220)	*OsARF19* (LOC_Os06g48950)	Positively regulating leaf angle and *OsGH3-5* and *OsBRI1*, and affecting the elongation of rice basal internodes and leaves by regulating cell elongation.	[60]
GA	*AtSPY* (AT3G11540)	*OsSPY* (LOC_Os08g44510)	Negatively regulating leaf angle and GA. Involving in BR signal transduction and control of the suppression of *SLR1*.	[64]
	AtGAI (AT1G14920)	*OsSLR1* (LOC_Os03g49990)	Negatively regulating leaf angle and GA signal transduction.	[64,66]
	AtGASA4 (AT5G15230)	*OsGSR1* (LOC_Os06g15620)	Positively regulating leaf angle and induced by *OsSLR1*. Directly binding to OsBRD2 to enhance the BR biosynthesis.	[64,66]
	OsmiR396	NA	Positively regulating leaf angle and BR-mediated signaling pathway.	[68]
	AtARF5 (AT3G13960)	*OsGRF4* (LOC_Os02g47280)	Negatively regulating leaf angle and cell enlargement and number, and being the *OsmiR396* target	[68]
	AtGFP1 (AT2G22840)	*OsGRF6* (LOC_Os03g51970)	Negatively regulating leaf angle and being the *OsmiR396* target.	[68]
	AtGID1C (AT5G27320)	*OsGID1* (LOC_Os05g33730)	Positively regulating leaf angle and mediating GA signaling in rice.	[89]
JA	*AtJAR1/AtGH3.11* (AT2G46370)	*OsGH3.5/OsJAR1* (LOC_Os05g50890)	Positively regulating leaf angle and being regulated by OsARF19. Redundant with other OsGH3.	[71]
Ethylene	*AtACS6* (AT4G11280)	*ZmACS7* (GRMZM5G894619)	Positively regulating leaf angle.	[76]
CK	*AtBBC1* (AT3G49010)	*ZmBBC1* (GRMZM2G145280)	Positively regulating leaf angle and cell division.	[85]
	AtAPS3 (AT4G14680)	*ZmAS1* (GRMZM2G149952)	Positively regulating leaf angle and cell division.	[85]

Table 1. *Cont.*

Pathway	Arabidopsis	Maize/Rice	Functions in Leaf Angle [1]	Refs
SLs	*AtMAX2* (AT2G42620)	*OsD3* (LOC_Os06g06050)	Negatively regulating leaf angle and related to the SL signaling in rice.	[9]
	AtD14 (AT3G03990)	*OsD14* (LOC_Os03g10620)	Negatively regulating leaf angle and dual function as a receptor and deactivator of bioactive SLs, related to the SL signaling in rice.	[78]
	AtCCD8 (AT4G32810)	*OsD10* (LOC_Os01g54270)	Negatively regulating leaf angle and encode carotenoid cleavage dioxygenase (CCD) 8 related to the SL biosynthesis in rice.	[78]
	AtCCD7 (AT2G44990)	*OsD17* (LOC_Os04g46470)	Negatively regulating leaf angle and encode carotenoid cleavage dioxygenase (CCD) 7 related to the SL biosynthesis in rice.	[78]
	AtD27 (AT1G03055)	*OsD27* (Os11g0587000)	Negatively regulating leaf angle and encode β-carotene isomerase, related to the SL biosynthesis in rice.	[78]

[1] Role of the corresponding gene in regulating leaf angle in rice or maize. NA, none available.

7. Future Perspectives

Leaf angle directly influences the shape of plant architecture, consequentially affecting yield. To feed the ever-increasing global population, demand of higher crop yield has triggered the breeders to breed new cultivars that maintain sustainable productivity and yield by dense planting with limited arable lands. Therefore, how to genetically manipulating leaf angle has become one of the most important tasks to be tackled in crop genetic improvement. Until now, the regulatory mechanism underlying leaf angle has been extensively elucidated. However, a few issues are still beyond understanding, preventing the application of specific gene resource in term of crop improvement. To address these issues, the relationship among these genes still needs to be further uncovered, in particular the interaction network. On another hand, knockout or overexpression of these genes generally caused pleiotropic effects on plant growth and development in addition to the LA, such as plant dwarfism and smaller grain. Therefore, identification of interest single nucleotide polymorphisms (SNPs) and/or haplotypes of LA genes could not only be an efficient strategy to extent our knowledge about the relevant regulatory network, but also provide suitable alleles for marker selection breeding. The studies of *ZmLG1* provide an excellent case for bridging the basic research and application. *ZmLG1* was initially identified as a key factor for formation of ligule and auricle in maize and rice, and then regard to be an important player during rice domestication [47,90,91]. Further large-scale genetic analysis revealed that *ZmLG1* was the major QTL controlling leaf angle in maize [92–95]. Based on these studies, a recent study demonstrated that genetic manipulation of *ZmLG1* gene can significantly increase photosynthetic efficiency and maize yield at higher planting density [96]. Therefore, genome resequencing of larger accessions in crop indeed facilitates deciphering interest haplotypes of LA regulators. Alternatively, generation of novel alleles by CRISPR-mediated gene editing is also a powerful and efficient approach regarding this issue. Nevertheless, it would be fascinating to ask if there are novel regulators involved in LA regulation independent on the phytohormones pathways, which might only modulate LA without pleiotropic effects on other traits. On the other hand, it is also attractive to see whether and how LA coordinates or integrates with other agronomic traits in promoting yield, since yield consists of numerous components, such as plant height and tiller number. Furthermore, manipulating the spatial and temporal expression pattern of LA genes may be also important for timely shaping the plant architecture to achieve optimal photosynthesis, as well as eliminating shade avoidance and neighbor interference.

In conclusion, BR signaling pathway performs along with other plant hormones to form a complex signaling crosstalk to coordinate LA architecture under regular growth condition and in response to environmental stimuli. The well-established regulatory network for LA would provide vast promising targets to be manipulated, no matter through traditional molecular marker-assisted breeding or the gene editing technology, for crop architecture improvement to pursue high production.

Author Contributions: Writing—original draft preparation, X.L., R.S. and Q.X.; writing—review and editing, X.L., P.W., Y.L., Z.Z., R.S. and Q.X.; visualization, X.L., P.W., Y.L., Z.Z. and S.G.; supervision, Q.X.; project administration, S.G. and Q.X.; funding acquisition, R.S. and Q.X. All authors have read and agreed to the published version of the manuscript.

Funding: This work was supported by the Major Program of Guangdong Basic and Applied Research (2019B030302006), the National Natural Science Foundation of China (31971920) and the support from the "Top Young Scientist of the Pearl River Talent Plan" (No. 20170104) to Qingjun Xie, as well as the National Natural Science Foundation of China (31771739) and the Natural Science Foundation of Guangdong Province-Guangzhou City Collaborative Key Project (2019B1515120061) to Rongxin Shen.

Acknowledgments: We are grateful to the supports of experimental platform and funding from the Guangdong Provincial Key Laboratory of Plant Molecular Breeding (No. GPKLPMB201804). We apologize in advance to colleagues whose valuable work was not cited due to article length considerations.

Conflicts of Interest: The authors declare no conflict of interest.

Abbreviations

LA	Leaf angle
BR	Brassinosteroid
IAA	Indoleacetic acid
GA	Gibberellin
SLs	Strigolactones
JA	Jasmonic acid
ABA	Abscisic acid

References

1. Duvick, D.N. The Contribution of breeding to yield advances in maize (*Zea mays* L.). *Adv. Agron.* **2005**, *86*, 83–145. [CrossRef]
2. Duvick, D.N. Genetic progress in yield of United States maize (*Zea mays* L.). *Maydica* **2005**, *50*, 193–202.
3. Leivar, P.; Tepperman, J.M.; Cohn, M.M.; Monte, E.; Al-Sady, B.; Erickson, E.; Quail, P.H. Dynamic antagonism between phytochromes and pif family basic helix-loop-helix factors induces selective reciprocal responses to light and shade in a rapidly responsive transcriptional network in arabidopsis. *Plant Cell* **2012**, *24*, 1398–1419. [CrossRef] [PubMed]
4. Khush, G.S. Green revolution: Preparing for the 21st century. *Genome* **1999**, *42*, 646–655. [CrossRef] [PubMed]
5. Sasaki, A.; Ashikari, M.; Ueguchi-Tanaka, M.; Itoh, H.; Nishimura, A.; Swapan, D.; Ishiyama, K.; Saito, T.; Kobayashi, M.; Khush, G.S.; et al. Green revolution: A mutant gibberellin-synthesis gene in rice. *Nature* **2002**, *416*, 701–702. [CrossRef] [PubMed]
6. Wu, K.; Wang, S.; Song, W.; Zhang, J.; Wang, Y.; Liu, Q.; Yu, J.; Ye, Y.; Li, S.; Chen, J.; et al. Enhanced sustainable green revolution yield via nitrogen-responsive chromatin modulation in rice. *Science* **2020**, *367*, eaaz2046. [CrossRef] [PubMed]
7. Sharman, B.C. Developmental anatomy of the shoot of *Zea mays* L. *Ann. Bot.* **1942**, *6*, 245–282. [CrossRef]
8. Sylvester, A.W.; Cande, W.Z.; Freeling, M. Division and differentiation during normal and liguleless-1 maize leaf development. *Development* **1990**, *110*, 985–1000.
9. Mantilla-Perez, M.B.; Fernandez, M.G.S. Differential manipulation of leaf angle throughout the canopy: Current status and prospects. *J. Exp. Bot.* **2017**, *68*, 5699–5717. [CrossRef]
10. Duvick, D.N.; Cassman, K. Post-green revolution trends in yield potential of temperate maize in the north-central united states. *Crop. Sci.* **1999**, *39*, 1622–1630. [CrossRef]
11. Sakamoto, T.; Morinaka, Y.; Ohnishi, T.; Sunohara, H.; Fujioka, S.; Ueguchi-Tanaka, M.; Mizutani, M.; Sakata, K.; Takatsuto, S.; Yoshida, S.; et al. Erect leaves caused by brassinosteroid deficiency increase biomass production and grain yield in rice. *Nat. Biotechnol.* **2005**, *24*, 105–109. [CrossRef] [PubMed]
12. Sinclair, T.R. Erect leaves and photosynthesis in rice. *Science* **1999**, *283*, 1456. [CrossRef]

13. Kong, F.; Zhang, T.; Liu, J.; Heng, S.; Shi, Q.; Zhang, H.; Wang, Z.; Ge, L.; Li, P.; Lu, X.; et al. Regulation of leaf angle by auricle development in maize. *Mol. Plant* **2017**, *10*, 516–519. [CrossRef]
14. Luo, X.; Zheng, J.; Huang, R.; Huang, Y.; Wang, H.; Jiang, L.; Fang, X. Phytohormones signaling and crosstalk regulating leaf angle in rice. *Plant Cell Rep.* **2016**, *35*, 2423–2433. [CrossRef] [PubMed]
15. Zhou, L.; Xiao, L.; Xue, H.-W. Dynamic cytology and transcriptional regulation of rice lamina joint development. *Plant Physiol.* **2017**, *174*, 1728–1746. [CrossRef]
16. Hong, Z.; Ueguchi-Tanaka, M.; Matsuoka, M. Brassinosteroids and rice architecture. *J. Pestic. Sci.* **2004**, *29*, 184–188. [CrossRef]
17. Duan, K.; Li, L.; Hu, P.; Xu, S.-P.; Xu, Z.-H.; Xue, H.-W. A brassinolide-suppressed rice MADS-box transcription factor, OsMDP1, has a negative regulatory role in BR signaling. *Plant J.* **2006**, *47*, 519–531. [CrossRef]
18. Cao, H.; Chen, S. Brassinosteroid-induced rice lamina joint inclination and its relation to indole-3-acetic acid and ethylene. *Plant Growth Regul.* **1995**, *16*, 189–196. [CrossRef]
19. Wang, Z.-Y.; Bai, M.-Y.; Oh, E.; Zhu, J.-Y. Brassinosteroid signaling network and regulation of photomorphogenesis. *Annu. Rev. Genet.* **2012**, *46*, 701–724. [CrossRef]
20. Zhang, C.; Bai, M.-Y.; Chong, K. Brassinosteroid-mediated regulation of agronomic traits in rice. *Plant Cell Rep.* **2014**, *33*, 683–696. [CrossRef]
21. Tong, H.; Chu, C. Functional specificities of brassinosteroid and potential utilization for crop improvement. *Trends Plant Sci.* **2018**, *23*, 1016–1028. [CrossRef] [PubMed]
22. Wang, W.; Bai, M.-Y.; Wang, Z.-Y. The brassinosteroid signaling network — a paradigm of signal integration. *Curr. Opin. Plant Biol.* **2014**, *21*, 147–153. [CrossRef]
23. Wu, Y.; Fu, Y.; Zhao, S.; Gu, P.; Zhu, Z.; Sun, C.; Tan, L. Clustered primary branch 1, a new allele of DWARF11, controls panicle architecture and seed size in rice. *Plant Biotechnol. J.* **2015**, *14*, 377–386. [CrossRef] [PubMed]
24. Morinaka, Y.; Sakamoto, T.; Inukai, Y.; Agetsuma, M.; Kitano, H.; Ashikari, M.; Matsuoka, M. Morphological alteration caused by brassinosteroid insensitivity increases the biomass and grain production of rice1. *Plant Physiol.* **2006**, *141*, 924–931. [CrossRef] [PubMed]
25. Tanabe, S.; Ashikari, M.; Fujioka, S.; Takatsuto, S.; Yoshida, S.; Yano, M.; Yoshimura, A.; Kitano, H.; Matsuoka, M.; Fujisawa, Y.; et al. A novel cytochrome p450 is implicated in brassinosteroid biosynthesis via the characterization of a rice dwarf mutant, dwarf11, with reduced seed length. *Plant Cell* **2005**, *17*, 776–790. [CrossRef]
26. Hong, Z.; Ueguchi-Tanaka, M.; Fujioka, S.; Takatsuto, S.; Yoshida, S.; Hasegawa, Y.; Ashikari, M.; Kitano, H.; Matsuoka, M. The rice brassinosteroid-deficient dwarf2 mutant, defective in the rice homolog of arabidopsis DIMINUTO/DWARF1, is rescued by the endogenously accumulated alternative bioactive brassinosteroid, dolichosterone. *Plant Cell* **2005**, *17*, 2243–2254. [CrossRef]
27. Mori, M.; Nomura, T.; Ooka, H.; Ishizaka, M.; Yokota, T.; Sugimoto, K.; Okabe, K.; Kajiwara, H.; Satoh, K.; Yamamoto, K.; et al. Isolation and characterization of a rice dwarf mutant with a defect in brassinosteroid biosynthesis1. *Plant Physiol.* **2002**, *130*, 1152–1161. [CrossRef]
28. Hong, Z.; Ueguchi-Tanaka, M.; Shimizu-Sato, S.; Inukai, Y.; Fujioka, S.; Shimada, Y.; Takatsuto, S.; Agetsuma, M.; Yoshida, S.; Watanabe, Y.; et al. Loss-of-function of a rice brassinosteroid biosynthetic enzyme, C-6 oxidase, prevents the organized arrangement and polar elongation of cells in the leaves and stem. *Plant J.* **2002**, *32*, 495–508. [CrossRef]
29. Wang, L.; Xu, Y.-Y.; Ma, Q.-B.; Li, D.; Xu, Z.-H.; Chong, K. Heterotrimeric G protein α subunit is involved in rice brassinosteroid response. *Cell Res.* **2006**, *16*, 916–922. [CrossRef]
30. Sun, Y.; Fan, X.-Y.; Cao, D.-M.; Tang, W.; He, K.; Zhu, J.-Y.; He, J.-X.; Bai, M.-Y.; Zhu, S.; Oh, E.; et al. Integration of brassinosteroid signal transduction with the transcription network for plant growth regulation in Arabidopsis. *Dev. Cell* **2010**, *19*, 765–777. [CrossRef]
31. Wang, H.; Yang, C.; Zhang, C.; Wang, N.; Lu, D.; Wang, J.; Zhang, S.; Wang, Z.-X.; Ma, H.; Wang, X. Dual Role of BKI1 and 14-3-3 s in Brassinosteroid Signaling to Link Receptor with Transcription Factors. *Dev. Cell* **2011**, *21*, 825–834. [CrossRef] [PubMed]
32. Tang, W.; Yuan, M.; Wang, R.; Yang, Y.; Wang, C.; Oses-Prieto, J.A.; Kim, T.-W.; Zhou, H.-W.; Deng, Z.; Gampala, S.S.; et al. PP2A activates brassinosteroid-responsive gene expression and plant growth by dephosphorylating BZR1. *Nat. Cell Biol.* **2011**, *13*, 124–131. [CrossRef] [PubMed]

33. Yamamuro, C.; Ihara, Y.; Wu, X.; Noguchi, T.; Fujioka, S.; Takatsuto, S.; Ashikari, M.; Kitano, H.; Matsuoka, M. Loss of function of a rice brassinosteroid insensitive1 homolog prevents internode elongation and bending of the lamina joint. *Plant Cell* **2000**, *12*, 1591–1605. [CrossRef] [PubMed]
34. Li, D.; Wang, L.; Wang, M.; Xu, Y.-Y.; Luo, W.; Liu, Y.-J.; Xu, Z.-H.; Li, J.; Chong, K. EngineeringOsBAK1gene as a molecular tool to improve rice architecture for high yield. *Plant Biotechnol. J.* **2009**, *7*, 791–806. [CrossRef] [PubMed]
35. Bai, M.-Y.; Zhang, L.-Y.; Gampala, S.S.; Zhu, S.-W.; Song, W.-Y.; Chong, K.; Wang, Z.-Y. Functions of OsBZR1 and 14-3-3 proteins in brassinosteroid signaling in rice. *Proc. Natl. Acad. Sci. USA* **2007**, *104*, 13839–13844. [CrossRef] [PubMed]
36. Zhang, C.; Xu, Y.; Guo, S.; Zhu, J.; Huan, Q.; Liu, H.; Wang, L.; Luo, G.; Wang, X.; Chong, K. Dynamics of brassinosteroid response modulated by negative regulator LIC in rice. *PLoS Genet.* **2012**, *8*, e1002686. [CrossRef]
37. Qiao, S.; Sun, S.; Wang, L.; Wu, Z.; Li, C.; Li, X.; Wang, T.; Leng, L.; Tian, W.; Lu, T.; et al. The RLA1/SMOS1 transcription factor functions with OsBZR1 to regulate brassinosteroid signaling and rice architecture. *Plant Cell* **2017**, *29*, 292–309. [CrossRef]
38. Hirano, K.; Yoshida, H.; Aya, K.; Kawamura, M.; Hayashi, M.; Hobo, T.; Sato-Izawa, K.; Kitano, H.; Ueguchi-Tanaka, M.; Matsuoka, M. Small organ size 1 and small organ size 2/dwarf and low-tillering form a complex to integrate auxin and brassinosteroid signaling in rice. *Mol. Plant* **2017**, *10*, 590–604. [CrossRef]
39. Tanaka, A.; Nakagawa, H.; Tomita, C.; Shimatani, Z.; Ohtake, M.; Nomura, T.; Jiang, C.-J.; Dubouzet, J.G.; Kikuchi, S.; Sekimoto, H.; et al. Brassinosteroid upregulated1, encoding a Helix-Loop-Helix Protein, Is a novel gene involved in brassinosteroid signaling and controls bending of the lamina joint in rice1. *Plant Physiol.* **2009**, *151*, 669–680. [CrossRef]
40. Zhang, L.-Y.; Bai, M.-Y.; Wu, J.; Zhu, J.-Y.; Wang, H.; Zhang, Z.; Wang, W.; Sun, Y.; Zhao, J.; Sun, X.; et al. Antagonistic HLH/bHLH transcription factors mediate brassinosteroid regulation of cell elongation and plant development in rice and Arabidopsis. *Plant Cell* **2009**, *21*, 3767–3780. [CrossRef]
41. Wang, K.; Li, M.-Q.; Chang, Y.-P.; Zhang, B.; Zhao, Q.-Z.; Zhao, W.-L. The basic helix-loop-helix transcription factor OsBLR1 regulates leaf angle in rice via brassinosteroid signalling. *Plant Mol. Biol.* **2020**, *102*, 589–602. [CrossRef] [PubMed]
42. Seo, H.; Kim, S.-H.; Lee, B.-D.; Lim, J.-H.; Lee, S.-J.; An, G.; Paek, N.-C. The rice basic Helix–Loop–Helix 79 (OsbHLH079) determines leaf angle and grain shape. *Int. J. Mol. Sci.* **2020**, *21*, 2090. [CrossRef] [PubMed]
43. Chen, L.; Xiong, G.; Cui, X.; Yan, M.; Xu, T.; Qian, Q.; Xue, Y.; Li, J.; Wang, Y. OsGRAS19 may be a novel component involved in the brassinosteroid signaling pathway in rice. *Mol. Plant* **2013**, *6*, 988–991. [CrossRef]
44. Lin, Z.; Yan, J.; Su, J.; Liu, H.; Hu, C.; Li, G.; Wang, F.; Lin, Y. Novel OsGRAS19 mutant, D26, positively regulates grain shape in rice (*Oryza sativa*). *Funct. Plant Biol.* **2019**, *46*, 857. [CrossRef]
45. Jiang, F.; Guo, M.; Yang, F.; Duncan, K.; Jackson, D.; Rafalski, A.; Wang, S.; Li, B. Mutations in an AP2 transcription factor-like gene affect internode length and leaf shape in maize. *PLoS ONE* **2012**, *7*, e37040. [CrossRef]
46. Kir, G.; Ye, H.; Nelissen, H.; Neelakandan, A.; Kusnandar, A.S.; Luo, A.; Inzé, D.; Sylvester, A.W.; Yin, Y.; Becraft, P.W. RNA interference knockdown of brassinosteroid insensitive1 in maize reveals novel functions for brassinosteroid signaling in controlling plant architecture. *Plant Physiol.* **2015**, *169*, 826–839. [CrossRef] [PubMed]
47. Moreno, M.A.; Harper, E.C.; Krueger, R.W.; Dellaporta, S.L.; Freeling, M. liguleless1 encodes a nuclear-localized protein required for induction of ligules and auricles during maize leaf organogenesis. *Genes Dev.* **1997**, *11*, 616–628. [CrossRef]
48. Lee, J.; Park, J.; Kim, S.L.; Yim, J.; An, G. Mutations in the rice liguleless gene result in a complete loss of the auricle, ligule, and laminar joint. *Plant Mol. Biol.* **2007**, *65*, 487–499. [CrossRef]
49. Tian, J.; Wang, C.; Xia, J.; Wu, L.; Xu, G.; Wu, W.; Li, D.; Qin, W.; Han, X.; Chen, Q.; et al. Teosinte ligule allele narrows plant architecture and enhances high-density maize yields. *Science* **2019**, *365*, 658–664. [CrossRef]
50. Liu, K.; Cao, J.; Yu, K.; Liu, X.; Gao, Y.; Chen, Q.; Zhang, W.; Peng, H.; Du, J.; Xin, M.; et al. Wheat TaSPL8 modulates leaf angle through auxin and brassinosteroid signaling. *Plant Physiol.* **2019**, *181*, 179–194. [CrossRef]

51. Ren, Z.; Wu, L.; Ku, L.; Wang, H.; Zeng, H.; Su, H.; Wei, L.; Dou, D.; Liu, H.; Cao, Y.; et al. ZmILI1 regulates leaf angle by directly affecting liguleless1 expression in maize. *Plant Biotechnol. J.* **2019**, *18*, 881–883. [CrossRef]
52. Fellner, M.; Horton, L.A.; Cocke, A.E.; Stephens, N.R.; Ford, E.D.; Van Volkenburgh, E. Light interacts with auxin during leaf elongation and leaf angle development in young corn seedlings. *Planta* **2003**, *216*, 366–376. [CrossRef]
53. Mockaitis, K.; Estelle, M. Auxin receptors and plant development: A new signaling paradigm. *Annu. Rev. Cell Dev. Biol.* **2008**, *24*, 55–80. [CrossRef]
54. Hagen, G. Auxin signal transduction. *Essays Biochem.* **2015**, *58*, 1–12. [CrossRef] [PubMed]
55. Vanneste, S.; Friml, J. Auxin: A trigger for change in plant development. *Cell* **2009**, *136*, 1005–1016. [CrossRef] [PubMed]
56. Lavy, M.; Estelle, M. Mechanisms of auxin signaling. *Development* **2016**, *143*, 3226–3229. [CrossRef]
57. Yoshikawa, T.; Ito, M.; Sumikura, T.; Nakayama, A.; Nishimura, T.; Kitano, H.; Yamaguchi, I.; Koshiba, T.; Hibara, K.-I.; Nagato, Y.; et al. The rice FISH BONE gene encodes a tryptophan aminotransferase, which affects pleiotropic auxin-related processes. *Plant J.* **2014**, *78*, 927–936. [CrossRef] [PubMed]
58. Zhao, S.-Q.; Xiang, J.-J.; Xue, H.-W. Studies on the rice leaf inclination1 (lc1), an iaa–amido synthetase, reveal the effects of auxin in leaf inclination control. *Mol. Plant* **2013**, *6*, 174–187. [CrossRef] [PubMed]
59. Bian, H.; Xie, Y.; Guo, F.; Han, N.; Ma, S.; Zeng, Z.; Wang, J.; Yang, Y.; Zhu, M. Distinctive expression patterns and roles of the miRNA393/TIR1 homolog module in regulating flag leaf inclination and primary and crown root growth in rice (*Oryza sativa*). *New Phytol.* **2012**, *196*, 149–161. [CrossRef] [PubMed]
60. Liu, X.; Yang, C.Y.; Miao, R.; Zhou, C.L.; Cao, P.H.; Lan, J.; Zhu, X.J.; Mou, C.L.; Huang, Y.S.; Liu, S.J.; et al. DS1/OsEMF1 interacts with OsARF11 to control rice architecture by regulation of brassinosteroid signaling. *Rice* **2018**, *11*, 46. [CrossRef] [PubMed]
61. Chen, S.-H.; Zhou, L.; Xu, P.; Xue, H.-W. SPOC domain-containing protein Leaf inclination3 interacts with LIP1 to regulate rice leaf inclination through auxin signaling. *PLoS Genet.* **2018**, *14*, e1007829. [CrossRef] [PubMed]
62. Wei, L.; Zhang, X.; Zhang, Z.; Liu, H.; Lin, Z. A new allele of the Brachytic2 gene in maize can efficiently modify plant architecture. *Heredity* **2018**, *121*, 75–86. [CrossRef] [PubMed]
63. Li, P.; Wei, J.; Wang, H.; Fang, Y.; Yin, S.; Xu, Y.; Liu, J.; Yang, Z.; Xu, C. Natural variation and domestication selection of zmpgp1 affects plant architecture and yield-related traits in maize. *Genes* **2019**, *10*, 664. [CrossRef] [PubMed]
64. Shimada, A.; Ueguchi-Tanaka, M.; Sakamoto, T.; Fujioka, S.; Takatsuto, S.; Yoshida, S.; Sazuka, T.; Ashikari, M.; Matsuoka, M. The riceSPINDLYgene functions as a negative regulator of gibberellin signaling by controlling the suppressive function of the DELLA protein, SLR1, and modulating brassinosteroid synthesis. *Plant J.* **2006**, *48*, 390–402. [CrossRef] [PubMed]
65. Zentella, R.; Sui, N.; Barnhill, B.; Hsieh, W.-P.; Hu, J.; Shabanowitz, J.; Boyce, M.; Olszewski, N.E.; Zhou, P.; Hunt, N.F.; et al. The Arabidopsis O-fucosyltransferase spindly activates nuclear growth repressor DELLA. *Nat. Methods* **2017**, *13*, 479–485. [CrossRef]
66. Wan, H.; Wang, J.; Liang, S.; Fang, H.; Xiao, Z. Estimating leaf area index by fusing MODIS and MISR data. In Proceedings of the 2006 IEEE International Symposium on Geoscience and Remote Sensing, Denver, CO, USA, 31 July–4 August 2006; Volume 29, pp. 1820–1823. [CrossRef]
67. Wei, L.; Gu, L.; Song, X.; Cui, X.; Lu, Z.; Zhou, M.; Wang, L.; Hu, F.; Zhai, J.; Meyers, B.C.; et al. Dicer-like 3 produces transposable element-associated 24-nt siRNAs that control agricultural traits in rice. *Proc. Natl. Acad. Sci. USA* **2014**, *111*, 3877–3882. [CrossRef]
68. Tang, Y.; Liu, H.; Guo, S.; Wang, B.; Li, Z.; Chong, K.; Xu, Y. OsmiR396d affects gibberellin and brassinosteroid signaling to regulate plant architecture in rice. *Plant Physiol.* **2017**, *176*, 946–959. [CrossRef]
69. Attia, K.A.; Abdelkhalik, A.F.; Ammar, M.H.; Wei, C.; Yang, J.; Lightfoot, D.A.; El-Sayed, W.M.; El-Shemy, H.A. Antisense phenotypes reveal a functional expression of OsARF1, an auxin response factor, in transgenic rice. *Curr. Issues Mol. Biol.* **2009**, *11*.
70. Song, Y.; You, J.; Xiong, L. Characterization of OsIAA1 gene, a member of rice Aux/IAA family involved in auxin and brassinosteroid hormone responses and plant morphogenesis. *Plant Mol. Biol.* **2009**, *70*, 297–309. [CrossRef]

71. Zhang, S.; Wang, S.; Xu, Y.; Yu, C.; Shen, C.; Qian, Q.; Geißler, M.; Jiang, D.A.; Qi, Y. The auxin response factor, OsARF19, controls rice leaf angles through positively regulating OsGH 3-5 and OsBRI 1. *Plant Cell Environ.* **2014**, *38*, 638–654. [CrossRef]
72. Ferrero-Serrano, Á.; Assmann, S.M. The α-subunit of the rice heterotrimeric G protein, RGA1, regulates drought tolerance during the vegetative phase in the dwarf rice mutant d1. *J. Exp. Bot.* **2016**, *67*, 3433–3443. [CrossRef] [PubMed]
73. Oki, K.; Inaba, N.; Kitagawa, K.; Fujioka, S.; Kitano, H.; Fujisawa, Y.; Kato, H.; Iwasaki, Y. Function of the subunit of rice heterotrimeric g protein in brassinosteroid signaling. *Plant Cell Physiol.* **2008**, *50*, 161–172. [CrossRef]
74. Jang, S.; An, G.; Li, H.-Y. Rice leaf angle and grain size are affected by the OsBUL1 transcriptional activator complex. *Plant Physiol.* **2016**, *173*, 688–702. [CrossRef] [PubMed]
75. Tong, H.; Xiao, Y.; Liu, D.; Gao, S.; Liu, L.; Yin, Y.; Jin, Y.; Qian, Q.; Chu, C. Brassinosteroid regulates cell elongation by modulating gibberellin metabolism in rice. *Plant Cell* **2014**, *26*, 4376–4393. [CrossRef] [PubMed]
76. Li, H.; Wang, L.; Liu, M.; Dong, Z.; Li, Q.; Fei, S.; Xiang, H.; Liu, B.; Jin, W. Maize plant architecture is regulated by the ethylene biosynthetic gene ZmACS7. *Plant Physiol.* **2020**. [CrossRef]
77. Li, X.; Sun, S.; Li, C.; Qiao, S.; Wang, T.; Leng, L.; Shen, H.; Wang, X. The Strigolactone-related mutants have enhanced lamina joint inclination phenotype at the seedling stage. *J. Genet. Genom.* **2014**, *41*, 605–608. [CrossRef]
78. Shindo, M.; Yamamoto, S.; Shimomura, K.; Umehara, M. Strigolactones Decrease leaf angle in response to nutrient deficiencies in rice. *Front. Plant Sci.* **2020**, *11*, 135. [CrossRef]
79. Gan, L.; Wu, H.; Wu, D.; Zhang, Z.; Guo, Z.; Yang, N.; Xia, K.; Zhou, X.; Oh, K.; Matsuoka, M.; et al. Methyl jasmonate inhibits lamina joint inclination by repressing brassinosteroid biosynthesis and signaling in rice. *Plant Sci.* **2015**, *241*, 238–245. [CrossRef]
80. Hu, Y.; Yu, D. Brassinosteroid insensitive2 interacts with abscisic acid insensitive5 to mediate the antagonism of brassinosteroids to abscisic acid during seed germination in Arabidopsis. *Plant Cell* **2014**, *26*, 4394–4408. [CrossRef]
81. Zhao, X.; Dou, L.; Gong, Z.; Wang, X.; Liu, X. BES 1 hinders abscisic acid insensitive 5 and promotes seed germination in Arabidopsis. *New Phytol.* **2018**, *221*, 908–918. [CrossRef]
82. Wang, H.; Tang, J.; Liu, J.; Hu, J.; Liu, J.; Chen, Y.; Cai, Z.; Wang, X. Abscisic acid signaling inhibits brassinosteroid signaling through dampening the dephosphorylation of BIN2 by ABI1 and ABI2. *Mol. Plant* **2018**, *11*, 315–325. [CrossRef] [PubMed]
83. Li, Q.-F.; Lu, J.; Zhou, Y.; Wu, F.; Tong, H.; Wang, J.-D.; Yu, J.-W.; Zhang, C.-Q.; Fan, X.-L.; Liu, Q. Abscisic acid represses rice lamina joint inclination by antagonizing brassinosteroid biosynthesis and signaling. *Int. J. Mol. Sci.* **2019**, *20*, 4908. [CrossRef]
84. Dou, D.; Han, S.; Ku, L.; Liu, H.; Su, H.; Ren, Z.; Zhang, D.; Zeng, H.; Dong, Y.; Liu, Z.; et al. ZmCLA4 regulates leaf angle through multiple plant hormone-mediated signal pathways in maize 2020. *bioRxiv* **2020**. [CrossRef]
85. Cao, Y.; Zeng, H.; Ku, L.; Ren, Z.; Han, Y.; Su, H.; Dou, D.; Liu, H.; Dong, Y.; Zhu, F.; et al. ZmIBH1-1 regulates plant architecture in maize. *J. Exp. Bot.* **2020**, *71*, 2943–2955. [CrossRef] [PubMed]
86. Bao, F.; Shen, J.; Brady, S.R.; Muday, G.K.; Asami, T.; Yang, Z. Brassinosteroids interact with auxin to promote lateral root development in Arabidopsis. *Plant Physiol.* **2004**, *134*, 1624–1631. [CrossRef] [PubMed]
87. Yuldashev, R.; Avalbaev, A.; Bezrukova, M.; Vysotskaya, L.; Khripach, V.; Shakirova, F. Cytokinin oxidase is involved in the regulation of cytokinin content by 24-epibrassinolide in wheat seedlings. *Plant Physiol. Biochem.* **2012**, *55*, 1–6. [CrossRef]
88. Peleg, Z.; Reguera, M.; Tumimbang, E.; Walia, H.; Blumwald, E. Cytokinin-mediated source/sink modifications improve drought tolerance and increase grain yield in rice under water-stress. *Plant Biotechnol. J.* **2011**, *9*, 747–758. [CrossRef]
89. Yoshida, H.; Tanimoto, E.; Hirai, T.; Miyanoiri, Y.; Mitani, R.; Kawamura, M.; Takeda, M.; Takehara, S.; Hirano, K.; Kainosho, M.; et al. Evolution and diversification of the plant gibberellin receptor GID1. *Proc. Natl. Acad. Sci. USA* **2018**, *115*, E7844–E7853. [CrossRef]
90. Ishii, T.; Numaguchi, K.; Miura, K.; Yoshida, K.; Thanh, P.T.; Htun, T.M.; Yamasaki, M.; Komeda, N.; Matsumoto, T.; Terauchi, R.; et al. OsLG1 regulates a closed panicle trait in domesticated rice. *Nat. Genet.* **2013**, *45*, 462–465. [CrossRef]

91. Zhu, Z.; Tan, L.; Fu, Y.; Liu, F.; Cai, H.; Xie, D.; Wu, F.; Wu, J.; Matsumoto, T.; Sun, C. Genetic control of inflorescence architecture during rice domestication. *Nat. Commun.* **2013**, *4*, 2200. [CrossRef]
92. Ding, J.; Zhang, L.; Chen, J.; Li, X.; Li, Y.; Cheng, H.; Huang, R.; Zhou, B.; Li, Z.; Wang, J.; et al. Genomic dissection of leaf angle in maize (*Zea mays* L.) using a four-way cross mapping population. *PLoS ONE* **2015**, *10*, e0141619. [CrossRef] [PubMed]
93. Li, C.; Li, Y.; Shi, Y.; Song, Y.; Zhang, D.; Buckler, E.S.; Zhang, Z.; Wang, T.; Li, Y. Genetic control of the leaf angle and leaf orientation value as revealed by ultra-high density maps in three connected maize populations. *PLoS ONE* **2015**, *10*, e0121624. [CrossRef]
94. Tian, F.; Bradbury, P.J.; Brown, P.J.; Hung, H.; Sun, Q.; Flint-Garcia, S.; Rocheford, T.R.; McMullen, M.D.; Holland, J.B.; Buckler, E.S. Genome-wide association study of leaf architecture in the maize nested association mapping population. *Nat. Genet.* **2011**, *43*, 159–162. [CrossRef] [PubMed]
95. Wang, Y.; Xu, J.; Deng, D.; Ding, H.; Bian, Y.; Yin, Z.; Wu, Y.; Zhou, B.; Zhao, Y. A comprehensive meta-analysis of plant morphology, yield, stay-green, and virus disease resistance QTL in maize (*Zea mays* L.). *Planta* **2015**, *243*, 459–471. [CrossRef] [PubMed]
96. Li, C.; Liu, C.; Qi, X.; Wu, Y.; Fei, X.; Mao, L.; Cheng, B.; Li, X.; Xie, C. RNA-guided Cas9 as an in vivo desired-target mutator in maize. *Plant Biotechnol. J.* **2017**, *15*, 1566–1576. [CrossRef] [PubMed]

International Journal of
Molecular Sciences

MDPI

Review

Roles of ASYMMETRIC LEAVES2 (AS2) and Nucleolar Proteins in the Adaxial–Abaxial Polarity Specification at the Perinucleolar Region in Arabidopsis

Hidekazu Iwakawa [1], Hiro Takahashi [2], Yasunori Machida [3,*] and Chiyoko Machida [1,*]

1 Graduate School of Bioscience and Biotechnology, Chubu University, 1200, Matsumoto-cho, Kasugai, Aichi 487-8501, Japan; iwakawa@isc.chubu.ac.jp
2 Graduate School of Medical Sciences, Kanazawa University, Kakuma-machi, Kanazawa, Ishikawa 920-1192, Japan; takahasi@p.kanazawa-u.ac.jp
3 Division of Biological Science, Graduate School of Science, Nagoya University, Furo-cho, Chikusa-ku, Nagoya, Aichi 464-8602, Japan
* Correspondence: yas@bio.nagoya-u.ac.jp (Y.M.); cmachida@isc.chubu.ac.jp (C.M.); Tel.: +81-52-789-2502 (Y.M.); +81-568-51-6276 (C.M.)

Received: 21 August 2020; Accepted: 29 September 2020; Published: 3 October 2020

check for updates

Abstract: Leaves of Arabidopsis develop from a shoot apical meristem grow along three (proximal–distal, adaxial–abaxial, and medial–lateral) axes and form a flat symmetric architecture. *ASYMMETRIC LEAVES2* (*AS2*), a key regulator for leaf adaxial–abaxial partitioning, encodes a plant-specific nuclear protein and directly represses the abaxial-determining gene *ETTIN/AUXIN RESPONSE FACTOR3* (*ETT/ARF3*). How AS2 could act as a critical regulator, however, has yet to be demonstrated, although it might play an epigenetic role. Here, we summarize the current understandings of the genetic, molecular, and cellular functions of AS2. A characteristic genetic feature of *AS2* is the presence of a number of (about 60) modifier genes, mutations of which enhance the leaf abnormalities of *as2*. Although genes for proteins that are involved in diverse cellular processes are known as modifiers, it has recently become clear that many modifier proteins, such as NUCLEOLIN1 (NUC1) and RNA HELICASE10 (RH10), are localized in the nucleolus. Some modifiers including ribosomal proteins are also members of the small subunit processome (SSUP). In addition, AS2 forms perinucleolar bodies partially colocalizing with chromocenters that include the condensed inactive 45S ribosomal RNA genes. AS2 participates in maintaining CpG methylation in specific exons of *ETT/ARF3*. *NUC1* and *RH10* genes are also involved in maintaining the CpG methylation levels and repressing *ETT/ARF3* transcript levels. AS2 and nucleolus-localizing modifiers might cooperatively repress *ETT/ARF3* to develop symmetric flat leaves. These results raise the possibility of a nucleolus-related epigenetic repression system operating for developmental genes unique to plants and predict that AS2 could be a molecule with novel functions that cannot be explained by the conventional concept of transcription factors.

Keywords: *ASYMMETRIC LEAVES2*; AS2/LOB domain; adaxial–abaxial polarity; *ETTIN/AUXIN RESPONSE FACTOR3* (*ETT/ARF3*); AS2 body; nucleolus; gene body methylation; ribosomal DNA (rDNA)

---

1. Leaf Developments in Arabidopsis

Leaves develop from a shoot apical meristem (SAM) as lateral organs along three axes: proximal–distal, adaxial–abaxial, and medial–lateral [1–7]. Initially, groups of cells on the peripheral zone of the SAM are specified in leaf primordia (P0, Figure 1) and grow along the proximal–distal axis (P1). Then, adaxial–abaxial structures are differentiated (P2). Subsequently, cells proliferate along the medial–lateral axis leading to flat and symmetric leaves (Figure 1) [2,3,8]. To date, numerous genes involved in adaxial–abaxial determination have been reported in *Arabidopsis thaliana* [2,9]. The *ASYMMETRIC LEAVES2* (*AS2*) and *ASYMMETRIC LEAVES1* (*AS1*), which encode a protein with the plant-specific AS2/LOB domain and a protein with the MYB (SANT) domain, respectively, were originally identified as factors involved in symmetric leaf lamina formation [10–13]. Recent studies have revealed, however, that AS2 and AS1 regulate proper morphology along all three axes of leaves. The *Rough Sheath2* (*RS2*) gene of maize, an ortholog of *PHANTASTICA* (*PHAN*) of *Antirrhinum majus* and *AS1* of Arabidopsis, is involved in the proximal–distal patterning of maize leaves through the repression of class 1 *KNOX* genes [10,14,15]. The *PHAN* gene is involved in growth and the adaxial–abaxial determination of lateral organs. In addition, its activity is required early in the growth of leaves in the direction of the proximal–distal axis [16,17]. The ectopic expression of class 1 *Knotted1-like homeobox* (*KNOX*) genes in *as1* and *as2* mutant plants results in reductions in the growth of leaf blades and petioles in Arabidopsis, and these phenotypes are suppressed by mutations of the class 1 *KNOX* genes, *brevipedicellus* (*bp*), *knat2*, and *knat6*. These results indicate that the *AS1* and *AS2* genes of Arabidopsis are involved in the establishment of the proximal–distal axis through the repression of the class 1 *KNOX* genes [18]. In addition, the formation of shorter petioles and leaf blades in *as1* and *as2* is due to repression of gibberellin-synthetic genes by the upregulation of *BP/KNAT1, KNAT2,* and *KNAT6* [18]. AS1, acting together with AS2, directly represses the expression of the *BP* and *KNAT2* genes [19]. In this review, we focus on the establishment of leaf adaxial–abaxial polarity.

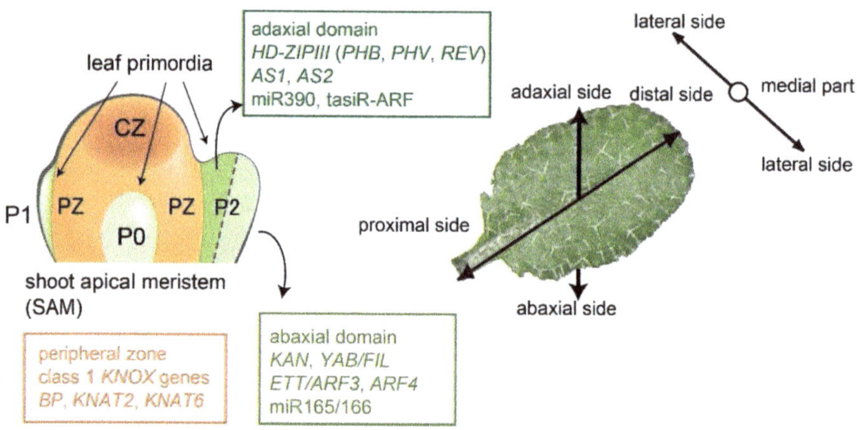

Figure 1. The leaf structure develops along three axes. Developmental compartments in the shoot apex around the apical meristem and the three structural leaf axes are schematically shown on the left and right sides, respectively (see details in text). CZ, central zone; PZ, peripheral zone; p0, primordium 0; p1, primordium 1; p2, primordium 2. Schematic representations are modified from ref. [2].

The *PHABULOSA* (*PHB*), *PHAVOLUTA* (*PHV*), and *REVOLUTA* (*REV*) genes encode class III homeodomain-leucine zipper (HD-ZIPIII) proteins, which determine adaxial cell fate [20–22]. Small RNAs play critical roles in specifying adaxial–abaxial polarity [23,24]. Micro RNAs miR165/166 promote the degradation of *HD-ZIPIII* transcripts in the abaxial domain, which results in the accumulation of HD-ZIPIII in the adaxial domain [24]. Members of the *KANADI* (*KAN*) gene family, which encode proteins with the GARP domain, determine abaxial cell fate [22,25]. The Arabidopsis genome contains six *YABBY* genes, which encode transcription factors with a zinc-finger domain and an HMG-related domain with a helix–loop–helix structure. The three *YABBY* genes, *FILAMENTOUS FLOWER* (*FIL*), *YABBY3* (*YAB3*), and *YAB2* are expressed in the abaxial domains of all leaf-derived organs, including cotyledons, leaves, and floral organs [26–30]. Furthermore, genetic analyses have shown that four *YABBY* genes (*FIL, YAB3, YAB2, YAB5*) govern embryo patterning and the growth of leaf lamina along the abaxial–adaxial boundary [30].

ETTIN/AUXIN RESPONSE FACTOR3 (*ETT/ARF3*) and *ARF4* also specify both abaxial cell fate and the lateral growth of leaf lamina [31]. Transcripts of both *ETT/ARF3* and *ARF4* are specifically degraded by the small RNA tasiR-ARF, which is generated through a miR390 pathway in the presumptive adaxial domain and contributes to the determination of the adaxial cell fate [23]. Because a loss of adaxial–abaxial polarity is often accompanied by a defect of leaf lamina expansion, it is suggested that the lateral growth of the lamina could be related to the determination of adaxial–abaxial identity, as previously proposed [2,32].

2. Roles of AS2–AS1 in the Development of Leaf Polarity

As described above, AS2 and AS1 proteins, which have AS2/LOB and R2R3 MYB (SANT) domains, respectively (Figure 2a), are identified by a yeast two-hybrid system, pull-down and gel-shift assays, and subcellular co-localization analyses. Because of their nature, these experiments indicate that AS2 and AS1 are physically associated with each other in vitro [19,33–36], implying also that they form a protein complex in the nucleus. Transcripts of *AS2* and *AS1* genes accumulate throughout the entire leaf primordia at early stages, in which the AS2–AS1 complex might be formed, but the accumulation patterns change as the leaves develop [37]. *AS2* transcripts are detected in the adaxial domain, while *AS1* transcripts are detected in the central region between the adaxial and abaxial domains of leaf primordia and the vasculature regions in more developed leaf primordia [10,37]. The plant-specific AS2/LOB domain includes a CXXC-type zinc-finger (ZF) motif, a leucine-zipper-like (LZL) region, and the internal-conserved-glycine (IcG) region between ZF and LZL (Figure 2a). The AS2/LOB domain is highly conserved in the AS2/LOB family, which consists of 42 members including AS2 in Arabidopsis [12,38,39]. Since the amino acid sequences outside of the domains are diverged among members and the transcription patterns of these genes differ for each gene, the roles of these genes in Arabidopsis development seem to be distinct. Members that might retain functions similar to those of the *AS2* gene do not appear to exist in this family, because the substitution of the AS2/LOB domain of AS2 with those of other members disrupts its function [39]. Considering similarities among the AS2/LOB domains, it is, however, undeniable that these family members may retain partially overlapping functions at the molecular level. They have often been described as transcription factors [40–47]. Recent results on AS2, however, suggest that the term "transcription factor" is not appropriate for a member of this family; they are better described as novel functional factors that could play a role in gene expression.

Transcriptome analyses of *as2* and *as1* mutants reveal that accumulations of *ETT/ARF3, KAN2,* and *YAB5* transcripts, all of which are related to the abaxial cell fate, are increased in *as2* and *as1*, whereas those of the adaxial domain-determining *HD-ZIPIII* are not changed [37,48]. A subsequent systematic analysis has revealed that *ETT/ARF3* is a direct target of the AS2–AS1 complex [49,50]. AS2–AS1 directly represses *ETT/ARF3* by binding to the upstream region of *ETT/ARF3*. Furthermore, AS2–AS1 indirectly represses *ETT/ARF3* via the tasiR-ARF pathway. AS2–AS1 induce the accumulation of miR390 involved in the generation of tasiR-ARF. Subsequently, both the *ETT/ARF3* and *ARF4* transcripts are degraded (Figure 2b). Therefore, the AS2–AS1 complex represses the expression of *ETT/ARF3* in the dual pathway [49]. Several phenotypes in *as2*, including defects of development along the adaxial–abaxial axis, are suppressed by the *ett arf4* double mutations. Consistent with these results, an overexpression of a tasiR-ARF-insensitive ETT/ARF3 cDNA produces *as2*-like leaves [51]. Similarly, lamina phenotypes of *as1* are also suppressed by the *ett arf4* double mutation. These results suggest that the elevated *ETT/ARF3* and *ARF4* expression in *as2* and *as1* cause several leaf phenotypes, including defects of adaxial–abaxial polarity in these mutants. The importance of the repression of these *ARFs* by AS2–AS1 is further confirmed by the analysis of modifier mutations of *as2* and *as1*, which are described in the next section. Increased expression levels of *KAN2* and *YAB5* in *as2* and *as1* are caused by indirect regulation by AS2–AS1 [49].

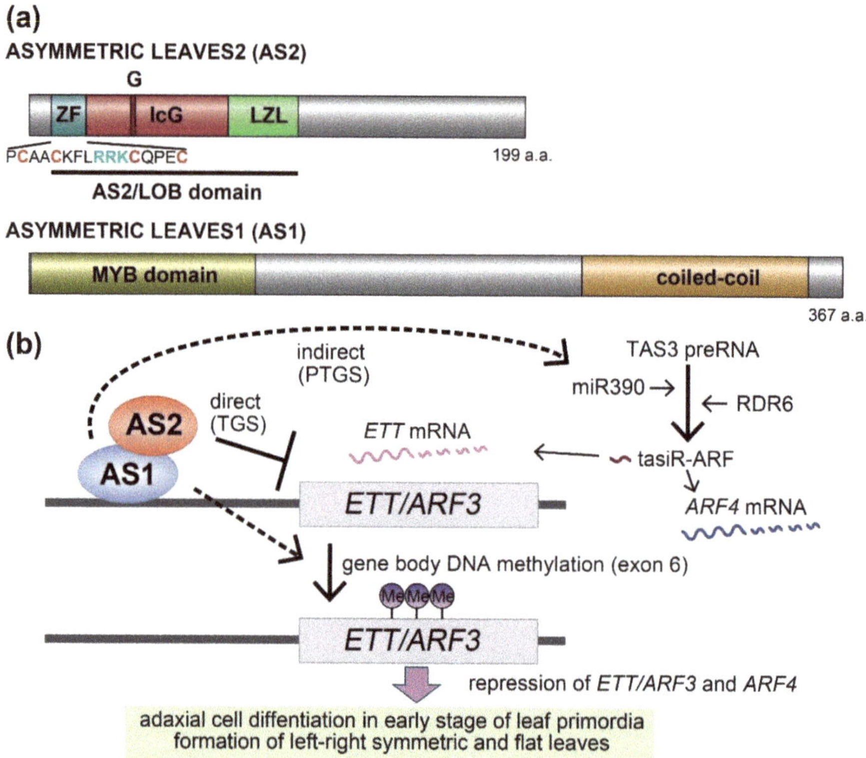

Figure 2. (**a**) Motif and domain organization of AS2 and AS1 proteins. The ZF motif, IcG, and LZL regions of AS2 and the MYB domain and coiled-coil structure of AS1 are shown. (**b**) Dual regulation of *ETT/ARF3* gene expression, including that by the possibly epigenetic system of AS2–AS1. The AS2–AS1 complex represses *ETT/ARF3* directly by binding to its promoter and represses *ETT/ARF3* and *ARF4* indirectly via stimulation of the miR390 and tasiR-ARF pathway. In addition, AS1 and AS2 maintain gene body DNA methylation of the *ETT/ARF3* gene. Solid lines indicate direct regulation and dashed black lines indicate indirect regulation. Schemes of (b) are modified from ref. [2].

3. Modifier Mutations That Enhance Defects of *AS2* and *AS1* in Leaf Adaxial–Abaxial Polarity

Various mutations (about 60) that markedly enhance the defects of adaxial leaf development in *as2* or *as1* have been reported [2]. The genes responsible for these mutations are considered as "modifiers" or modifier genes, which affect the phenotypic expression of other genes. Double mutants generate abaxialized filamentous (needle-like, pin-shaped, pointed) leaves that have lost the adaxial domain (Figure 3). Causative mutations occur in genes that are involved in chromatin modification, biogenesis of small RNAs, and DNA replication [2,52]. Mutations in genes encoding ribosomal proteins are also identified as modifiers in *as2* or *as1* [2,53–58]. In addition, mutations in genes encoding nucleolar proteins, such as *RNA HELICASE10* (*RH10*), *NUCLEOLIN1* (*NUC1*), *ROOT INITIATION DEFECTIVE2* (*RID2*), and *APUM23* are involved in ribosome biosynthesis, and enhance the phenotypes of *as2* and *as1* [59–63]. Mutations in *HDT1* and *HDT2* for nucleolar histone deacetylases (HDACs), which localize to the nucleolus, also act as modifiers of the *as2* and *as1* phenotypes [35]. We especially focus on the roles of nucleolar proteins in this review (Table 1).

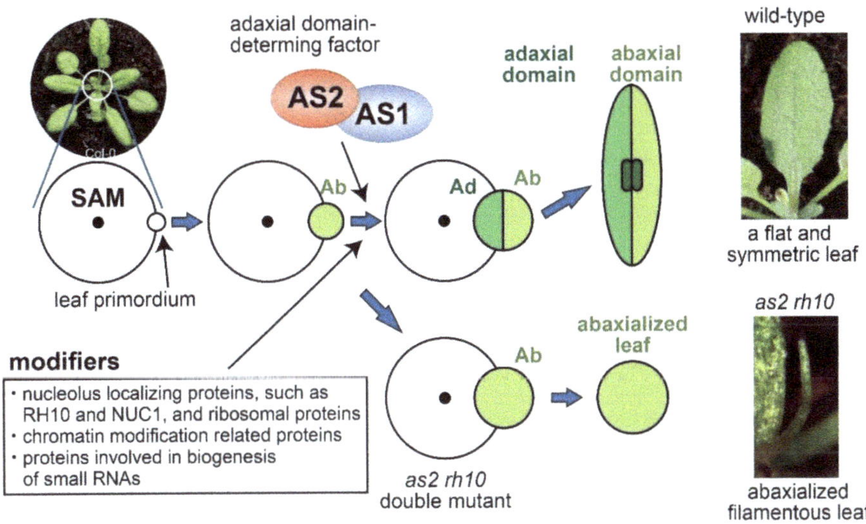

Figure 3. Development of leaves along with three axes. Top views of SAM are schematically shown by open circle. Dot indicates the center of the SAM. Adaxial and abaxial domains are shown by green and light green, respectively. AS2-AS1 contributes to the determination of adaxial domain followed by the medial-lateral growth of leaves with vasculature (indicated by dark green rectangles). Modifiers act cooperatively with AS2–AS1 at leaf primordia to develop the adaxial (Ad) domain from abaxialized (Ab) leaf primordia and to generate leaves with a flat and symmetric structure. The double mutation into AS2 (or AS1) and modifiers results in the production of abaxialized filamentous leaves. Photograph of wild-type leaf is modified from ref. [64].

Transcript levels of several abaxial-determining genes (*KAN2*, *YAB5*, *ETT/ARF3*, and *ARF4*) are slightly upregulated in the *as2-1* single mutant and each of the modifier single mutants and are markedly increased in the *as2-1* and modifier double mutants (for example, *as2-1 rh10-1*). When the double mutations of *ETT/ARF3* and *ARF4* are introduced to double mutants with *as2-1* and one of the modifier mutations, such as *as2-1 nuc1-1* or *as2-1 rh10-1*, the abaxialized filamentous leaves phenotype (e.g., *as2 rh10* leaves in Figure 3) is restored to the expanded shapes [59,64,65]. These results show that the upregulation of these *ARF* genes in the double mutants is responsible for the disappearance of their adaxial specification in their filamentous leaves. These genetic observations suggest that the repression of these *ARF* genes by the synergistic action of AS2–AS1 and modifier proteins is critical for the proper development of the adaxial domain. These results suggest that modifier proteins act cooperatively with AS2–AS1 to generate flat and symmetric leaves (Figure 3). The modifier genes that encode nucleolar proteins are summarized below.

Nucleoli are membrane-less organelles that appear to assemble through the phase separation of their molecular components [66]. The nucleoli contain internal subcompartments of ribosome biogenesis such as rDNA transcription, the processing of the precursor rRNA to generate mature rRNAs, assembly of these rRNAs, and many ribosomal proteins to generate each of small and large subunits of ribosomes. Genomic regions positioned in close proximity to the nucleolus are known as nucleolus-associated domains (NADs). Recent analyses of DNA sequencing that have been purified along with the nucleolus suggests that NADs in both animal and plant cells are enriched in regions displaying heterochromatic signatures [67,68].

***NUCLEOLIN1* (*NUC1*) gene**: Nucleolin, one of the most abundant non-ribosomal proteins in the nucleolus, has been described in a large variety of organisms [69]. The Arabidopsis genome encodes two nucleolin-like proteins—NUC1 and NUC2. Only the NUC1 gene, however, is ubiquitously expressed under normal growth conditions [61].

The single mutant *nuc1-1* exhibits a pointed narrow leaf shape, which is often observed in other modifier mutations [59,61,70,71]. In *nuc1-1* plants, nucleolar disorganization is observed and accumulated levels of pre-rRNA precursors are detected, indicating that NUC1 is involved in the processing of pre-rRNAs [61,72–74]. An analysis of high-throughput sequencing of DNA purified from the nucleoli of the *NUC1* mutant revealed that NUC1 is required for global genomic organization and stability [67,75]. In addition, human nucleolin is reported to be an assembly intermediate of the SSUP and its candidate components [62,76,77]. The *as2-1 nuc1-1* and *as1-1 nuc1-1* double mutant plants generate filamentous leaves. These mutant phenotypes are partially suppressed by the mutation in *ETT/ARF3*, indicating a role in the repression of *ETT/ARF3* gene expression for the formation of flat symmetric leaves in the wild-type plants [65].

***RNA HELICASE10* (*RH10*) gene**: The mutation of *rh10* was isolated as a modifier of *as2* and *as1*. Transcript levels of the abaxial genes, such as *ETT/ARF3* and *ARF4*, are elevated in *as2-1 rh10-1*, generating abaxialized filamentous leaves. This phenotype is suppressed by the *ett/arf3 arf4* double mutation, indicating a role in the repression of *ETT/ARF3* and *ARF4* gene expression for the formation of flat symmetric leaves in the wild-type plants [59]. RH10 is localized to the nucleolus in leaf primordia cells and is an ortholog of budding yeast Rrp3 and human DDX47, which belong to the DEAD-box RNA helicase family, a component of the nucleolar protein complex designated as the small subunit (SSUP) involved in 18S rRNA biogenesis [77,78]. It is reported that the DEAD-box RNA helicase family has an indispensable role in gene regulation through RNA metabolism [77–80]. DDX47 is necessary for maintaining the pluripotency of mouse stem cells [81]. In *rh10-1*, various defects are detected in SSUP-related events, such as the accumulation of 35S/33S rRNA precursors and a reduction in the 18S/25S ratio [59]. Nucleoli are enlarged in the *rh10-1* mutant [59]. RH10 may be involved in the early stages of processing reactions of the precursors of ribosomal RNAs.

***ROOT INITIATION DEFECTIVE2* (*RID2*) gene**: *RID2* encodes an evolutionarily conserved methyltransferase-like protein, an orthologous protein of the budding yeast, Bud23, which exhibits tight functional and physical interactions with some of the SSUP components [82–84]. The RID2 protein is localized in nuclei and accumulates mainly within nucleoli [60]. RID2 is involved in the processing of pre-rRNAs at various early stages [85,86]. Nucleolar enlargement is also observed in the *rid2* mutant. A mutation in the *RID2* gene has an effect on the adaxial–abaxial organization of leaves on the *as2* background, generating filamentous leaves and upregulating *ARF3/ETT* and *ARF4* as found in other modifier mutants and *as2-1* [59].

***APUM23* gene**: *APUM23*, which encodes a protein that is a member of the Pumilio/PUF domain protein family with its pumilio-like RNA-binding repeats, is localized to the nucleolus and is involved in the processing of 35S pre-rRNA [63,87]. The *apum23-1* mutant has enlarged nucleoli [63]. The double mutants *apum23-3 as2-2* and *apum23-3 as1-1* produce filamentous leaves, suggestive of the involvement of APUM23 in leaf development, similarly as with other nucleolar modifiers.

Ribosomal protein genes: It is worth noting that Arabidopsis double mutants of the ribosomal protein gene *rps6a-1*, which has a 9 bp deletion in the coding region of *RPS6A*, and *as2-1* exhibit strong adaxial leaf defects, as indicated by the fact that 80% of the double mutant leaves are filamentous [56]. Rps6 of budding yeast is one of five small-ribosomal-subunit proteins (Rps4, Rps6, Rps7, Rps9, and Rps14) that are components of the SSUP, which is a large ribonucleoprotein required for the biogenesis of the 18S rRNA [88]. Genetic interactions between *AS2/AS1* and homologues of *Rps4*, *Rps7*, *Rps9*, and Rps14 in Arabidopsis have yet to be examined. Three other *RPS* mutants (*rps23B*, *rps23B*, and *rps23B*) and fourteen *RPL* genes for ribosomal proteins in the large subunit also enhance the leaf phenotype in *as2* and/or *as1* (Table 1). It would also be intriguing to examine the relationships between the adaxial defects and ribosomal protein genes for such ribonucleoprotein complexes as a large subunit processome [89] in the nucleolus of Arabidopsis [53–57]. Therefore, the wild-type *AS2* gene, which is specific in plants, may appear to attenuate defects resulting from mutations in the ribosomal protein gene.

***HDT1* and *HDT2* genes**: *HDT1*, which encodes plant-specific nucleolar histone deacetylases (HDACs), is one of the factors responsible for gene silencing of megabase-scale rRNA loci and gene dosage control in nucleolar dominance [90,91], which are achieved by a highly condensed heterochromatic state that is associated with H3K9me2 and 5-methylcytosine enrichment in the promoter regions of rDNA genes [91]. Knockdown of the Arabidopsis genes *HDT1* and *HDT2* for nucleolar histone deacetylases (HDACs) enhances the leaf adaxial defects of *as2* and *as1* to generate severely abaxialized filamentous leaves, as seen in *as2-1 rh10-1* [35]. Considering the role of HDT1 in an epigenetic silencing of rDNAs (in nucleolar dominance), such as in the allopolyploid hybrid *Arabidopsis suecica* between *A. thaliana* and *A. arenosa* [91], the cooperative repression of the abaxial genes by AS2 and epigenetic silencing system of rDNAs described above are involved in the development of flat symmetric leaves.

Table 1. Gene mutations that act as modifiers to enhance leaf adaxial–abaxial abnormalities in *as2* and *as1*.

1. Gene Name (Mutant Name)	2. AGI Code	3. Protein	4. Cellular Process and Status	5.References
I. Genes involved in rRNA processing				
NUCLEOLIN1 (*nuc1*)	AT1G48920	NUCLEOLIN	rRNA processing and ribosome biogenesis Components of SSUP-like complex	[59,61,70,71]
RNA HELICASE10 (*rh10*)	AT5G60990	DEAD-box RNA helicase family protein	pre-rRNA processing Components of SSUP-like complex	[59]
ROOT INITIATION DEFECTIVE2 (*rid2*)	AT5G57280	RNA methyltransferase-like protein	pre-rRNA processing	[59,60]
APUM23 (*apum23*)	AT1G72320	Pumillio protein containing PUF domain	pre-rRNA processing and rRNA maturation	[63]
II. Genes for ribosomal proteins				
RPL4D (*rpl4d*)	AT5G02870	Ribosomal proteins	Subunits of ribosome; components of pre-rRNA-protein complex	[53–57]
RPL5A (*pgy3/ae6/oli5/rpl5a*)	AT3G25520			
RPL5B (*rpl5b/oli7*)	AT5G39740			
RPL7B (*rpl7b*)	AT2G01250			
RPL9C (*rpl9c/pgy2*)	AT1G33140			
RPL10aB (*rpl10ab/pgy1*)	AT2G27530			
RPL18C (*rpl18c*)	AT5G27850			
RPL24b (*stv1*)	AT3G53020			
RPL27ac (*rpl27ac*)	AT1G70600			
RPL28A (*ae5/rpl28a*)	AT2G19730			
PRL36aB (*api2*)	AT4G14320			
RPL36aA (*rpl36aa*)	AT3G59540			
RPL38B (*rpl38b*)	AT4G31985			
RPL39C (*rpl39c*)	AT3G23390			
RPS6A * (*rps6a*)	AT4G31700			
RPS21B (*rps21b*)	AT3G53890			
RPS24B (*rps24b*)	AT5G28060			
RPS28B (*rps28b*)	AT5G03850			
III. Genes involved in histone modification				
HDT1 (*hdt1/hd2a/hda3*)	AT3G44750	Histone deacetylase (plant-specific class)	Deacetylation of nucleosomal histone H3, transcription of rDNAs	[35,90,92]
HDT2 (*hdt2/hd2b*)	AT5G22650	Histone deacetylase (plant-specific class)	Deacetylation of nucleosomal histone H3, transcription of rDNAs	[35,90,92]

* Rps6 of budding yeast is one of the proteins that was identified as a bona fide component of the SSUP.

4. AS2 Bodies: Perinucleolar Granules Co-Localized Partially with the Chromocenter

The AS2-fused YFP (Yellow Fluorescent Protein) was used to investigate subnuclear localization of AS2 protein. The AS2 protein is localized to perinucleolar bodies known as AS2 bodies as well as to the nucleoplasm in the leaf cells of Arabidopsis and some interphase cells of a cultured tobacco cell line BY-2 (Figure 4) [35,93]. As mentioned in Section 2, AS2 has the AS2/LOB domain that includes ZF, IcG, and LZL regions (Figure 2a), which are essential for the formation of AS2 bodies at the perinucleolar regions [94]; the carboxyl-terminal half of AS2 is nonessential for the body formation, but essential for the developmental function of AS2 [12,93]; AS1 co-localizes with AS2 in the cell bodies (Figure 4) [35].

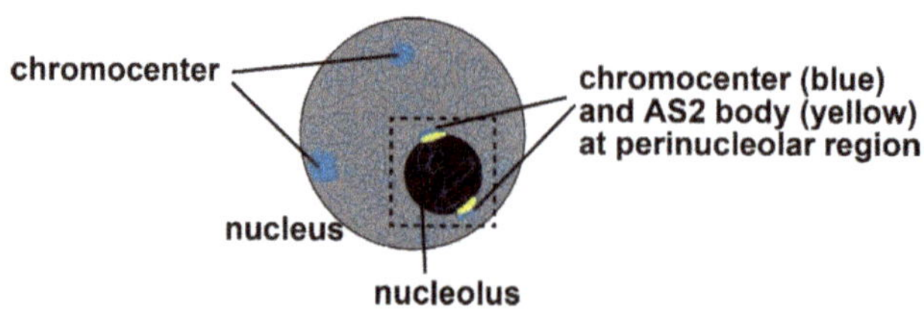

Figure 4. Schematic representation of the nucleus is shown. Chromosomes and AS2 bodies are indicated by blue and yellow, respectively. AS1 and AS2 are co-localized on AS2 bodies.

In addition, the amino acid residues that are highly conserved within and adjacent to the ZF motifs of all the AS2 family members are critically important for the body formation: four cysteine residues; proline and alanine residues next to the first cysteine residue; the RRK cluster (Figure 2). The RRK sequence is found within proposed nucleolar localization signals (NoLSs) [95–98] and it is likely that this cluster participates in the perinucleolar localization of AS2. These amino acid residues and three regions (ZF, IcG, and LZL) in the AS2/LOB domain are also required for the ability of AS2 to complement the *as2* mutation and to bind to the coding sequence of the target *ARF3/ETT* gene, showing that the formation of AS2 bodies is related to the genetic functions of AS2 in leaf formation. The AS2 bodies appear to be located to the peripheral regions of nucleoli and are partially overlapped with perinucleolar chromocenters with condensed chromatin-containing ribosomal RNA genes (45S rDNA repeats), suggesting that AS2 bodies interact with 45S rDNA repeats (Figure 4) [94].

It should be noted that the proportions of cells in which AS2 bodies are generated in plants differ from those in cultured cells. AS2 bodies are detected in only a few percentages of interphase cells of the tobacco-cultured cell line BY-2 and the Arabidopsis-cultured cell line MM2d transformed with the AS2-fused YFP constructs [93,94]. AS2 bodies are, however, detected in almost all interphase cells of the adaxial domain in leaf primordia of the Arabidopsis plants with the AS2-fused YFP construct [94]. The average number of AS2 bodies per YFP-positive cells at interphase (and/or the G0 stage) in leaf primordia was calculated as 1.9 [94]. In contrast, AS2 bodies are formed in all M phase cells of both cell lines, MM2d and BY-2 and in all M phase cells of leaf primordia; AS2 bodies are separated into daughter cells during the M phase progression [93,94]. These observations imply that the formation and distribution of AS2 bodies might be modulated developmentally in plants and in a cell-cycle-dependent manner.

The subcellular localization of AS2 appears to be subject to multiple controls, since AS2 was exported to the cytoplasm via the action of the geminivirus-encoded nuclear shuttle protein [12,94,99].

Although mechanisms for the formation of AS2 bodies and their roles in repressing the target genes for leaf development have yet to be discovered, the identification of AS2 body components and investigations of how these molecules interact within the nucleolus would provide answers for these questions.

5. AS2–AS1 Binds to Exon 1 of the Target Gene *ETT/ARF3*, and Is Involved in Maintaining CpG Methylation in Exon 6

Four mechanisms have been investigated for the repression of target gene *ETT/ARF3* expression by AS2–AS1: (1) direct binding of the AS1–AS2 complex to the 5′-upstream regions of *ETT/ARF3* to reduce the expression activity of *ETT/ARF3* (Figure 5) [49]; (2) indirect activation of miR390-dependent post-transcriptional gene silencing to negatively regulate both *ETT/ARF3* and *ARF4* (Figure 2b) [49,50]; (3) direct binding of AS2 to the synthetic GCGGCG-containing nucleotides [47,50], and exon 1 of the *ETT/ARF3* gene containing the CGCCGC (Figure 5) [65]; (4) maintenance of the status of gene body (CpG) methylation in exon 6 of ETT/ARF3 (Figure 5) [49,65,100]. In the present review, we focus on the last two topics.

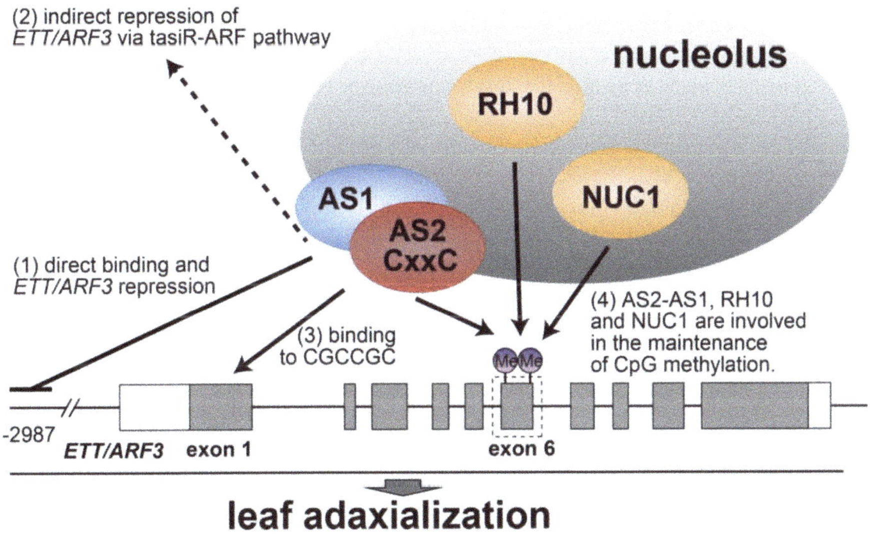

Figure 5. Molecular relationships between AS2–AS1 and the target gene *ETT/ARF3*. AS1–AS2 directly binds the *ETT/ARF3* regulatory region and represses *ETT/ARF3* expression. AS2 binds to the specific CGC repeat sequence in exon 1. AS2–AS1, RH10, and NUC1 are involved in the maintenance of CpG methylation in exon 6. RH10 and NUC1 are proteins localized in the nucleolus.

Several protein members of the AS2/LOB family, including AS2, bind synthetic double-stranded DNAs containing the GCGGCG sequence [101]. AS2, specifically, also binds in vitro the double-stranded CGCCGC sequence in exon 1 of the target gene *ETT/ARF3* [65]. The zinc-finger motif containing the RRK (Arg-Arg-Lys) sequence in AS2 is essential for this binding [65], the formation of AS2 bodies and functions in the development of leaves with normal shapes [94]. Modes of molecular interactions between the amino acid residues in RRK and each of the deoxyribonucleotides in GCGGCG have recently been proposed based on the results of SEC–SAXS (size exclusion chromatography–small angle X-ray scattering) experiments [47]. Since 32 out of 42 members of the family harbor the RRK and/or RRR sequence in the ZF motifs [12], it should be informative to investigate the possible roles of the clusters of these basic amino acid residues in other members in plant physiology, development, and growth [102].

AS2 and AS1 play a role in maintaining cytosine methylation mediated by METHYLTRANSFERASE1 (MET1) in six CpG dinucleotides in exon 6 of *ETT/ARF3* (Figure 5) [49]. Because levels of CpG methylation are inversely related to the *ETT/ARF3* transcript levels, AS2 and AS1 possibly regulate the transcriptional repression of *ETT/ARF3* through CpG methylation in the recruitment of methylation activity and/or inhibition of demethylation activity at exon 6 [103]. As described in Section 3, mutations in the *RH10*, *NUC1*, and *RID2* genes for nucleolar proteins enhance defects in leaf morphology in the *as2* mutant and, in parallel with this observation, result in an increase in the transcript level of target genes *ETT/ARF3* and *ARF4*. The levels of CpG methylation at some of the CpG dinucleotides in exon 6 of *ETT/ARF3* decrease in *rh10* and *nuc1* mutants, and further decrease in *rh10 as2* and *nuc1 as2*, suggesting that these nucleolar proteins, in addition to AS2, also take part in maintaining the cytosine methylation of CpG dinucleotides in exon 6 of *ETT/ARF3* [65,103].

How can AS2 be involved in maintaining MET1-regulated CpG methylation in exon 6 of the *ETT/ARF3* gene? MET1 is an ortholog of the Dnmt1 of vertebrates and acts as DNA methyltransferase, which methylates hemimethylated CpG, converting it to fully methylated CpG during DNA replication [104,105]. MET1 is part of a putative protein complex involved in the maintenance of DNA methylation in Arabidopsis [106–111]. MET1 is similar to Dnmt1, in terms of the domain organization [109,112], except that MET1 has no amino acid sequence for the ZF-CxxC motif. If AS2 forms a protein complex with MET1, AS2 provides the ZF motif, which has DNA binding activity, as described above in this section, to the MET1-containing putative protein complex. The promoter regions of inactive 45S rDNAs in Arabidopsis are highly methylated by MET1 and their chromatin states are highly condensed at perinucleolar regions [113,114]. MET1 requires NUC1, one of the AS2 modifiers, and nucleolar histone deacetylase HDA6 for this methylation [113,114] and directly interacts with HDA6 [115,116], which is also associated with AS2 and AS1 [36]. The CpG methylation system for the 45S rDNA might be also involved in CpG methylation in the *ETT/ARF3* gene around perinucleolar areas; the *ETT/ARF3* gene might be recruited to such an area by an action of AS2 (Figure 6).

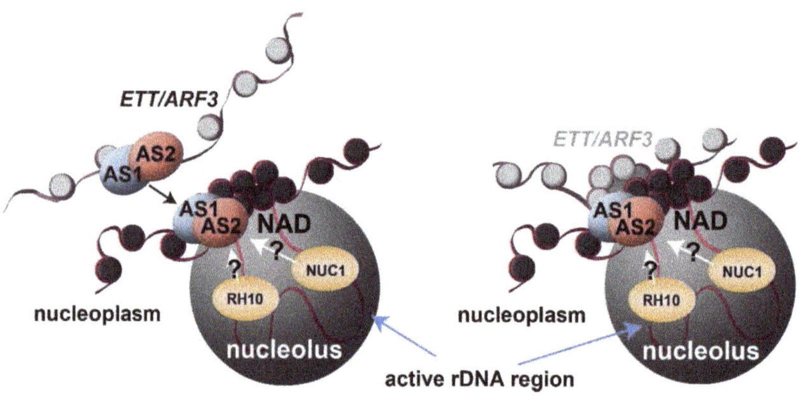

Figure 6. Models of the roles of nucleolar proteins in AS2–AS1 involved in epigenetic regulation of the *ETT/ARF3* gene. Nucleolus and surrounding structures are shown. Nucleosomes are indicated by coiled structure composed of red lines (DNA) and light/dark grey circle (histone octamer). Different genomes are distinguished by different darkness of nucleosomes. NUC1 affected the localization patterns of AS2 bodies at the peripheral region of the nucleolus, which are required for leaf development. AS2 bodies are partially overlapped with chromocenters, represented by dense nucleosome at the peripheral region of the nucleolus. The 45S rDNA repeat loci include transcriptionally active and inactive regions, which chromosomal status are loosened in the nucleolus and condensed on the peri-nucleolus (overlapping with chromocenter), respectively, suggestive of an interaction of AS2 bodies with inactive 45S rDNA. RH10 and NUC1 in addition to AS2 are involved in the maintenance of CpG methylation in exon 6 of *ETT/ARF3* in the nucleoplasm (left panel) or in AS2 bodies (right panel). NUC1 is involved in the maintenance of CpG methylation in 45S rDNA.

The AS2–AS1 complex is also involved in the establishment of leaf proximal–distal polarity to repress the class 1 *KNOX* homeobox genes *BREVIPEDICELLUS* (*BP*), *KNAT2*, and *KNAT6* (Figure 1) [18]. The AS2–AS1 complex physically interacts with CURLY LEAF (CLF), the polycomb repressive complex 2

(PRC2) core component, and LIKE-HETEROCHROMATIN PROTEIN1 (LHP1), the PRC1 component, and recruits PRC2 to the homeobox genes *BP* and *KNAT2* [117,118]. AS2–AS1 interacts with the *BP* promoter, likely through the recruitment of the chromatin-remodeling factor HIRA (histone-regulator A) and forms a repressive chromatin state [19]. AS2–AS1 also interacts with LEAF FLOWER RELATED (LFR) in the chromatin remodeling complex and is associated with H3K27me3 in the *BP* gene, but not with the *ETT/ARF3* gene [119]. AS2–AS1 is required for the correct temporal repression of *ETT/ARF3*, which involves a PRC2-independent mechanism [50]. Despite their pivotal role, the means by which AS2–AS1 epigenetically represses *ETT/ARF3* in the establishment of leaf adaxial–abaxial polarity remains unsolved.

6. Subcellular localization of AS2

Although it is often reported that AS2 and other members of the AS2/LOB family are nuclear proteins [2,12,35,38,93,120], mechanisms of the nuclear localization of AS2 protein are poorly understood. The RRK sequence in the zinc-finger motif of AS2 (Figure 2a) is only a basic amino acid cluster, which is thought to be critical for nuclear localization. The examination of subcellular localization of the mutant as2 (as2-RRK/3A) with the alanine replacement at the RRK sequence in the zinc-finger motif (Figure 2a) with the alanine cluster, however, shows that the mutant protein is still present in the nucleoplasm; it is not exported to the cytoplasm and does not form the AS2 bodies at the perinucleolar region [94]. These observations show that the RRK sequence is not involved in the nuclear localization of AS2. This result is consistent with the previous finding [121]: that is, the mutant proteins of ASL18/LBD16, another member of the AS2/LOB family, from which the RRK sequence is deleted are still localized to the nucleus and nuclear localization signals are proposed to be present in the coiled-coil sequence and the carboxyl-terminal region. Furthermore, the mutant as2 protein that lacks the ICG region is exclusively localized to the cytoplasm [94]. It is also reported that AS2 was exported to the cytoplasm via the action of the geminivirus-encoded nuclear shuttle protein and localized to the plant P-body complex [99]. Therefore, the AS2 protein might be subject to multiple subcellular localization controls, depending on its interactions with other proteins and other unknown cellular conditions.

As described above, 32 among the 42 members of the AS2/LOB family harbor ZF motifs, the amino acid sequences of which include RRK or RRR sequences [12,38]. The observation that the as2 mutant lacking RRK is localized to the nucleoplasm, but does not form AS2 bodies at the nucleolar periphery, suggests that RRK appears to be involved in the transport of AS2 to the perinucleolar region and/or to the formation of AS2 bodies. The RRK sequence is present within the proposed NoLSs [95–98]. It is intriguing to test whether this cluster of the basic amino acid residues in the zinc-finger could be directly involved in the transfer to the peripheral region of the nucleoli from the nucleoplasm and/or the formation of AS2 bodies by a phase separation mechanism, because the nucleolus and many nucleolar bodies are proposed to be formed through such a physico-chemical molecular interaction [122].

7. Possible Roles of AS2 Bodies in Epigenetic Repression of *ETT/ARF3*

As described in Section 3, the level of the leaf abaxial gene *ETT/ARF3* expression is influenced by modifier proteins that are localized to the nucleolus. For example, the AS2–AS1 complex binds directly to the upstream region of the *ETT/ARF3* gene to repress its transcription [49,64]. Furthermore, the *ETT/ARF3* transcriptional level is altered by mutations in various genes for nucleolus-localized proteins, such as RH10, RID2, and NUC1, which affect the biogenesis of ribosomal RNAs and the formation of the nucleolus with a properly organized morphology [59–61]. Perturbation of rRNA biogenesis correlates with structural disorders of the nucleolus, such as nucleolar enlargement in plant cellsand in animal cells [60,85,86,123]. It is, however, still unknown how structural disorders of the nucleolus affects leaf development mediated by AS2–AS1.

Perinucleolar regions might provide the molecular architectures, such as nucleolus-associated chromatin domains (NADs), which correspond to regions of low transcriptional levels [67,68,124,125]. In Arabidopsis, many of 45S rRNA genes are condensed as heterochromatin and silenced by epigenetic mechanisms that include DNA methylation and histone modification at the periphery of the nucleolus [114,126–128]. MET1, HDA6, and chromatin assembly factor (CAF-1) are all involved in the formation of such an epigenetic state in the perinucleolar subdomain [114,126,127]. AS2 and AS1, which are associated with HDA6 [36], are colocalized to AS2 bodies in the peripheral region of the nucleolus (Figures 4 and 6) [35,93,94]. The *ETT/ARF3* gene undergoes MET1-dependent CpG methylation in exon 6 [49]. As described in Sections 4 and 6, mutant proteins of AS2 (as2-RRK/3A, Figure 2a) that do not form AS2 bodies are not functional in leaf morphogenesis [94]. Mutations in RH10, RID2, and NUC1 might affect the integrity of nucleolar morphology [59–61], which would then alter the transcriptional patterns of *ETT/ARF3*, a target gene of AS2, although the subnucleolar localization of the target gene to the peripheral subdomain (Figure 6) has yet to be demonstrated. It is an interesting problem to elucidate how the *ETT/ARF3* gene transcribed by RNA polymerase II is regulated in the nucleolus or its peripheral region.

Recently, AS2 was shown to bind to DNAs other than *ETT/ARF3* [129]. Since AS2 is a plant-specific DNA binding protein, elucidation of the interaction between AS2 and these DNA molecules should uncover a more global and novel regulatory system mediated by AS2 and the nucleolus in plant cells [19,101].

Author Contributions: H.I. prepared the manuscript related to Sections 1, 2, 4 and 6. H.T. prepared the manuscript related to Section 5. Y.M. prepared the manuscript related to Sections 4 and 6. C.M. supervised all the research in this manuscript and prepared the manuscript related to Sections 1–3, 5 and 7. All authors have read and agreed to the published version of the manuscript.

Funding: This work was supported by the Japan Society for the Promotion of Science (JSPS) KAKENHI (grant numbers JP18H03330, JP18K06297, JP19K22892, JP20K06702 and 16H06279 (PAGS)) and the Research Foundation for the Electrotechnology of Chubu.

Acknowledgments: The authors are grateful to Takanori Suzuki at Ishihara Sangyo Kaisha LTD, Yuki Sakamoto at Osaka University, Sachihiro Matsunaga and Tetsuya Higashiyama at University of Tokyo, Daisuke Kurihara and Shogo Matsumoto at Nagoya University, Michiko Sasabe at Hirosaki University, and Shoko Kojima and Sayuri Ando at Chubu University for their helpful discussions.

Conflicts of Interest: The authors declare no conflict of interest.

Abbreviations

FISH	Fluorescence in situ Hybridization
GFP	Green Fluorescent Protein
BY-2	Bright Yellow-2
tasiR	trans-acting small interfering RNA

References

1. Bar, M.; Ori, N. Compound leaf development in model plant species. *Curr. Opin. Plant. Biol.* **2015**, *23*, 61–69. [CrossRef] [PubMed]
2. Machida, C.; Nakagawa, A.; Kojima, S.; Takahashi, H.; Machida, Y. The complex of ASYMMETRIC LEAVES (AS) proteins plays a central role in antagonistic interactions of genes for leaf polarity specification in Arabidopsis. *Wiley Interdiscip. Rev. Dev. Biol.* **2015**, *4*, 655–671. [CrossRef] [PubMed]
3. Fouracre, J.P.; Poethig, R.S. The role of small RNAs in vegetative shoot development. *Curr. Opin. Plant. Biol.* **2016**, *29*, 64–72. [CrossRef] [PubMed]
4. Du, F.; Guan, C.; Jiao, Y. Molecular Mechanisms of Leaf Morphogenesis. *Mol. Plant* **2018**, *11*, 1117–1134. [CrossRef] [PubMed]
5. Alvarez, J.P.; Furumizu, C.; Efroni, I.; Eshed, Y.; Bowman, J.L. Active suppression of a leaf meristem orchestrates determinate leaf growth. *eLife* **2016**, *5*. [CrossRef] [PubMed]
6. Conklin, P.A.; Strable, J.; Li, S.; Scanlon, M.J. On the mechanisms of development in monocot and eudicot leaves. *New Phytol.* **2019**, *221*, 706–724. [CrossRef]
7. Kuhlemeier, C.; Timmermans, M.C. The Sussex signal: Insights into leaf dorsiventrality. *Development* **2016**, *143*, 3230–3237. [CrossRef]

8. Satterlee, J.W.; Scanlon, M.J. Coordination of Leaf Development Across Developmental Axes. *Plants* **2019**, *8*, 433. [CrossRef]
9. Nakata, M.T.; Tameshige, T.; Takahara, M.; Mitsuda, N.; Okada, K. The functional balance between the WUSCHEL-RELATED HOMEOBOX1 gene and the phytohormone auxin is a key factor for cell proliferation in Arabidopsis seedlings. *Plant. Biotechnol.* **2018**, *35*, 141–154. [CrossRef]
10. Byrne, M.E.; Barley, R.; Curtis, M.; Arroyo, J.M.; Dunham, M.; Hudson, A.; Martienssen, R.A. Asymmetric leaves 1 mediates leaf patterning and stem cell function in Arabidopsis. *Nature* **2000**, *408*, 967–971. [CrossRef]
11. Semiarti, E.; Ueno, Y.; Tsukaya, H.; Iwakawa, H.; Machida, C.; Machida, Y. The ASYMMETRIC LEAVES2 gene of Arabidopsis thaliana regulates formation of a symmetric lamina, establishment of venation and repression of meristem-related homeobox genes in leaves. *Development* **2001**, *128*, 1771–1783. [PubMed]
12. Iwakawa, H.; Ueno, Y.; Semiarti, E.; Onouchi, H.; Kojima, S.; Tsukaya, H.; Hasebe, M.; Soma, T.; Ikezaki, M.; Machida, C.; et al. The ASYMMETRIC LEAVES2 gene of Arabidopsis thaliana, required for formation of a symmetric flat leaf lamina, encodes a member of a novel family of proteins characterized by cysteine repeats and a leucine zipper. *Plant Cell Physiol.* **2002**, *43*, 467–478. [CrossRef] [PubMed]
13. Lin, W.C.; Shuai, B.; Springer, P.S. The Arabidopsis LATERAL ORGAN BOUNDARIES-domain gene ASYMMETRIC LEAVES2 functions in the repression of KNOX gene expression and in adaxial-abaxial patterning. *Plant Cell* **2003**, *15*, 2241–2252. [CrossRef] [PubMed]
14. Timmermans, M.C.; Hudson, A.; Becraft, P.W.; Nelson, T. ROUGH SHEATH2: A Myb protein that represses knox homeobox genes in maize lateral organ primordia. *Science* **1999**, *284*, 151–153. [CrossRef]
15. Tsiantis, M.; Schneeberger, R.; Golz, J.F.; Freeling, M.; Langdale, J.A. The maize rough sheath2 gene and leaf development programs in monocot and dicot plants. *Science* **1999**, *284*, 154–156. [CrossRef]
16. Waites, R.; Selvadurai, H.R.; Oliver, I.R.; Hudson, A. The PHANTASTICA gene encodes a MYB transcription factor involved in growth and dorsoventrality of lateral organs in Antirrhinum. *Cell* **1998**, *93*, 779–789. [CrossRef]
17. Waites, R.; Hudson, A. The Handlebars gene is required with Phantastica for dorsoventral asymmetry of organs and for stem cell activity in Antirrhinum. *Development* **2001**, *128*, 1923–1931.
18. Ikezaki, M.; Kojima, M.; Sakakibara, H.; Kojima, S.; Ueno, Y.; Machida, C.; Machida, Y. Genetic networks regulated by ASYMMETRIC LEAVES1 (AS1) and AS2 in leaf development in Arabidopsis thaliana: KNOX genes control five morphological events. *Plant J.* **2010**, *61*, 70–82. [CrossRef]
19. Guo, M.; Thomas, J.; Collins, G.; Timmermans, M.C. Direct repression of KNOX loci by the ASYMMETRIC LEAVES1 complex of Arabidopsis. *Plant Cell* **2008**, *20*, 48–58. [CrossRef]
20. McConnell, J.R.; Emery, J.; Eshed, Y.; Bao, N.; Bowman, J.; Barton, M.K. Role of PHABULOSA and PHAVOLUTA in determining radial patterning in shoots. *Nature* **2001**, *411*, 709–713. [CrossRef]
21. Emery, J.F.; Floyd, S.K.; Alvarez, J.; Eshed, Y.; Hawker, N.P.; Izhaki, A.; Baum, S.F.; Bowman, J.L. Radial patterning of Arabidopsis shoots by class III HD-ZIP and KANADI genes. *Curr. Biol.* **2003**, *13*, 1768–1774. [CrossRef] [PubMed]
22. Merelo, P.; Paredes, E.B.; Heisler, M.G.; Wenkel, S. The shady side of leaf development: The role of the REVOLUTA/KANADI1 module in leaf patterning and auxin-mediated growth promotion. *Curr. Opin. Plant. Biol.* **2017**, *35*, 111–116. [CrossRef] [PubMed]
23. Montgomery, T.A.; Howell, M.D.; Cuperus, J.T.; Li, D.; Hansen, J.E.; Alexander, A.L.; Chapman, E.J.; Fahlgren, N.; Allen, E.; Carrington, J.C. Specificity of ARGONAUTE7-miR390 interaction and dual functionality in TAS3 trans-acting siRNA formation. *Cell* **2008**, *133*, 128–141. [CrossRef] [PubMed]
24. Tatematsu, K.; Toyokura, K.; Miyashima, S.; Nakajima, K.; Okada, K. A molecular mechanism that confines the activity pattern of miR165 in Arabidopsis leaf primordia. *Plant J.* **2015**, *82*, 596–608. [CrossRef]
25. Kerstetter, R.A.; Bollman, K.; Taylor, R.A.; Bomblies, K.; Poethig, R.S. KANADI regulates organ polarity in Arabidopsis. *Nature* **2001**, *411*, 706–709. [CrossRef]
26. Sawa, S.; Ito, T.; Shimura, Y.; Okada, K. FILAMENTOUS FLOWER controls the formation and development of arabidopsis inflorescences and floral meristems. *Plant Cell* **1999**, *11*, 69–86. [CrossRef]
27. Siegfried, K.R.; Eshed, Y.; Baum, S.F.; Otsuga, D.; Drews, G.N.; Bowman, J.L. Members of the YABBY gene family specify abaxial cell fate in Arabidopsis. *Development* **1999**, *126*, 4117–4128.
28. Watanabe, K.; Okada, K. Two discrete cis elements control the Abaxial side-specific expression of the FILAMENTOUS FLOWER gene in Arabidopsis. *Plant Cell* **2003**, *15*, 2592–2602. [CrossRef]

29. Golz, J.F.; Roccaro, M.; Kuzoff, R.; Hudson, A. GRAMINIFOLIA promotes growth and polarity of Antirrhinum leaves. *Development* **2004**, *131*, 3661–3670. [CrossRef]
30. Sarojam, R.; Sappl, P.G.; Goldshmidt, A.; Efroni, I.; Floyd, S.K.; Eshed, Y.; Bowman, J.L. Differentiating Arabidopsis shoots from leaves by combined YABBY activities. *Plant Cell* **2010**, *22*, 2113–2130. [CrossRef]
31. Sessions, A.; Nemhauser, J.L.; McColl, A.; Roe, J.L.; Feldmann, K.A.; Zambryski, P.C. ETTIN patterns the Arabidopsis floral meristem and reproductive organs. *Development* **1997**, *124*, 4481–4491. [PubMed]
32. Shi, J.; Dong, J.; Xue, J.; Wang, H.; Yang, Z.; Jiao, Y.; Xu, L.; Huang, H. Model for the role of auxin polar transport in patterning of the leaf adaxial-abaxial axis. *Plant J.* **2017**, *92*, 469–480. [CrossRef] [PubMed]
33. Phelps-Durr, T.L.; Thomas, J.; Vahab, P.; Timmermans, M.C. Maize rough sheath2 and its Arabidopsis orthologue ASYMMETRIC LEAVES1 interact with HIRA, a predicted histone chaperone, to maintain knox gene silencing and determinacy during organogenesis. *Plant Cell* **2005**, *17*, 2886–2898. [CrossRef] [PubMed]
34. Yang, J.Y.; Iwasaki, M.; Machida, C.; Machida, Y.; Zhou, X. Chua NH: BetaC1, the pathogenicity factor of TYLCCNV, interacts with AS1 to alter leaf development and suppress selective jasmonic acid responses. *Genes Dev.* **2008**, *22*, 2564–2577. [CrossRef]
35. Ueno, Y.; Ishikawa, T.; Watanabe, K.; Terakura, S.; Iwakawa, H.; Okada, K.; Machida, C.; Machida, Y. Histone deacetylases and ASYMMETRIC LEAVES2 are involved in the establishment of polarity in leaves of Arabidopsis. *Plant Cell* **2007**, *19*, 445–457. [CrossRef]
36. Luo, M.; Yu, C.W.; Chen, F.F.; Zhao, L.; Tian, G.; Liu, X.; Cui, Y.; Yang, J.Y.; Wu, K. Histone deacetylase HDA6 is functionally associated with AS1 in repression of KNOX genes in arabidopsis. *PLoS Genet* **2012**, *8*, e1003114. [CrossRef]
37. Iwakawa, H.; Iwasaki, M.; Kojima, S.; Ueno, Y.; Soma, T.; Tanaka, H.; Semiarti, E.; Machida, Y.; Machida, C. Expression of the ASYMMETRIC LEAVES2 gene in the adaxial domain of Arabidopsis leaves represses cell proliferation in this domain and is critical for the development of properly expanded leaves. *Plant J.* **2007**, *51*, 173–184. [CrossRef]
38. Shuai, B.; Reynaga-Peña, C.G.; Springer, P.S. The lateral organ boundaries gene defines a novel, plant-specific gene family. *Plant. Physiol.* **2002**, *129*, 747–761. [CrossRef]
39. Matsumura, Y.; Iwakawa, H.; Machida, Y.; Machida, C. Characterization of genes in the ASYMMETRIC LEAVES2/LATERAL ORGAN BOUNDARIES (AS2/LOB) family in Arabidopsis thaliana, and functional and molecular comparisons between AS2 and other family members. *Plant J.* **2009**, *58*, 525–537. [CrossRef]
40. Song, B.; Tang, Z.; Li, X.; Li, J.; Zhang, M.; Zhao, K.; Liu, H.; Zhang, S.; Wu, J. Mining and evolution analysis of lateral organ boundaries domain (LBD) genes in Chinese white pear (*Pyrus bretschneideri*). *BMC Genom.* **2020**, *21*, 644. [CrossRef]
41. Guo, B.J.; Wang, J.; Lin, S.; Tian, Z.; Zhou, K.; Luan, H.Y.; Lyu, C.; Zhang, X.Z.; Xu, R.G. A genome-wide analysis of the ASYMMETRIC LEAVES2/LATERAL ORGAN BOUNDARIES (AS2/LOB) gene family in barley (Hordeum vulgare L.). *J. Zhejiang Univ. Sci. B* **2016**, *17*, 763–774. [CrossRef] [PubMed]
42. Guo, Z.; Xu, H.; Lei, Q.; Du, J.; Li, C.; Wang, C.; Yang, Y.; Sun, X. The Arabidopsis transcription factor LBD15 mediates ABA signaling and tolerance of water-deficit stress by regulating ABI4 expression. *Plant J.* **2020**. [CrossRef] [PubMed]
43. Zhang, Y.; Li, Z.; Ma, B.; Hou, Q.; Wan, X. Phylogeny and Functions of LOB Domain Proteins in Plants. *Int. J. Mol. Sci.* **2020**, *21*, 2278. [CrossRef] [PubMed]
44. Yu, J.; Xie, Q.; Li, C.; Dong, Y.; Zhu, S.; Chen, J. Comprehensive characterization and gene expression patterns of LBD gene family in Gossypium. *Planta* **2020**, *251*, 81. [CrossRef]
45. Liu, H.; Cao, M.; Chen, X.; Ye, M.; Zhao, P.; Nan, Y.; Li, W.; Zhang, C.; Kong, L.; Kong, N.; et al. Genome-Wide Analysis of the Lateral Organ Boundaries Domain (LBD) Gene Family in. *Int. J. Mol. Sci.* **2019**, *20*, 5360. [CrossRef]
46. Zhang, Z.; Zhao, H.; Li, W.; Wu, J.; Zhou, Z.; Zhou, F.; Chen, H.; Lin, Y. Genome-wide association study of callus induction variation to explore the callus formation mechanism of rice. *J. Integr. Plant. Biol.* **2019**, *61*, 1134–1150. [CrossRef]
47. Chen, W.F.; Wei, X.B.; Rety, S.; Huang, L.Y.; Liu, N.N.; Dou, S.X.; Xi, X.G. Structural analysis reveals a "molecular calipers" mechanism for a LATERAL ORGAN BOUNDARIES DOMAIN transcription factor protein from wheat. *J. Biol. Chem.* **2019**, *294*, 142–156. [CrossRef]

48. Takahashi, H.; Iwakawa, H.; Nakao, S.; Ojio, T.; Morishita, R.; Morikawa, S.; Machida, Y.; Machida, C.; Kobayashi, T. Knowledge-based fuzzy adaptive resonance theory and its application to the analysis of gene expression in plants. *J. Biosci. Bioeng.* **2008**, *106*, 587–593. [CrossRef]
49. Iwasaki, M.; Takahashi, H.; Iwakawa, H.; Nakagawa, A.; Ishikawa, T.; Tanaka, H.; Matsumura, Y.; Pekker, I.; Eshed, Y.; Vial-Pradel, S.; et al. Dual regulation of ETTIN (ARF3) gene expression by AS1-AS2, which maintains the DNA methylation level, is involved in stabilization of leaf adaxial-abaxial partitioning in Arabidopsis. *Development* **2013**, *140*, 1958–1969. [CrossRef]
50. Husbands, A.Y.; Benkovics, A.H.; Nogueira, F.T.; Lodha, M.; Timmermans, M.C. The ASYMMETRIC LEAVES Complex Employs Multiple Modes of Regulation to Affect Adaxial-Abaxial Patterning and Leaf Complexity. *Plant Cell* **2015**, *27*, 3321–3335. [CrossRef]
51. Hunter, C.; Willmann, M.R.; Wu, G.; Yoshikawa, M.; de la Luz Gutiérrez-Nava, M.; Poethig, S.R. Trans-acting siRNA-mediated repression of ETTIN and ARF4 regulates heteroblasty in Arabidopsis. *Development* **2006**, *133*, 2973–2981. [CrossRef] [PubMed]
52. Luong, T.Q.; Keta, S.; Asai, T.; Kojima, S.; Nakagawa, A.; Micol, J.L.; Xia, S.; Machida, Y.; Machida, C. A genetic link between epigenetic repressor AS1-AS2 and DNA replication factors in establishment of adaxial-abaxial leaf polarity of Arabidopsis. *Plant. Biotechnol.* **2018**, *35*, 39–49. [CrossRef] [PubMed]
53. Pinon, V.; Etchells, J.P.; Rossignol, P.; Collier, S.A.; Arroyo, J.M.; Martienssen, R.A.; Byrne, M.E. Three PIGGYBACK genes that specifically influence leaf patterning encode ribosomal proteins. *Development* **2008**, *135*, 1315–1324. [CrossRef] [PubMed]
54. Yao, Y.; Ling, Q.; Wang, H.; Huang, H. Ribosomal proteins promote leaf adaxial identity. *Development* **2008**, *135*, 1325–1334. [CrossRef]
55. Szakonyi, D.; Byrne, M.E. Ribosomal protein L27a is required for growth and patterning in Arabidopsis thaliana. *Plant J.* **2011**, *65*, 269–281. [CrossRef]
56. Horiguchi, G.; Mollá-Morales, A.; Pérez-Pérez, J.M.; Kojima, K.; Robles, P.; Ponce, M.R.; Micol, J.L.; Tsukaya, H. Differential contributions of ribosomal protein genes to Arabidopsis thaliana leaf development. *Plant J.* **2011**, *65*, 724–736. [CrossRef]
57. Casanova-Sáez, R.; Candela, H.; Micol, J.L. Combined haploinsufficiency and purifying selection drive retention of RPL36a paralogs in Arabidopsis. *Sci. Rep.* **2014**, *4*, 4122. [CrossRef]
58. Kojima, K.; Tamura, J.; Chiba, H.; Fukada, K.; Tsukaya, H.; Horiguchi, G. Two Nucleolar Proteins, GDP1 and OLI2, Function as Ribosome Biogenesis Factors and Are Preferentially Involved in Promotion of Leaf Cell Proliferation without Strongly Affecting Leaf Adaxial-Abaxial Patterning in Arabidopsis thaliana. *Front. Plant Sci.* **2017**, *8*, 2240. [CrossRef]
59. Matsumura, Y.; Ohbayashi, I.; Takahashi, H.; Kojima, S.; Ishibashi, N.; Keta, S.; Nakagawa, A.; Hayashi, R.; Saéz-Vásquez, J.; Echeverria, M.; et al. A genetic link between epigenetic repressor AS1-AS2 and a putative small subunit processome in leaf polarity establishment of Arabidopsis. *Biol. Open* **2016**, *5*, 942–954. [CrossRef]
60. Ohbayashi, I.; Konishi, M.; Ebine, K.; Sugiyama, M. Genetic identification of Arabidopsis RID2 as an essential factor involved in pre-rRNA processing. *Plant J.* **2011**, *67*, 49–60. [CrossRef]
61. Pontvianne, F.; Matía, I.; Douet, J.; Tourmente, S.; Medina, F.J.; Echeverria, M.; Sáez-Vásquez, J. Characterization of AtNUC-L1 reveals a central role of nucleolin in nucleolus organization and silencing of AtNUC-L2 gene in Arabidopsis. *Mol. Biol. Cell* **2007**, *18*, 369–379. [CrossRef] [PubMed]
62. Sáez-Vásquez, J.; Delseny, M. Ribosome Biogenesis in Plants: From Functional 45S Ribosomal DNA Organization to Ribosome Assembly Factors. *Plant Cell* **2019**, *31*, 1945–1967. [CrossRef] [PubMed]
63. Abbasi, N.; Kim, H.B.; Park, N.I.; Kim, H.S.; Kim, Y.K.; Park, Y.I.; Choi, S.B. APUM23, a nucleolar Puf domain protein, is involved in pre-ribosomal RNA processing and normal growth patterning in Arabidopsis. *Plant J.* **2010**, *64*, 960–976. [CrossRef] [PubMed]
64. Takahashi, H.; Iwakawa, H.; Ishibashi, N.; Kojima, S.; Matsumura, Y.; Prananingrum, P.; Iwasaki, M.; Takahashi, A.; Ikezaki, M.; Luo, L.; et al. Meta-analyses of microarrays of Arabidopsis asymmetric leaves1 (as1), as2 and their modifying mutants reveal a critical role for the ETT pathway in stabilization of adaxial-abaxial patterning and cell division during leaf development. *Plant Cell Physiol.* **2013**, *54*, 418–431. [CrossRef] [PubMed]

65. Vial-Pradel, S.; Keta, S.; Nomoto, M.; Luo, L.; Takahashi, H.; Suzuki, M.; Yokoyama, Y.; Sasabe, M.; Kojima, S.; Tada, Y.; et al. Arabidopsis Zinc-Finger-Like Protein ASYMMETRIC LEAVES2 (AS2) and Two Nucleolar Proteins Maintain Gene Body DNA Methylation in the Leaf Polarity Gene ETTIN (ARF3). *Plant Cell Physiol.* **2018**, *59*, 1385–1397. [CrossRef]
66. Feric, M.; Vaidya, N.; Harmon, T.S.; Mitrea, D.M.; Zhu, L.; Richardson, T.M.; Kriwacki, R.W.; Pappu, R.V.; Brangwynne, C.P. Coexisting Liquid Phases Underlie Nucleolar Subcompartments. *Cell* **2016**, *165*, 1686–1697. [CrossRef]
67. Pontvianne, F.; Carpentier, M.C.; Durut, N.; Pavlištová, V.; Jaške, K.; Schořová, Š.; Parrinello, H.; Rohmer, M.; Pikaard, C.S.; Fojtová, M.; et al. Identification of Nucleolus-Associated Chromatin Domains Reveals a Role for the Nucleolus in 3D Organization of the A. thaliana Genome. *Cell Rep.* **2016**, *16*, 1574–1587. [CrossRef]
68. Bersaglieri, C.; Santoro, R. Genome Organization in and around the Nucleolus. *Cells* **2019**, *8*, 579. [CrossRef]
69. Ginisty, H.; Sicard, H.; Roger, B.; Bouvet, P. Structure and functions of nucleolin. *J. Cell Sci.* **1999**, *112 Pt 6*, 761–772.
70. Kojima, H.; Suzuki, T.; Kato, T.; Enomoto, K.; Sato, S.; Tabata, S.; Sáez-Vasquez, J.; Echeverría, M.; Nakagawa, T.; Ishiguro, S.; et al. Sugar-inducible expression of the nucleolin-1 gene of Arabidopsis thaliana and its role in ribosome synthesis, growth and development. *Plant J.* **2007**, *49*, 1053–1063. [CrossRef]
71. Petricka, J.J.; Nelson, T.M. Arabidopsis nucleolin affects plant development and patterning. *Plant Physiol.* **2007**, *144*, 173–186. [CrossRef] [PubMed]
72. Micol-Ponce, R.; Sarmiento-Mañús, R.; Ruiz-Bayón, A.; Montacié, C.; Sáez-Vasquez, J.; Ponce, M.R. Arabidopsis RIBOSOMAL RNA PROCESSING7 Is Required for 18S rRNA Maturation. *Plant Cell* **2018**, *30*, 2855–2872. [CrossRef] [PubMed]
73. Durut, N.; Sáez-Vásquez, J. Nucleolin: Dual roles in rDNA chromatin transcription. *Gene* **2015**, *556*, 7–12. [CrossRef] [PubMed]
74. Montacié, C.; Durut, N.; Opsomer, A.; Palm, D.; Comella, P.; Picart, C.; Carpentier, M.C.; Pontvianne, F.; Carapito, C.; Schleiff, E.; et al. Nucleolar Proteome Analysis and Proteasomal Activity Assays Reveal a Link between Nucleolus and 26S Proteasome in. *Front. Plant Sci.* **2017**, *8*, 1815. [CrossRef]
75. Picart, C.; Pontvianne, F. Plant nucleolar DNA: Green light shed on the role of Nucleolin in genome organization. *Nucleus* **2017**, *8*, 11–16. [CrossRef]
76. Turner, A.J.; Knox, A.A.; Prieto, J.L.; McStay, B.; Watkins, N.J. A novel small-subunit processome assembly intermediate that contains the U3 snoRNP, nucleolin, RRP5, and DBP4. *Mol. Cell Biol.* **2009**, *29*, 3007–3017. [CrossRef]
77. Phipps, K.R.; Charette, J.; Baserga, S.J. The small subunit processome in ribosome biogenesis—Progress and prospects. *Wiley Interdiscip Rev. RNA* **2011**, *2*, 1–21. [CrossRef]
78. Granneman, S.; Bernstein, K.A.; Bleichert, F.; Baserga, S.J. Comprehensive mutational analysis of yeast DEXD/H box RNA helicases required for small ribosomal subunit synthesis. *Mol. Cell Biol.* **2006**, *26*, 1183–1194. [CrossRef]
79. Dragon, F.; Gallagher, J.E.; Compagnone-Post, P.A.; Mitchell, B.M.; Porwancher, K.A.; Wehner, K.A.; Wormsley, S.; Settlage, R.E.; Shabanowitz, J.; Osheim, Y.; et al. A large nucleolar U3 ribonucleoprotein required for 18S ribosomal RNA biogenesis. *Nature* **2002**, *417*, 967–970. [CrossRef]
80. Feng, J.M.; Tian, H.F.; Wen, J.F. Origin and evolution of the eukaryotic SSU processome revealed by a comprehensive genomic analysis and implications for the origin of the nucleolus. *Genome Biol. Evol.* **2013**, *5*, 2255–2267. [CrossRef]
81. You, K.T.; Park, J.; Kim, V.N. Role of the small subunit processome in the maintenance of pluripotent stem cells. *Genes Dev.* **2015**, *29*, 2004–2009. [CrossRef]
82. Sardana, R.; White, J.P.; Johnson, A.W. The rRNA methyltransferase Bud23 shows functional interaction with components of the SSU processome and RNase MRP. *RNA* **2013**, *19*, 828–840. [CrossRef] [PubMed]
83. Sardana, R.; Zhu, J.; Gill, M.; Johnson, A.W. Physical and functional interaction between the methyltransferase Bud23 and the essential DEAH-box RNA helicase Ecm16. *Mol. Cell Biol.* **2014**, *34*, 2208–2220. [CrossRef] [PubMed]
84. Zhu, J.; Liu, X.; Anjos, M.; Correll, C.C.; Johnson, A.W. Utp14 Recruits and Activates the RNA Helicase Dhr1 to Undock U3 snoRNA from the Preribosome. *Mol. Cell Biol.* **2016**, *36*, 965–978. [CrossRef] [PubMed]

85. Ohbayashi, I.; Lin, C.Y.; Shinohara, N.; Matsumura, Y.; Machida, Y.; Horiguchi, G.; Tsukaya, H.; Sugiyama, M. Evidence for a Role of ANAC082 as a Ribosomal Stress Response Mediator Leading to Growth Defects and Developmental Alterations in Arabidopsis. *Plant Cell* **2017**, *29*, 2644–2660. [CrossRef]
86. Ohbayashi, I.; Sugiyama, M. Plant Nucleolar Stress Response, a New Face in the NAC-Dependent Cellular Stress Responses. *Front. Plant Sci.* **2017**, *8*, 2247. [CrossRef]
87. Huang, K.C.; Lin, W.C.; Cheng, W.H. Salt hypersensitive mutant 9, a nucleolar APUM23 protein, is essential for salt sensitivity in association with the ABA signaling pathway in Arabidopsis. *BMC Plant Biol.* **2018**, *18*, 40. [CrossRef]
88. Bernstein, K.A.; Gallagher, J.E.; Mitchell, B.M.; Granneman, S.; Baserga, S.J. The small-subunit processome is a ribosome assembly intermediate. *Eukaryot. Cell* **2004**, *3*, 1619–1626. [CrossRef]
89. McCann, K.L.; Charette, J.M.; Vincent, N.G.; Baserga, S.J. A protein interaction map of the LSU processome. *Genes Dev.* **2015**, *29*, 862–875. [CrossRef]
90. Lawrence, R.J.; Earley, K.; Pontes, O.; Silva, M.; Chen, Z.J.; Neves, N.; Viegas, W.; Pikaard, C.S. A concerted DNA methylation/histone methylation switch regulates rRNA gene dosage control and nucleolar dominance. *Mol. Cell* **2004**, *13*, 599–609. [CrossRef]
91. Pontes, O.; Lawrence, R.J.; Silva, M.; Preuss, S.; Costa-Nunes, P.; Earley, K.; Neves, N.; Viegas, W.; Pikaard, C.S. Postembryonic establishment of megabase-scale gene silencing in nucleolar dominance. *PLoS ONE* **2007**, *2*, e1157. [CrossRef]
92. Zhou, C.; Labbe, H.; Sridha, S.; Wang, L.; Tian, L.; Latoszek-Green, M.; Yang, Z.; Brown, D.; Miki, B.; Wu, K. Expression and function of HD2-type histone deacetylases in Arabidopsis development. *Plant J.* **2004**, *38*, 715–724. [CrossRef] [PubMed]
93. Luo, L.; Ando, S.; Sasabe, M.; Machida, C.; Kurihara, D.; Higashiyama, T.; Machida, Y. Arabidopsis ASYMMETRIC LEAVES2 protein required for leaf morphogenesis consistently forms speckles during mitosis of tobacco BY-2 cells via signals in its specific sequence. *J. Plant Res.* **2012**, *125*, 661–668. [CrossRef] [PubMed]
94. Luo, L.; Ando, S.; Sakamoto, Y.; Suzuki, T.; Takahashi, H.; Ishibashi, N.; Kojima, S.; Kurihara, D.; Higashiyama, T.; Yamamoto, K.T.; et al. The formation of perinucleolar bodies is important for normal leaf development and requires the zinc-finger DNA-binding motif in Arabidopsis ASYMMETRIC LEAVES2. *Plant J.* **2020**, *101*, 1118–1134. [CrossRef] [PubMed]
95. Song, Z.; Wu, M. Identification of a novel nucleolar localization signal and a degradation signal in Survivin-deltaEx3: A potential link between nucleolus and protein degradation. *Oncogene* **2005**, *24*, 2723–2734. [CrossRef] [PubMed]
96. Musinova, Y.R.; Lisitsyna, O.M.; Golyshev, S.A.; Tuzhikov, A.I.; Polyakov, V.Y.; Sheval, E.V. Nucleolar localization/retention signal is responsible for transient accumulation of histone H2B in the nucleolus through electrostatic interactions. *Biochim. Biophys. Acta* **2011**, *1813*, 27–38. [CrossRef] [PubMed]
97. de Melo, I.S.; Jimenez-Nuñez, M.D.; Iglesias, C.; Campos-Caro, A.; Moreno-Sanchez, D.; Ruiz, F.A.; Bolívar, J. NOA36 protein contains a highly conserved nucleolar localization signal capable of directing functional proteins to the nucleolus, in mammalian cells. *PLoS ONE* **2013**, *8*, e59065. [CrossRef]
98. Earley, L.F.; Kawano, Y.; Adachi, K.; Sun, X.X.; Dai, M.S.; Nakai, H. Identification and characterization of nuclear and nucleolar localization signals in the adeno-associated virus serotype 2 assembly-activating protein. *J. Virol.* **2015**, *89*, 3038–3048. [CrossRef]
99. Ye, J.; Yang, J.; Sun, Y.; Zhao, P.; Gao, S.; Jung, C.; Qu, J.; Fang, R.; Chua, N.H. Geminivirus Activates ASYMMETRIC LEAVES 2 to Accelerate Cytoplasmic DCP2-Mediated mRNA Turnover and Weakens RNA Silencing in Arabidopsis. *PLoS Pathog.* **2015**, *11*, e1005196. [CrossRef]
100. Vial-Pradel, S.; Hasegawa, Y.; Nakagawa, A.; Miyaki, S.; Machida, Y.; Kojima, S.; Machida, C.; Takahashi, H. SIMON: Simple methods for analyzing DNA methylation by targeted bisulfite next-generation sequencing. *Plant. Biotechnol.* **2019**, *36*, 213–222. [CrossRef]
101. Husbands, A.; Bell, E.M.; Shuai, B.; Smith, H.M.; Springer, P.S. LATERAL ORGAN BOUNDARIES defines a new family of DNA-binding transcription factors and can interact with specific bHLH proteins. *Nucleic Acids Res.* **2007**, *35*, 6663–6671. [CrossRef] [PubMed]
102. Ohashi-Ito, K.; Iwamoto, K.; Fukuda, H. LOB DOMAIN-CONTAINING PROTEIN 15 Positively Regulates Expression of VND7, a Master Regulator of Tracheary Elements. *Plant Cell Physiol.* **2018**, *59*, 989–996. [CrossRef] [PubMed]

103. Nishimura, T. Gene Body Methylation Involved in Leaf Development. *Plant Cell Physiol.* **2018**, *59*, 1288–1289. [CrossRef] [PubMed]
104. Lyko, F. The DNA methyltransferase family: A versatile toolkit for epigenetic regulation. *Nat. Rev. Genet.* **2018**, *19*, 81–92. [CrossRef]
105. Song, J.; Rechkoblit, O.; Bestor, T.H.; Patel, D.J. Structure of DNMT1-DNA complex reveals a role for autoinhibition in maintenance DNA methylation. *Science* **2011**, *331*, 1036–1040. [CrossRef]
106. Long, H.K.; Blackledge, N.P.; Klose, R.J. ZF-CxxC domain-containing proteins, CpG islands and the chromatin connection. *Biochem. Soc. Trans.* **2013**, *41*, 727–740. [CrossRef]
107. Nishiyama, A.; Yamaguchi, L.; Sharif, J.; Johmura, Y.; Kawamura, T.; Nakanishi, K.; Shimamura, S.; Arita, K.; Kodama, T.; Ishikawa, F.; et al. Uhrf1-dependent H3K23 ubiquitylation couples maintenance DNA methylation and replication. *Nature* **2013**, *502*, 249–253. [CrossRef]
108. Song, J.; Du, Z.; Ravasz, M.; Dong, B.; Wang, Z.; Ewing, R.M. A Protein Interaction between β-Catenin and Dnmt1 Regulates Wnt Signaling and DNA Methylation in Colorectal Cancer Cells. *Mol. Cancer Res.* **2015**, *13*, 969–981. [CrossRef]
109. Zhang, Z.M.; Liu, S.; Lin, K.; Luo, Y.; Perry, J.J.; Wang, Y.; Song, J. Crystal Structure of Human DNA Methyltransferase 1. *J. Mol. Biol.* **2015**, *427*, 2520–2531. [CrossRef]
110. Du, J. Structure and Mechanism of Plant DNA Methyltransferases. *Adv. Exp. Med. Biol.* **2016**, *945*, 173–192. [CrossRef]
111. Ferry, L.; Fournier, A.; Tsusaka, T.; Adelmant, G.; Shimazu, T.; Matano, S.; Kirsh, O.; Amouroux, R.; Dohmae, N.; Suzuki, T.; et al. Methylation of DNA Ligase 1 by G9a/GLP Recruits UHRF1 to Replicating DNA and Regulates DNA Methylation. *Mol. Cell* **2017**, *67*, 550–565.e555. [CrossRef] [PubMed]
112. Ryazanova, A.Y.; Abrosimova, L.A.; Oretskaya, T.S.; Kubareva, E.A. *Diverse Domains of (Cytosine-5)-DNA Methyltransferases: Structural and Functional Characterization, Methylation-From DNA, RNA and Histones to Diseases and Treatment*; Dricu, A., Ed.; InTech: Rijeka, Croatia, 2012; pp. 29–69.
113. Pontvianne, F.; Abou-Ellail, M.; Douet, J.; Comella, P.; Matia, I.; Chandrasekhara, C.; Debures, A.; Blevins, T.; Cooke, R.; Medina, F.J.; et al. Nucleolin is required for DNA methylation state and the expression of rRNA gene variants in Arabidopsis thaliana. *PLoS Genet* **2010**, *6*, e1001225. [CrossRef] [PubMed]
114. Pontvianne, F.; Blevins, T.; Chandrasekhara, C.; Mozgová, I.; Hassel, C.; Pontes, O.M.; Tucker, S.; Mokros, P.; Muchová, V.; Fajkus, J.; et al. Subnuclear partitioning of rRNA genes between the nucleolus and nucleoplasm reflects alternative epiallelic states. *Genes Dev.* **2013**, *27*, 1545–1550. [CrossRef] [PubMed]
115. To, T.K.; Kim, J.M.; Matsui, A.; Kurihara, Y.; Morosawa, T.; Ishida, J.; Tanaka, M.; Endo, T.; Kakutani, T.; Toyoda, T.; et al. Arabidopsis HDA6 regulates locus-directed heterochromatin silencing in cooperation with MET1. *PLoS Genet* **2011**, *7*, e1002055. [CrossRef]
116. Liu, X.; Yu, C.W.; Duan, J.; Luo, M.; Wang, K.; Tian, G.; Cui, Y.; Wu, K. HDA6 directly interacts with DNA methyltransferase MET1 and maintains transposable element silencing in Arabidopsis. *Plant Physiol.* **2012**, *158*, 119–129. [CrossRef]
117. Lodha, M.; Marco, C.F.; Timmermans, M.C. The ASYMMETRIC LEAVES complex maintains repression of KNOX homeobox genes via direct recruitment of Polycomb-repressive complex2. *Genes Dev.* **2013**, *27*, 596–601. [CrossRef]
118. Li, Z.; Li, B.; Liu, J.; Guo, Z.; Liu, Y.; Li, Y.; Shen, W.H.; Huang, Y.; Huang, H.; Zhang, Y.; et al. Transcription factors AS1 and AS2 interact with LHP1 to repress KNOX genes in Arabidopsis. *J. Integr. Plant Biol.* **2016**, *58*, 959–970. [CrossRef]
119. Lin, X.; Gu, D.; Zhao, H.; Peng, Y.; Zhang, G.; Yuan, T.; Li, M.; Wang, Z.; Wang, X.; Cui, S. LFR is functionally associated with AS2 to mediate leaf development in Arabidopsis. *Plant J.* **2018**, *95*, 598–612. [CrossRef]
120. Xu, C.; Luo, F.; Hochholdinger, F. LOB Domain Proteins: Beyond Lateral Organ Boundaries. *Trends Plant Sci.* **2016**, *21*, 159–167. [CrossRef]
121. Kim, M.J.; Kim, J. Identification of nuclear localization signal in ASYMMETRIC LEAVES2-LIKE18/LATERAL ORGAN BOUNDARIES DOMAIN16 (ASL18/LBD16) from Arabidopsis. *J. Plant Physiol.* **2012**, *169*, 1221–1226. [CrossRef]
122. Correll, C.C.; Bartek, J.; Dundr, M. The Nucleolus: A Multiphase Condensate Balancing Ribosome Synthesis and Translational Capacity in Health, Aging and Ribosomopathies. *Cells* **2019**, *8*, 869. [CrossRef] [PubMed]

123. Nishimura, K.; Kumazawa, T.; Kuroda, T.; Katagiri, N.; Tsuchiya, M.; Goto, N.; Furumai, R.; Murayama, A.; Yanagisawa, J.; Kimura, K. Perturbation of ribosome biogenesis drives cells into senescence through 5S RNP-mediated p53 activation. *Cell Rep.* **2015**, *10*, 1310–1323. [CrossRef] [PubMed]
124. Padeken, J.; Heun, P. Nucleolus and nuclear periphery: Velcro for heterochromatin. *Curr. Opin. Cell Biol.* **2014**, *28*, 54–60. [CrossRef] [PubMed]
125. Carpentier, M.C.; Picart-Picolo, A.; Pontvianne, F. A Method to Identify Nucleolus-Associated Chromatin Domains (NADs). *Methods Mol. Biol.* **2018**, *1675*, 99–109. [CrossRef] [PubMed]
126. Chandrasekhara, C.; Mohannath, G.; Blevins, T.; Pontvianne, F.; Pikaard, C.S. Chromosome-specific NOR inactivation explains selective rRNA gene silencing and dosage control in Arabidopsis. *Genes Dev.* **2016**, *30*, 177–190. [CrossRef] [PubMed]
127. Mohannath, G.; Pontvianne, F.; Pikaard, C.S. Selective nucleolus organizer inactivation in Arabidopsis is a chromosome position-effect phenomenon. *Proc. Natl. Acad. Sci. USA* **2016**, *113*, 13426–13431. [CrossRef] [PubMed]
128. Pavlištová, V.; Dvořáčková, M.; Jež, M.; Mozgová, I.; Mokroš, P.; Fajkus, J. Phenotypic reversion in fas mutants of Arabidopsis thaliana by reintroduction of FAS genes: Variable recovery of telomeres with major spatial rearrangements and transcriptional reprogramming of 45S rDNA genes. *Plant J.* **2016**, *88*, 411–424. [CrossRef]
129. O'Malley, R.C.; Huang, S.C.; Song, L.; Lewsey, M.G.; Bartlett, A.; Nery, J.R.; Galli, M.; Gallavotti, A.; Ecker, J.R. Cistrome and Epicistrome Features Shape the Regulatory DNA Landscape. *Cell* **2016**, *165*, 1280–1292. [CrossRef]

International Journal of
Molecular Sciences

MDPI

Review

The Regulation of CIN-like TCP Transcription Factors

Jingqiu Lan [1] and Genji Qin [1,2,*]

1 State Key Laboratory of Protein and Plant Gene Research, School of Life Sciences, Peking University, Beijing 100871, China; lanjq@pku.edu.cn
2 School of Advanced Agricultural Sciences, Peking University, Beijing 100871, China
* Correspondence: qingenji@pku.edu.cn

Received: 30 May 2020; Accepted: 20 June 2020; Published: 24 June 2020

check for updates

Abstract: TEOSINTE BRANCHED1/CYCLOIDEA/PROLIFERATING CELL FACTOR 1 and 2 (TCP) family proteins are the plant-specific transcription factors extensively participating in diverse developmental processes by integrating external cues with internal signals. The roles of CINCINNATA (CIN)-like TCPs are conserved in control of the morphology and size of leaves, petal development, trichome formation and plant flowering. The tight regulation of CIN-like TCP activity at transcriptional and post-transcriptional levels are central for plant developmental plasticity in response to the ever-changing environmental conditions. In this review, we summarize recent progresses with regard to the function and regulation of CIN-like TCPs. CIN-like TCPs are regulated by abiotic and biotic cues including light, temperature and pathogens. They are also finely controlled by microRNA319 (miRNA319), chromatin remodeling complexes and auxin homeostasis. The protein degradation plays critical roles in tightly controlling the activity of CIN-like TCPs as well.

Keywords: CIN-like TCP transcription factors; regulation; light; high temperature; microRNA319; BRAHMA; TIE1 transcriptional repressors; TEAR1 E3 ligases

---

1. Introduction

Developmental plasticity is central for sessile plants in adaptation to the environmental conditions [1]. The molecular bases for plant developmental plasticity or the mechanisms by which plants translate the environmental cues into the internal signals to direct the optimal growth and development in different plant growing conditions are important for plant survival and are useful for crop improvement by molecular breeding. Since the discovery of the founding members of TEOSINTE BRANCHED1/CYCLOIDEA/PROLIFERATING CELL FACTOR 1 and 2 (TCP) protein family in plants more than twenty years ago [2–4], TCP proteins have emerged as a central hub for integrating the internal and external cues to control plant developmental plasticity.

TCP is an acronym of the name of founding genes isolated from three species, i.e., *TEOSINTE BRANCHED1* (*TB1*) from maize (*Zea mays*) [2,5], *CYCLOIDEA* (*CYC*) from snapdragon (*Antirrhinum majus*) [3], and *PROLIFERATING CELL FACTOR 1* and *2* (*PCF1* and *PCF2*) from rice (*Oryza sativa*) [4]. *TB1* is a famous maize domestication gene. TB1 represses the outgrowth of axillary branches and promotes the formation of female inflorescences in domesticated maize, while in teosinte—which is the wild ancestor of maize—the twice lower expression of *TB1* leads to a decrease of apical dominance and an increase of shoot branches [5]. The *CYC* gene was isolated from snapdragon. *CYC* is specifically expressed in the dorsal primordia and controls the flower zygomorphic trait. Disruption of both *CYC* and its close homolog *DICHOTOMA* (*DICH*) in snapdragon results in radially symmetric flowers [6]. Both TB1 and CYC play pivotal roles in shaping plant key morphologies. The rice PCF proteins were found to directly bind to the promoter region of *PROLIFERATING CELL NUCLEAR ANTIGEN* (*PCNA*) gene which encodes a protein acting as a DNA polymerase sliding clamp implicated in DNA replication and cell cycle regulation [4]. Further analysis of the protein sequences of TB1, CYC and PCF proteins found that they all contain a conserved region predicted to form a non-canonical basic helix-loop-helix (bHLH) structure named as the TCP domain [7]. Since PCF1 and PCF2 had DNA-binding activity, TCP proteins were deduced to act as transcription factors and the TCP domain was proposed to be responsible for DNA binding and protein-protein interaction [4,7,8].

According to the sequence differences in the TCP domain, TCPs are classified into class I and class II subfamilies [7] (Figure 1). The TCP domain of class II TCPs contains additional four-amino acid residues in the conserved basic region [7]. The class II TCPs are further divided into CINCINNATA (CIN)-like TCPs and CYC/TB1-like TCPs based on the additional sequence differences in the TCP domain [9]. The CYC/TB1 TCP subgroup also carries a conserved glutamic acid-cysteine-glutamic acid (ECE) motif outside the TCP domain [10]. The *CIN* gene was isolated from snapdragon by analyzing the *cin* mutant which produces abnormal leaves and petals with undulated edges [11,12] and is the founding member of the CIN-like TCP subgroup (Figure 1). CIN controls leaf flatness by tightly regulating cell proliferation and differentiation in the different areas of leaf blades [11]. In the model plant *Arabidopsis*, the CIN-like TCPs include eight members which are further grouped into two clades based on the existence of microRNA (miRNA) binding site outside the sequence encoding TCP domain. *TCP2, TCP3, TCP4, TCP10,* and *TCP24* have the miRNA binding sites and post-transcriptionally regulated by miR319 [13], while TCP5, TCP13 and TCP17 form a small clade named as TCP5-like CIN-TCPs that were proved to be important for plant thermomorphogenesis (Figure 2) [14].

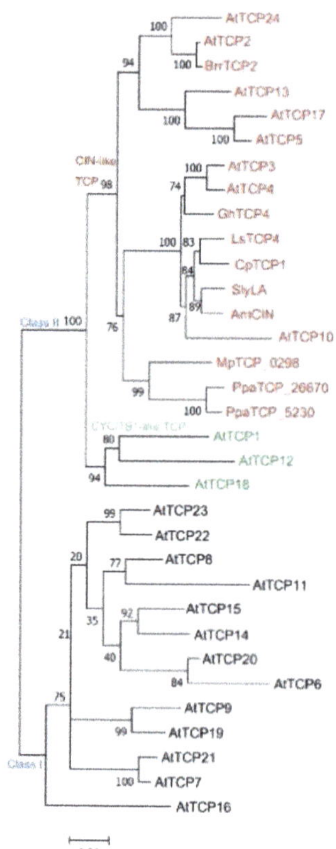

Figure 1. The phylogeny of TCP transcription factors, including all the TCP proteins in *Arabidopsis thaliana*

and the CIN-like TCPs of other species mentioned in this review. Multiple alignments of the full-length TCP proteins were conducted using MAFFT Version 7 [15] with L-INS-i iterative refinement methods. The phylogenetic tree was constructed with the Maximum Likelihood (ML) method using the IQ-tree2 software [16] with the VT+F+R4 model with 1000 bootstrap replications. The subfamilies and subclasses (Class I, Class II, CIN-like TCP and CYC-like TCPs) are indicated above the divergent branches. The proteins in red words are the CIN-like TCPs which are mainly discussed in this review. The prefixes of TCP proteins are indicated the species. At: *Arabidopsis thaliana*; Brr: *Brassica rapa*; Gh: *Gossypium hirsutum*; Ls: *Lactuca sativa*; Cp: *Cyclamen persicum*; Sly: *Solanum lycopersium*; Am: *Antirrhinum majus*; Mp: *Marchantia polymorpha*; Ppa: *Physcomitrella patens*. The bootstrap support is indicated above the branches. The scale bar denotes the branch length.

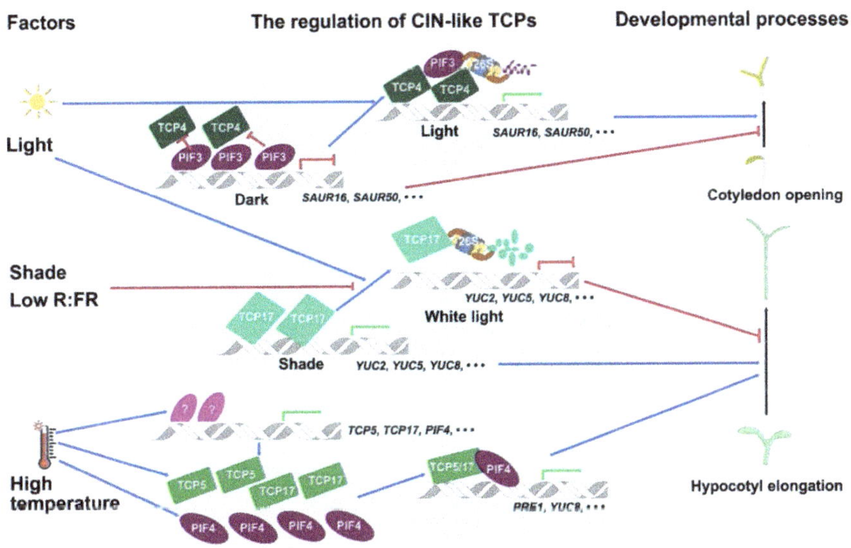

Figure 2. An overview of the regulation mechanisms of CIN-like TCP transcription factors by light and temperature during cotyledon opening and hypocotyl elongation processes. The external stimuli including light and high temperature are summarized at the left column. The schematic diagram includes the regulation mechanisms of CIN-like TCPs at the transcriptional and the protein levels. The arrows directly pointing on the double helix symbols indicate transcriptional regulations. The arrows pointing to the proteins indicate the regulations of protein stabilities. The proteins related with the "26S" symbols indicate protein degradation through the ubiquitin-26S proteasome pathway. The blue arrows represent the positive regulation, and the red arrows with dash-headed ends indicate the negative regulation. The green arrows and red dash-headed ends at the double-helix icons indicates the activation and repression of gene expression, respectively. All the unknown factors are indicated with question marks. R:FR, red light: far red light ratio; PIFs, PHYTOCHROME-INTERACTION FACTORs; SAURs, SMALL AUXIN UPREGULATED RNAs; YUCs, YUCCAs.

TCP transcription factors constitute a plant-specific protein family which is conserved in plant kingdom. TCP homologs are identified from diverse plant species [7]. It is proved that TCP proteins are existed in the early land plants during evolutionary history [17–20]. However, it is still in dispute whether they are present in pluricellular green algae [17,20]. The TCP protein family is significantly expanded in angiosperm species by gene or whole-genome duplication independently in basal angiosperm, magnoliids, basal eudicot, monocot, and many major groups within eudicot [10,17,20–25]. It is hard to distinguish whether class I or class II subfamily is the first to appear in plant kingdom, because the genome of liverwort *Marchantia polymorpha* contains the members belonging to both of the two families [19,26]. As for class II TCPs, the CIN-like TCP subgroup is predicted to be more ancestral than the CYC/TB1-like TCPs, since the class II TCPs all belong to CIN-like TCP subgroup in the non-vascular plants [18,19,26,27]. The CYC/TB1-like TCP group is proposed to originate in angiosperm species and to evolve independently in basal eudicot groups and monocot species [23,24,28–31].

TCP family transcription factors governs various key developmental processes during the life cycle of plants. TCPs regulate seed germination, leaf development, outgrowth of shoot branches, flowering, flower development, silique and ovule development, photomorphogenesis, thermomorphogenesis, circadian rhythms, defense responses and senescence [11,32–53]. The tight regulation of TCPs is very important for plant development and survival. Plants evolve many ways to tightly regulate TCP activity. The aim of this review is to give a comprehensive overview on current knowledge relevant to the roles of CIN-like TCPs in different species and the fine regulation of CIN-like TCP by external stimuli, miRNA and other proteins. To understand the detailed functions of TCPs in plants, the downstream targets regulated by TCPs, the regulation of CYC/TB1-like TCPs, please refer to the excellent recent reviews [39,54,55].

2. The Functions of CIN-Like TCP Transcription Factors in Different Species

One of the most prominent roles of CIN-like TCP transcription factors is that they play a conserved and central role in control of leaf flatness, size, shape and complexity. The loss of *CIN* function in snapdragon *cin* mutant disrupted leaf flatness and forms defective simple leaves with larger size and wavy margins [11,12,56]. In *Arabidopsis*, CIN-like TCPs have highly redundant and additive roles in regulating the morphogenesis of simple leaves (Figure 2). The *Arabidopsis tcp* single mutants produced leaves with no obvious differences from wild-type control. However, disruption of *TCP4* and *TCP10* had already led to larger and curled leaves. The high-order multiple *CIN*-like *tcp* mutants caused even severer leaf curvature and wavier leaf margins in a dose-dependent manner [53,57–59], indicating that the activity of CIN-like TCPs is pivotal for shaping leaf forms. The *CIN*-like *TCP* homolog in turnip (*Brassica rapa*), *BrrTCP2*, has conserved function in control of leaf size and morphology. Overexpression of *BrrTCP2* reduced the leaf size of wild-type *Arabidopsis* and restored the leaf morphology of the *Arabidopsis* multiple mutant *tcp2 tcp4 tcp10* [60]. In the regulation of leaf morphology, CIN-like TCPs repress the activity of leaf marginal meristem which determines leaf serrations in simple leaves or complexity of compound leaves in different plants. In lettuce (*Lactuca sativa*), the Empire type cultivars have more serrated leaves than the Salinas type cultivars. The molecular base is that Empire type cultivars carry a retrotransposable element inserted in the upstream of *LsTCP4* gene, causing lower expression level of *LsTCP4* than that in the Salina type cultivars. The downregulation of *LsTCP4* by the insertion led to the severer leaf serration in Empire type cultivars [61]. However, differential expression analysis between broad- and curly-leaved plants of *Cichorium endivia*, a close relative of *L. sativa* that also displayed wavy or serrated leaves, did not identify *TCP4-like* homologous genes as differentially expressed in leaves with different morphologies, and the two transcripts were abundant in both leaf types [62]. Tomato forms compound leaves regulated by *LACEOLATE* (*LA*) homologous to *CIN*-like *TCPs*. Downregulation of *LA* generated more and larger leaflets, causing super-compound leaves. On the contrary, overexpression of *LA* resulted in the compound leaves turning into simple leaves [63–65]. *CpTCP1* in cyclamen (*Cyclamen persicum*) is a homolog of *CIN*-like *TCPs*. Disruption of TCP function by a dominant repressor in which the ethylene-responsive element binding factor-associated amphiphilic repression (EAR) repression domain (SRDX) was fused to CpTCP1 caused irregular protrusions of acicular and branched shapes in the leaf margins [66]. CIN-like TCPs also regulate the leafy head of Chinese cabbage (*Brassica rapa*). Altering the spatio-temporal expression patterns of *BrpTCP4* led to a cylindrical head shape from a round one [67]. Furthermore, the genetic manipulation of CIN-like TCP activity resulted in different sizes and shapes of leaves in both simple and compound leaves [64,68]. These findings indicate that CIN-like TCPs are central regulators of leaf morphology and that the tight control of the spatio-temporal TCP activity is fundamental in determining diverse leaves in different species.

CIN-like TCPs also modulate the development of organs homologous to leaves such as petals. The *Arabidopsis* single mutant *tcp5* produces wider petals than the wild-type control [69]. Moreover, the *35S:miR-3TCP* transgenic plants in which an artificial miRNA targeting to *TCP5*, *TCP13*, and *TCP17* was expressed to knock down the three genes generate petals with even increased width from tip to base [69]. Besides *TCP5*-like *CIN-TCPs* which was identified to determine petal size, the other five *CIN*-like *TCP* genes targeted by miRNA319 also played vital roles in control of petal growth. The mutant carrying a loss-of-function mutation in *miR319a* (named as $MiR319a^{129}$) exhibited narrow petals and sterile anthers, indicating that CIN-like TCPs not only inhibit the growth of petal [70], but also play an essential role in plant fertility. The overexpression of the miR319-resistant form of *TCP4* by a petal-specific promoter rescued the narrow petals in $MiR319a^{129}$ mutant [70,71]. CIN-like TCPs also modify the morphology of petals besides petal sizes. Expression of a dominant repressor in which TCP3 was fused to an EAR motif to disrupt the function of TCPs resulted in curled petals in *Arabidopsis*. Expression of other CIN-like TCP chimeric repressors also caused curled petals [72]. The function of CIN-like TCPs in regulating petal development is conserved among different species. For examples, the introduction of chimeric repressors of *Arabidopsis* CIN-like TCPs in *Chrysanthemum morifolium* or *Ipomoea nil* also led to similar wavy and serrated petals [73,74]. Suppression of TCP functions by expression of chimeric repressors of CpTCP1 homologous to *Arabidopsis* TCP3 in *C. persicum* caused ruffled petals [66].

At the cellular level, CIN-like TCPs regulate cell proliferation, cell elongation or expansion and cell differentiation. During leaf and petal development, CIN-like TCPs inhibit cell proliferation and promote cell differentiation. Disruption of CIN-like TCPs prolong the leaf cell proliferation in the leaf blade with more rapid growth in the margin than in the center of blade, leading to the increased number of pavement cells and wavy margins [11,12,35,58,59,75]. As specialized epidermal cells, trichomes are also regulated by CIN-like TCP transcription factors. CIN-like TCPs suppress the trichome differentiation and subsequent trichome branching. The numbers of trichomes and trichome branches were both significantly increased in *jaw-D* and *tcp2 tcp4 tcp10* mutants, but were decreased in *TCP4* overexpression lines [76]. The function of CIN-like TCPs is also conserved in the regulation of trichome formation. Overexpression of *miR319a* in *Populus tomentosa* resulted in higher density of trichomes on the leaf surface when compared with that of wild-type control. When the functions of CIN-like TCPs were enhanced by inhibiting the roles of miR319, the number of trichomes was largely decreased [77]. Cotton fibers are specific trichome types on the seed epidermis. The constitutive overexpression of *GhTCP4* homologous to *Arabidopsis* TCP4 in upland cotton (*Gossypium hirsutum*) repressed the elongation of cotton fiber [78]. However, CIN-like TCPs positively regulate hypocotyl cell elongation in *Arabidopsis*. Induction of CIN-like TCPs using mTCP4-GR in which TCP4 fusion with rat glucocorticoid receptor (GR) by dexamethasone (DEX) treatment in transgenic lines significantly increased the length of hypocotyl cells (Figure 2) [79]. Overexpression of *TCP5*-like *CIN-TCPs* led to the significant increase of hypocotyl under shade, high temperature or under normal growth conditions (Figure 2) [14]. In consistence with the results, the *tcp5 tcp13 tcp17* triple mutant displayed short hypocotyls [14]. These findings demonstrate that CIN-like TCPs control cell proliferation, elongation and differentiation in a specific cell type-dependent manner at different context.

CIN-like TCPs are reported to be essential for regulating other biological processes. For examples, CIN-like TCPs facilitate the transition from vegetative to reproductive growth. The flowering time of *cin*-like *tcp* multiple mutants was significantly postponed, while overexpression of *TCP4* led to early flowering in *Arabidopsis* [80]. The tomato *LA* gene belonging to *CIN*-like *TCP* group controls flowering as well [81]. In addition, CIN-like TCPs participate in developmental plasticity in response to biotic stresses in *Arabidopsis* and rice. CIN-like TCPs are also implicated in the typical morphological alterations caused by infection of phytopathogens such as phytoplasmas in *Arabidopsis* [43–45]. Rice ragged stunt virus (RRSV) downregulated rice *TCP21* belonging to miR319-targeted CIN-like TCPs by up-regulating the expression of *miR319* gene. Overexpression of *TCP21* increased the rice resistance to RRSV [82].

3. Light Regulates CIN-Like TCP Transcription Factors

Light is a critical environmental stimulus affecting plant development and growth including cotyledon opening, hypocotyl elongation and flowering [83–86]. When seeds germinate in dark under soil and then the seedlings grow out with exposure to light in nature, plants undergo an important morphological change from skotomorphogenesis to photomorphogenesis including cotyledon opening and inhibition of hypocotyl elongation [83,87,88]. The bHLH transcription factors PHYTOCHROME-INTERACTING FACTORs (PIFs) including PIF3 are central regulators in promoting skotomorphogenesis by suppressing cotyledon opening and the elongation of hypocotyl [89,90]. However, the molecular mechanisms of light-induced cotyledon opening are not well-known. Recently, CIN-like TCPs have been identified to participate in controlling light-induced cotyledon opening during photomorphogenesis (Figure 2). Interestingly, *CIN*-like *TCP* genes including *TCP3, TCP4* and *TCP10* are predominantly expressed in cotyledons under both light and dark growth conditions [13]. Why do CIN-like TCPs promote cotyledon opening in the light but not affect cotyledon closing in dark? Chromatin immunoprecipitation sequencing (ChIP-seq) and RNA sequencing (RNA-seq) analyses showed that TCP4 directly bind to the promoter regions of *SMALL AUXIN UPREGULATED RNA* (*SAUR*) genes including *SAUR16* and *SAUR50*. The promoter regions of the *SAUR* genes are also directly targeted by PHYTOCHROME INTERACTING FACTOR3 (PIF3), a key component inhibiting cotyledon opening. The molecular mechanism is that the accumulated PIF3 in the dark represses the transactivation activity of TCP4 possibly by competing the binding to the promoter regions of *SAUR* genes with TCP4 in the dark, while in the light PIFs are rapidly degraded and causing more TCP4 proteins to bind to the promoters of the *SAUR* genes and to upregulate their expression to promote cotyledon opening (Figure 2) [86]. However, PIF3 does not interact with TCP4 in this process. The exact mechanism by which PIF3 inhibits TCP4 binding to the promoter regions of *SAUR* genes is still an open question.

In addition to controlling the light-regulated cotyledon opening in plant photomorphogenesis, CIN-like TCPs also participate in the regulation of light-regulated hypocotyl elongation under shade. The shade avoidance syndrome (SAS) of plants caused by neighboring shade or low ratio of red light to far red light (R:FR) includes long hypocotyl, elongated leaf petiole, reduced shoot branches and early flowering [91]. It is known that shade or low R:FR upregulates the expression level of *BRC1* or *TB1* belonging to *CYC/TB1*-like *TCP* subgroup [37], while recently TCP5-like CIN-TCPs has been reported to regulate the rapid growth of hypocotyl in response to shade (Figure 2) [92]. The hypocotyl elongation of the triple mutant *tcp5 tcp13 tcp17* was insensitive to shade, while overexpression of *TCP17* led to longer hypocotyls under shade or white light. TCP17 is an unstable protein which is stabilized by shade. When plants were transferred from shade to white light, TCP17 was degraded and the degradation were inhibited by treatment with the 26S proteasome inhibitor MG132 [92]. This result indicates that white light promotes the degradation of TCP17 via the 26S proteasome, while shade inhibits the process (Figure 2). Interestingly, the transcriptional level of *TCP17* was rapidly downregulated by shade in reverse, indicating accumulation of TCP17 under shade is dependent on the post-transcriptional regulation [92]. It will be very interesting to identify the E3 ligase mediating the degradation of TCP17 under white light and the molecular mechanisms of suppression of the TCP17 degradation machinery by shade.

4. High Temperature Regulates CIN-Like TCP Transcription Factors

Ambient temperature is one of the most important environmental factors governing plant behavior. Plants adopt a series of morphological changes called thermomorphogenesis in adaptation to high temperature [93,94]. Thermomorphogenesis includes leaf hyponastic growth, petiole elongation and hypocotyl elongation [93]. TCP5-like CIN-TCPs have recently been identified to act as key factors in positively regulating plant thermomorphogenesis. High temperature not only induces the expression of *TCP5, TCP13* and *TCP17* genes at the transcriptional level, but also stabilizes the protein of TCP5-like CIN-TCPs at the post-transcriptional level in *Arabidopsis* (Figure 2) [14,95].

Interestingly, high temperature treatment regulates both the expression level and the expression pattern of *TCP5*. When TCP5pro-GUS transgenic lines in which *GUS* reporter gene was driven by *TCP5* promoter was treated under high temperature, the GUS staining was strengthened in the hypocotyls and cotyledons, and at the same time was shifted from the leaf blades to petioles, in consistence with the leaf trait of thermomorphogenesis with elongated petioles and reduced areas of blades [14]. High temperature also up-regulates the expression of *PIF4* which is the first key factor identified in control of plant thermomorphogenesis [96,97]. TCP5 protein not only directly bound to the promoter region of *PIF4* gene to increase its expression level [14], but also interacted with PIF4 at the protein level [14]. Moreover, TCP17 protein interacted with the blue light receptor CRYTOCHROME1 (CRY1) at lower temperature to block the activity of TCP17. High ambient temperature increased the protein stability of TCP17 and led to the release of TCP17 from TCP17-CRY1 complex, promoting the interactions between TCP17 and PIF4 [93]. The interactions between PIF4 with TCP5 or TCP17 synergistically promoted the expression of a lot of common downstream genes including *PRE1* and *YUC8*, thus enhancing plant thermomorphogenesis (Figure 2) [14,95]. Accordingly, overexpression of *CIN-like TCP5* gene led to constitutive thermomorphogenesis, while the hypocotyls and petioles of *tcp5 tcp13 tcp17* were shorter than that of wild-type control under normal temperature or high temperature [14,95]. It is worth mentioning that although PIF4 is homologous to PIF3 which is a key regulator in photomorphogenesis [86], they use different mechanisms to regulate the activity of CIN-like TCPs. PIF3 do not interact with TCP4, but inhibits the binding activity of TCP4 to the promoter of their downstream genes in an unknown way under dark [86]. Adversely, PIF4 interacts with TCP5-like CIN-TCPs and obviously strengthened their transactivation activity in activating the downstream genes [14,95]. These results demonstrate that high temperature regulates the function of TCP5-like CIN-TCPs which positively regulate plant thermomorphogenesis by a different mechanism underlying the regulation of cotyledon opening by TCP4 in *Arabidopsis*. However, the transcription factors and E3 ligases that are responsible for regulating the expression of *TCP5*-like *CIN-TCPs* and the stability of their products under different ambient temperatures need to be further identified.

5. Phytoplasmas Regulate CIN-Like TCP Transcription Factors

Phytoplasmas are phytopathogens transmitted by insects and infect a wide range of plant species, causing great economic losses in agriculture [47,98]. Like the most pathogens, phytoplasmas produce effectors to alter the host pathogen interface in facilitating their growth during infection [47]. The effectors cause some typical changes of plant morphology including overgrowth of lateral branches, altered leaf shape and sterile flowers [98]. The aster yellows phytoplasma witches' broom (AY-WB) strain infect a wide range of dicot and monocot species [47,48]. The secreted AY-WB protein 11 (SAP11) is a virulence nuclear effector with a nuclear localization signal at its N-terminus. Overexpression of *SAP11* in *Arabidopsis* produced serrate and wavy leaves almost identical to those of *jaw-D* and the multiple *cin*-like *tcp* mutants [48]. SAP11 interacts with CIN-like TCP proteins [48], leading to the TCP degradation which is not inhibited by the 26S proteasome inhibitor epoxomicin or protease inhibitor cocktail (Figure 2). This indicates that the SAP11-mediating TCP protein degradation is not through ubiquitin-26S proteasome pathway [48]. Because CIN-like TCPs positively regulate the expression of *LOX2* gene by directly binding to its promoter [53], the overexpression of *SAP11* caused the downregulation of the *LOX2* gene and reduced the production of jasmonic acid (JA) in both *Arabidopsis* and tobacco (*Nicotiana benthamiana*), facilitating the infection of phytoplasmas [47,49]. The SAP11 protein homologs in different phytoplasmas strains displayed the varied abilities in control of the stability of CIN-like TCPs. These strains include AY-WB, onion yellow strain M (OY-M), peanut pupurea witches' broom (PnWB), *Candidatus* phytoplasmas mali (CaPM) [49]. When SAP11 homologs were co-expressed with CIN-like TCPs in tobacco, the abundance of TCP proteins were measured to determine the abilities of SAP11 proteins in mediating TCP degradation [49]. The results showed that $SAP11_{AYWB}$ had the strongest ability to mediate the degradation of TCP2, TCP3, TCP4, TCP5, TCP10 and TCP24, while $SAP11_{CaPM}$ only mediated the degradation of TCP2 and TCP10 with lower ability than $SAP11_{AYWB}$. $SAP11_{PnWB}$ and $SAP11_{OYM}$ only exhibited a weak ability to destabilize TCP2 [45]. The SAP11 homolog from the Maize Bushy Stunt Phytoplasmas (SAP_{MBSP}) have been shown to only interact with CYC/TB1-like TCPs, but not any members of CIN-like TCPs in maize [46]. Accordingly, the MBSP-infected maize showed overgrowth of tillers controlled by the CYC/TB1-like TCPs, but not had any effects on the morphology of leaves [46]. Similarly, SWP1 which is a SAP11-like phytoplasmas effector from wheat blue dwarf phytoplasma interacted with BRC1 and mediate the degradation of BRC1 when *SWP1* was overexpressed in *Arabidopsis* [50]. These findings indicate that the effector SAP11 proteins from different phytoplasmas strains have different specificity in promoting the degradation of TCPs. As the SAP11 protein have no protease activity, the mechanisms underlying SAP11-mediated TCP degradation remains to be further discovered [48].

6. miRNAs Regulate CIN-Like TCP Transcription Factors

miRNAs are small RNAs that recognize targeting mRNA via base pairing to the highly complementary binding sites and suppress the stability and translation of mRNAs [99,100]. A subset of *CIN*-like *TCP* genes contains a miR319-targeting sequence at the 3′-terminus of transcripts in almost all angiosperm groups [13,63]. The Arabidopsis *jagged and wavy-Dominant* (*jaw-D*) mutant was first identified from a collection of activation tagging mutants by forward genetics [13]. The mutant *jaw-D* displayed a predominant phenotype with the serrated and curved leaves [13,101]. Further analysis showed that T-DNA with four cauliflower mosaic virus (CaMV) 35S enhancer was inserted in neighboring region of MIRNA gene *MIR319a* in *jaw-D*. The expression of *miR319a* was activated and the target *CIN*-like *TCP* genes including *TCP2, TCP3, TCP4, TCP10* and *TCP24* were significantly downregulated in the mutant, suggesting that the transcript abundance of the corresponding *TCP* genes was regulated by miR319a (Figure 3) [13]. The overexpression of *miR319* also caused epinastic cotyledons, more trichomes, defective secondary cell wall biosynthesis and venation patterning, a modest delay in flowering, crinkled petals, short stamen, reduced male fertility and crinkled fruits by downregulating *CIN*-like *TCP* genes [42,70,102–104]. *Arabidopsis* genome contains three *MIR319* genes including *MIR319a, MIR319b,* and *MIR319c* which have highly redundant function in control of the abundance of *CIN*-like *TCP* transcripts [70]. However, the three *MIR319* genes also showed largely non-overlapping expression patterns revealed by GUS reporter analysis in plants, suggesting that they may have distinct roles in control of *TCP* abundance in a temporal and spatial manner during plant development [70]. During leaf development, the *MIR319a* gene is only expressed at the stipules, which is completely complementary to the expression pattern of *MIR319c* that the highest expression level is detected at the basal region of leaf primordia and young leaves, indicating the functional divergences between the two genes. *MIR319b* is only expressed in the sepal and stamen abscission zones of inflorescences at the reproductive stage [70]. *MIR319a* and *MIR319c* have partially spatiotemporal overlapping expression patterns during early inflorescence development [70]. Though the GUS activity for promoter analysis of *MIR319b* was not detected in leaves, the *mir319b* single mutant moderately reduced the size of leaf serrations, and *mir319a/b* double mutant almost entirely suppressed serration formation [59], suggesting that *MIR319b* is essential for leaf development with a possible low expression level in leaves.

miR319 is a conserved and ancient plant miRNA family and plays important roles in plant morphological adaptation to environmental conditions by targeting *TCP* for degradation. The miR319 and miR159 share highly similarity in mature miRNA sequence, secondary structure, conservation pattern and biogenesis in *Arabidopsis*. miR319 and miR159 are proposed to evolve from a common ancestor in land plants [105]. miR159 did not induce the cleavage of *TCP* mRNAs due to the specificity of sequences, while miR319 mediated the cleavage of *MYB33* and *MYB65* mRNA, which are pivotal targets of miR159 [101,105]. Two miR319 copies were identified in the genome of *M. polymorpha,* which also contains two *MpTCP* genes [106]. However, the two *MpTCP* genes have no possible miR319-targeting site and one target of miR319 was identified as *MpMYB33* [106,107]. In *Physcomitrella* and *Selaginella*, the *TCP* genes also have no miR319-targeting sites [108–110], indicating that miR319 regulation of *CIN*-like *TCP* possibly evolve after the divergence of lycophytes and euphyllophytes.

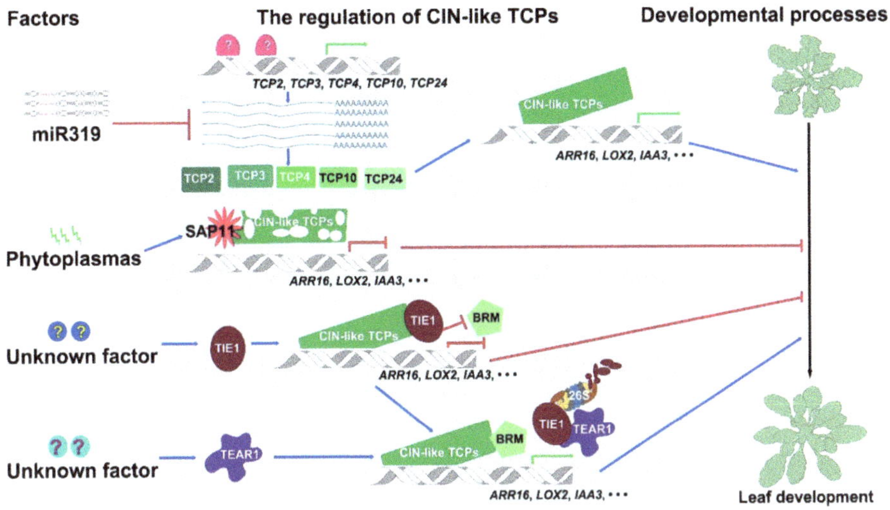

Figure 3. An overview of the regulation mechanisms of CIN-like TCP transcription factors during leaf development. The external stimuli and internal factors are summarized at the left column. The schematic diagram includes the regulation mechanisms of CIN-like TCPs at the transcriptional level, at the post-transcriptional level, and at the protein level. The arrows directly pointing on the double helix symbols indicate transcriptional regulations. The arrows pointing to the proteins indicate the regulations of protein stabilities or antagonistic functions. The proteins related with the "26S" symbols indicate protein degradation through the ubiquitin-26S proteasome pathway. The blue arrows represent the positive regulation, and the red arrows with dash-headed ends indicate the negative regulation. The green arrows and red dash-headed ends at the double-helix icons indicates the activation and repression of gene expression, respectively. All the unknown factors are indicated with question marks. R:FR, red light: far red light ratio; YUCs, YUCCAs; LOX2, LIPOXYGENASE 2; SAP11, SECRETED AY-WB PROTEIN 11; ARR16, ARABIDOPSIS RESPONSE REGULATOR 16; IAA3, INDOLE-3-ACETIC ACID INDUCIBLE 3; BRM, BRAHMA; TIE1, TCP INTERACTOR CONTAINING EAR MOTIF PROTEIN 1; TEAR1, TIE1-ASSOCIATED RING-TYPE E3 LIGASE 1.

7. Chromatin Remodeling Complexes Regulate the Activity of CIN-Like TCPs

The activity of CIN-like TCPs is controlled by chromatin remodeling complexes including SWITCH/SUCROSE NONFERMENTING (SWI/SNF) complex and TCP INTERACTOR CONTAINING EAR MOTIF PROTEIN 1 (TIE1)-TOPLESS (TPL)/TOPLESS-RELATED (TPR) complex at the protein level (Figure 3). SWI/SNF complexes use ATPase to provide the energy in deciding the nucleosome position conformation and thus determining the accessibility of chromatin [111]. *BRAHMA* (*BRM*) encodes a SWI/SNF ATPase in *Arabidopsis* [112–114]. The hypomorphic mutations in *BRM* suppressed the phenotypes including fewer trichomes and smooth margins in *TCP4* overexpression lines [114]. And the hypomorphic *brm* mutants produced curled leaves and delayed leaf maturation resembling the multiple *cin*-like *tcp* mutants, indicating that BRM promotes the activity of CIN-like TCPs (Figure 3) [114]. BRM interacts with TCP4 and together bind to the promoter region of type A *ARABIDOPSIS RESPONSE REGULATOR* (*ARR*) gene *ARR16* to promote the expression of *ARR16* (Figure 3) [115]. The modulation of CIN-like TCP activity by BRM provides a fine regulation of leaf sensitivity to the phytohormone cytokinin (CK) during leaf development.

Compared with the positive regulation of CIN-like TCP activity mediated by BRM, TIE1-TPL/TPR complexes repressed CIN-like TCP activity by recruiting histone deacetylases (HDA) (Figure 3) [116]. TIE1 was identified to regulate TCP activity by analyzing a gain-of-function mutant *tie1-D* obtained by screening a collection of activation tagging mutants for leaf-defective ones. Overexpression of *TIE1* in *tie1-D* or in transgenic plants using CaMV 35S promoter to drive *TIE1* all led to curled and serrated leaves that are observed in the multiple *cin*-like *tcp* mutants [116]. *TIE1* encodes a transcriptional repressor containing a typical EAR motif at the C-terminal end. Indeed, TIE1 has transcriptional repression activity and directly interacts with the corepressor TPL/TPRs through EAR motif. TIE1 also interacts with CIN-like TCPs via the N-terminal domain. Consequently, TIE1 suppresses the activity of CIN-like TCPs by acting as a bridge connecting corepressor TPL/TPRs with CIN-like TCPs during leaf development (Figure 3) [116]. Interestingly, TIE1 also interacted with BRC1 belonging to CYC/TB1-like TCP group [40]. *TIE1* had overlapping expression pattern with *BRC1* in young axillary buds and overexpression of *TIE1* resulted in excessive branches, indicating that *TIE1* also represses the activity of BRC1 during shoot branching [40]. The function of *TIE1* is conserved in controlling shoot branching in cotton (*Gossypium hirsutum*) [117]. GhTIE1 interacted with CYC subclade proteins GhBRC1, GhBRC2, and GhTCP13 *in vivo*. Silencing of *GhTIE1* in cotton seriously decreased shoot branching [117]. A similar mechanism in suppression of CIN-like TCP activity is mediated by SPOROCYTELESS/NOZZLE (SPL/NZZ) during ovule development [118]. SPL/NZZ is a key regulator responsible for promoting the differentiation of megasporocytes. No megasporocytes were formed in the ovules of *spl/nzz* mutants. SPL/NZZ also contains a typical EAR repressor motif at the C-terminal domain and has the transcriptional repression activity. SPL/NZZ uses C-terminal EAR motif to interact with TPL/TPRs and uses its N-terminal domain to interact with CIN-like TCPs [118]. Overexpression of *SPL* in T-DNA activation tagging mutant *spl-D* caused the defective ovule arrangement in ovaries resembling to that of the multiple *cin*-like *tcp* mutants. Consistently, overexpression of the *CIN*-like *TCPs* led to no megasporocytes resembling the phenotype of *spl* loss-of-function mutants [118]. These results indicate that SPL inhibits the activity of CIN-like TCPs in a way similar to TIE1 by connecting TPL/TPR corepressors with CIN-like TCPs.

The regulation of CIN-like TCP activity by TIE1, SPL or BRM during leaf or ovule development is parallel to the regulation of key regulators in auxin signaling. The EAR motif-containing AUXIN (AUX)/INDOLE-3-ACETIC ACID (IAA) repressors mediate auxin signaling by recruiting TPL/TPRs to suppress the activity of AUXIN RESPONSE FACTORS (ARFs) [119–121]. Auxin triggers the degradation of AUX/IAA via 26S proteasome, the released ARFs such as MONOPTEROS (MP) bind to SWI/SNF chromatin remodeling ATPases BRM to promote the accessibility of chromatin and the expression of downstream genes. Interestingly, TIE1 is also an unstable protein as AUX/IAA repressors and the degradation of TIE1 is mediated by an E3 ligase TIE1-ASSOCIATED RING-TYPE E3 LIGASE1 (TEAR1) (Figure 3) [122]. Disruption of *TEAR1* leads to serrated and curled leaves similar to that observed in the multiple *cin*-like *tcp* mutants and *tie1-D* [122]. These findings suggest that *TEAR1* indirectly regulates the activity of CIN-like TCPs by switching the interactors of CIN-like TCPs from TIE1 to BRM (Figure 3), thus changing the chromatin state to control leaf development. However, the signals triggering the TIE1 degradation to release the suppression of CIN-like TCPs by TEAR1 need to be further identified.

8. Concluding Remarks and Perspectives

CIN-like TCPs are key transcription factors essential for plant growth and development in response to environmental cues and internal signals. The temporal and spatial activity of CIN-like TCPs determines cell proliferation, expansion and differentiation of cells in different organs in shaping plant morphology at various developmental stages. Consequently, the fine-tuning of CIN-like TCP activity is critical for plant developmental plasticity. At the transcriptional level, *CIN*-like *TCPs* are dynamically and specifically expressed in organs and also are induced by environmental signals including light and temperature [14,92,95]. However, the upstream regulation which determines the dynamic expression pattern and induction of *CIN*-like *TCP* genes are insufficient. The transcriptional repressor RABBIT EARS (RBE) has been reported to decreased the expression of *TCP4*, *TCP5*, *TCP13* and *TCP17* in promoting petal growth and *TCP4* and *TCP5* are possibly direct targets of RBE [69,71]. More studies on detailed analysis of the promoter regions of *CIN*-like *TCPs* are necessary for elucidating other upstream regulators, especially the direct regulators. The truncated promoters can be used to drive reporters in determining the minimal regions required for the expression patterns of *CIN*-like *TCPs*. Transcription factors directly interacting with the promoters of *CIN*-like *TCPs* could be identified by yeast-one-hybrid screening.

CIN-like TCPs are central for regulating biosynthesis and signaling of different phytohormones including auxin, JA and brassinosteroid (BR) [53,123,124]. However, little is known about how phytohormones regulate CIN-like TCPs. It has been shown that auxin, gibberellin (GA), strigolactone (SL) and cytokinin (CK) regulate *BRC1* belonging to *CYC/TB1*-like *TCP* group of class II *TCPs* [36,125,126]. The decreased auxin level by overexpression of *IAA CARBOXYL METHYLTRANSFERASE1* (*IAMT1*) which converted IAA to methyl-IAA ester led to curly leaves and reduced the expression level of some *CIN*-like *TCPs* [127], indicating that auxin positively regulates *CIN*-like *TCPs* at the transcriptional level. Further studies are needed to determine whether other plant hormones and environmental signals except light and temperature could possibly regulate CIN-like TCPs and how these signals could be integrated to control the activity of CIN-like TCPs.

At the post-transcriptional level, the miR319-*TCP* regulation module is conserved and widely studied in several plant species [13,63]. Could the other miRNAs targeting *CIN*-like *TCPs* exist in different plant species? which are those transcription factors deciding the expression level and pattern of *MIR319* genes? These questions are still open.

At the protein level, we know little about the degradation mechanisms of CIN-like TCPs mediated by 26S proteasome or other protein degradation pathways. The regulation mechanisms of CIN-like TCPs by class I TCP transcription factors and other interacting proteins are still largely unknown. It is still a challenge to thoroughly understand the shaping of plant morphology controlled by the CIN-like TCP-centered network under various environmental and developmental conditions in *Arabidopsis* and the other plant species.

Author Contributions: J.L. and G.Q. wrote the manuscript; J.L. and G.Q. have read and agreed to publish the version of the manuscript. All authors have read and agreed to the published version of the manuscript.

Funding: This research was supported by the National Science Fund for Distinguished Young Scholars of China (Grant No. 31725005) and the Science Fund for the Creative Research Groups of the National Natural Science Foundation of China (Grant No. 31621001).

Acknowledgments: Because of space constraints, we apologize to all colleagues whose work and publications have not been mentioned and cited.

Conflicts of Interest: The authors declare no conflict of interest.

References

1. Gaillochet, C.; Lohmann, J.U. The never-ending story: From pluripotency to plant developmental plasticity. *Development* **2015**, *142*, 2237–2249. [CrossRef]
2. Doebley, J.; Stec, A.; Gustus, C. Teosinte Branched1 and the origin of maize - evidence for epistasis and the evolution of dominance. *Genetics* **1995**, *141*, 333–346. [PubMed]
3. Luo, D.; Carpenter, R.; Vincent, C.; Copsey, L.; Coen, E. Origin of floral asymmetry in Antirrhinum. *Nature* **1996**, *383*, 794–799. [CrossRef] [PubMed]
4. Kosugi, S.; Ohashi, Y. PCF1 and PCF2 specifically bind to cis elements in the rice proliferating cell nuclear antigen gene. *Plant Cell* **1997**, *9*, 1607–1619. [PubMed]
5. Doebley, J.; Stec, A.; Hubbard, L. The evolution of apical dominance in maize. *Nature* **1997**, *386*, 485–488. [CrossRef]
6. Luo, D.; Carpenter, R.; Copsey, L.; Vincent, C.; Clark, J.; Coen, E. Control of organ asymmetry in flowers of Antirrhinum. *Cell* **1999**, *99*, 367–376. [CrossRef]

7. Cubas, P.; Lauter, N.; Doebley, J.; Coen, E. The TCP domain: A motif found in proteins regulating plant growth and development. *Plant J.* **1999**, *18*, 215–222. [CrossRef]
8. Kosugi, S.; Ohashi, Y. DNA binding and dimerization specificity and potential targets for the TCP protein family. *Plant J.* **2002**, *30*, 337–348. [CrossRef]
9. Martin-Trillo, M.; Cubas, P. TCP genes: A family snapshot ten years later. *Trends Plant Sci.* **2010**, *15*, 31–39. [CrossRef]
10. Howarth, D.G.; Donoghue, M.J. Phylogenetic analysis of the "'ECE" (CYC/TB1) clade reveals duplications predating the core eudicots. *Proc. Natl. Acad. Sci. USA* **2006**, *103*, 9101–9106. [CrossRef]
11. Nath, U.; Crawford, B.C.W.; Carpenter, R.; Coen, E. Genetic control of surface curvature. *Science* **2003**, *299*, 1404–1407. [CrossRef] [PubMed]
12. Crawford, B.C.W.; Nath, U.; Carpenter, R.; Coen, E.S. CINCINNATA controls both cell differentiation and growth in petal lobes and leaves of antirrhinum. *Plant Physiol.* **2004**, *135*, 244–253. [CrossRef] [PubMed]
13. Palatnik, J.F.; Allen, E.; Wu, X.; Schommer, C.; Schwab, R.; Carrington, J.C.; Weigel, D. Control of leaf morphogenesis by microRNAs. *Nature* **2003**, *425*, 257–263. [CrossRef] [PubMed]
14. Han, X.; Yu, H.; Yuan, R.; Yang, Y.; An, F.; Qin, G. Arabidopsis transcription factor TCP5 controls plant thermomorphogenesis by positively regulating PIF4 activity. *Iscience* **2019**, *15*, 611–622. [CrossRef]
15. Katoh, K.; Rozewicki, J.; Yamada, K.D. MAFFT online service: Multiple sequence alignment, interactive sequence choice and visualization. *Brief. Bioinform.* **2019**, *20*, 1160–1166. [CrossRef]
16. Minh, B.Q.; Schmidt, H.A.; Chernomor, O.; Schrempf, D.; Woodhams, M.D.; von Haeseler, A.; Lanfear, R. IQ-TREE 2: New models and efficient methods for phylogenetic inference in the genomic era. *Mol. Biol. and Evol.* **2020**, *37*, 1530–1534. [CrossRef]
17. Liu, M.M.; Wang, M.M.; Yang, J.; Wen, J.; Guo, P.C.; Wu, Y.W.; Ke, Y.Z.; Li, P.F.; Li, J.N.; Du, H. Evolutionary and comparative expression analyses of TCP transcription factor gene family in land plants. *Int. J. Mol. Sci.* **2019**, *20*, 3591. [CrossRef]
18. Rensing, S.A.; Lang, D.; Zimmer, A.D.; Terry, A.; Salamov, A.; Shapiro, H.; Nishiyama, T.; Perroud, P.F.; Lindquist, E.A.; Kamisugi, Y.; et al. The *Physcomitrella* genome reveals evolutionary insights into the conquest of land by plants. *Science* **2008**, *319*, 64–69. [CrossRef]
19. Bowman, J.L.; Kohchi, T.; Yamato, K.T.; Jenkins, J.; Shu, S.Q.; Ishizaki, K.; Yamaoka, S.; Nishihama, R.; Nakamura, Y.; Berger, F.; et al. Insights into land plant evolution garnered from the Marchantia polymorpha genome. *Cell* **2017**, *171*, 287–304. [CrossRef]
20. Navaud, O.; Dabos, P.; Carnus, E.; Tremousaygue, D.; Herve, C. TCP transcription factors predate the emergence of land plants. *J. Mol. Evol.* **2007**, *65*, 23–33. [CrossRef]
21. Gubitz, T.; Caldwell, A.; Hudson, A. Rapid molecular evolution of CYCLOIDEA-like genes in Antirrhinum and its relatives. *Mol. Bio. Evol.* **2003**, *20*, 1537–1544. [CrossRef]
22. Chapman, M.A.; Leebens-Mack, J.H.; Burke, J.M. Positive selection and expression divergence following gene duplication in the sunflower CYCLOIDEA gene family. *Mol. Bio. Evol.* **2008**, *25*, 1260–1273. [CrossRef]
23. Mondragon-Palomino, M.; Trontin, C. High time for a roll call: Gene duplication and phylogenetic relationships of TCP-like genes in monocots. *Ann. Bot.* **2011**, *107*, 1533–1544. [CrossRef] [PubMed]
24. Horn, S.; Pabón-Mora, N.; Theuß, V.S.; Busch, A.; Zachgo, S. Analysis of the CYC/TB1 class of TCP transcription factors in basal angiosperms and magnoliids. *Plant J.* **2015**, *81*, 559–571. [CrossRef] [PubMed]
25. Citerne, H.L.; Reyes, E.; Le Guilloux, M.; Delannoy, E.; Simonnet, F.; Sauquet, H.; Weston, P.H.; Nadot, S.; Damerval, C. Characterization of CYCLOIDEA-like genes in Proteaceae, a basal eudicot family with multiple shifts in floral symmetry. *Ann. Bot.* **2017**, *119*, 367–378. [CrossRef] [PubMed]
26. Busch, A.; Deckena, M.; Almeida-Trapp, M.; Kopischke, S.; Kock, C.; Schuessler, E.; Tsiantis, M.; Mithoefer, A.; Zachgo, S. MpTCP1 controls cell proliferation and redox processes in *Marchantia polymorpha*. *New Phytol.* **2019**, *224*, 1627–1641. [CrossRef] [PubMed]
27. Floyd, S.K.; Bowman, J.L. The ancestral developmental tool kit of land plants. *Int. J. Plant Sci.* **2007**, *168*, 1–35. [CrossRef]
28. Koelsch, A.; Gleissberg, S. Diversification of CYCLOIDEA-like TCP genes in the basal eudicot families Fumariaceae and Papaveraceae s.str. *Plant Biol.* **2006**, *8*, 680–687. [CrossRef]
29. Damerval, C.; Le Guilloux, M.; Jager, M.; Charon, C. Diversity and evolution of CYCLOIDEA-like TCP genes in relation to flower development in Papaveraceae. *Plant Physiol.* **2007**, *143*, 759–772. [CrossRef]

30. Bartlett, M.E.; Specht, C.D. Changes in expression pattern of the *teosinte branched1*-like genes in the Zingiberales provide a mechanism for evolutionary shifts in symmetry across the order. *Am. J. Bot.* **2011**, *98*, 227–243. [CrossRef]
31. Jabbour, F.; Cossard, G.; Le Guilloux, M.; Sannier, J.; Nadot, S.; Damerval, C. Specific duplication and dorsoventrally asymmetric expression patterns of *Cycloidea*-like genes in zygomorphic species of Ranunculaceae. *PLoS ONE* **2014**, *9*, e95727. [CrossRef] [PubMed]
32. Resentini, F.; Felipo-Benavent, A.; Colombo, L.; Blazquez, M.A.; Alabadi, D.; Masiero, S. TCP14 and TCP15 mediate the promotion of seed germination by gibberellins in *Arabidopsis thaliana*. *Mol. Plant* **2015**, *8*, 482–485. [CrossRef]
33. Tatematsu, K.; Nakabayashi, K.; Kamiya, Y.; Nambara, E. Transcription factor AtTCP14 regulates embryonic growth potential during seed germination in *Arabidopsis thaliana*. *Plant J.* **2008**, *53*, 42–52. [CrossRef] [PubMed]
34. Alvarez, J.P.; Furumizu, C.; Efroni, I.; Eshed, Y.; Bowman, J.L. Active suppression of a leaf meristem or chestrates determinate leaf growth. *Elife* **2016**, *5*, e15023. [CrossRef] [PubMed]
35. Challa, K.R.; Rath, M.; Nath, U. The CIN-TCP transcription factors promote commitment to differentiation in Arabidopsis leaf pavement cells via both auxin-dependent and independent pathways. *PLoS Genet.* **2019**, *15*, e1007988. [CrossRef] [PubMed]
36. Aguilar-Martinez, J.A.; Poza-Carrion, C.; Cubas, P. Arabidopsis BRANCHED1 acts as an integrator of branching signals within axillary buds. *Plant Cell* **2007**, *19*, 458–472. [CrossRef]
37. Gonzalez-Grandio, E.; Poza-Carrion, C.; Sorzano, C.O.S.; Cubas, P. BRANCHED1 promotes axillary bud dormancy in response to shade in Arabidopsis. *Plant Cell* **2013**, *25*, 834–850. [CrossRef]
38. Gonzalez-Grandio, E.; Pajoro, A.; Franco-Zorrilla, J.M.; Tarancon, C.; Immink, R.G.H.; Cubas, P. Abscisic acid signaling is controlled by a BRANCHED1/HD-ZIP I cascade in Arabidopsis axillary buds. *Proc. Natl. Acad. Sci. USA* **2017**, *114*, E245–E254. [CrossRef]
39. Wang, M.; Le Moigne, M.-A.; Bertheloot, J.; Crespel, L.; Perez-Garcia, M.-D.; Oge, L.; Demotes-Mainard, S.; Hamama, L.; Daviere, J.-M.; Sakr, S. BRANCHED1: A key hub of shoot branching. *Front. Plant Sci.* **2019**, *10*, 76. [CrossRef]
40. Yang, Y.; Nicolas, M.; Zhang, J.; Yu, H.; Guo, D.; Yuan, R.; Zhang, T.; Yang, J.; Cubas, P.; Qin, G. The TIE1 transcriptional repressor controls shoot branching by directly repressing *BRANCHED1* in Arabidopsis. *PLoS Genet.* **2018**, *14*, e1007296. [CrossRef]
41. Ballester, P.; Navarrete-Gomez, M.; Carbonero, P.; Onate-Sanchez, L.; Ferrandiz, C. Leaf expansion in Arabidopsis is controlled by a TCP-NGA regulatory module likely conserved in distantly related species. *Physiol. Planta.* **2015**, *155*, 21–32. [CrossRef]
42. Bresso, E.; Chorostecki, U.; Rodriguez, R.E.; Palatnik, J.F.; Schommera, C. Spatial control of gene expression by miR319-regulated TCP transcription factors in leaf development. *Plant Physiol.* **2018**, *176*, 1694–1708. [CrossRef] [PubMed]
43. Balsemao-Pires, E.; Andrade, L.R.; Sachetto-Martins, G. Functional study of TCP23 in *Arabidopsis thaliana* during plant development. *Plant Physiol. Biochem.* **2013**, *67*, 120–125. [CrossRef] [PubMed]
44. Wu, J.-F.; Tsai, H.-L.; Joanito, I.; Wu, Y.-C.; Chang, C.-W.; Li, Y.-H.; Wang, Y.; Hong, J.C.; Chu, J.-W.; Hsu, C.-P.; et al. LWD-TCP complex activates the morning gene *CCA1* in Arabidopsis. *Nat. Commun.* **2016**, *7*, 13181. [CrossRef] [PubMed]
45. Chang, S.H.; Tan, C.M.; Wu, C.-T.; Lin, T.-H.; Jiang, S.-Y.; Liu, R.-C.; Tsai, M.-C.; Su, L.-W.; Yang, J.-Y. Alterations of plant architecture and phase transition by the phytoplasma virulence factor SAP11. *J. Exp. Bot.* **2018**, *69*, 5389–5401. [CrossRef]
46. Pecher, P.; Moro, G.; Canale, M.C.; Capdevielle, S.; Singh, A.; MacLean, A.; Sugio, A.; Kuo, C.-H.; Lopes, J.R.S.; Hogenhout, S.A. Phytoplasma SAP11 effector destabilization of TCP transcription factors differentially impact development and defence of Arabidopsis versus maize. *PLoS Pathog.* **2019**, *15*, e1008035. [CrossRef]
47. Sugio, A.; Kingdom, H.N.; MacLean, A.M.; Grieve, V.M.; Hogenhout, S.A. Phytoplasma protein effector SAP11 enhances insect vector reproduction by manipulating plant development and defense hormone biosynthesis. *Proc. Natl. Acad. Sci. USA* **2011**, *108*, E1254–E1263. [CrossRef]
48. Sugio, A.; MacLean, A.M.; Hogenhout, S.A. The small phytoplasma virulence effector SAP11 contains distinct domains required for nuclear targeting and CIN-TCP binding and destabilization. *New Phytol.* **2014**, *202*, 838–848. [CrossRef]

49. Tan, C.M.; Li, C.-H.; Tsao, N.-W.; Su, L.-W.; Lu, Y.-T.; Chang, S.H.; Lin, Y.Y.; Liou, J.-C.; Hsieh, L.-C.; Yu, J.-Z.; et al. Phytoplasma SAP11 alters 3-isobutyl-2-methoxypyrazine biosynthesis in *Nicotiana benthamiana* by suppressing NbOMT1. *J. Exp. Bot.* **2016**, *67*, 4415–4425. [CrossRef]
50. Wang, N.; Yang, H.; Yin, Z.; Liu, W.; Sun, L.; Wu, Y. Phytoplasma effector SWP1 induces witches' broom symptom by destabilizing the TCP transcription factor BRANCHED1. *Mol. Plant Pathol.* **2018**, *19*, 2623–2634. [CrossRef]
51. Danisman, S.; van der Wal, F.; Dhondt, S.; Waites, R.; de Folter, S.; Bimbo, A.; van Dijk, A.-J.; Muino, J.M.; Cutri, L.; Dornelas, M.C.; et al. Arabidopsis Class I and Class II TCP transcription factors regulate jasmonic acid metabolism and leaf development antagonistically. *Plant Physiol.* **2012**, *159*, 1511–1523. [CrossRef] [PubMed]
52. Koyama, T. A hidden link between leaf development and senescence. *Plant Sci.* **2018**, *276*, 105–110. [CrossRef] [PubMed]
53. Schommer, C.; Palatnik, J.F.; Aggarwal, P.; Chetelat, A.; Cubas, P.; Farmer, E.E.; Nath, U.; Weigel, D. Control of jasmonate biosynthesis and senescence by miR319 targets. *PLoS Biol.* **2008**, *6*, 1991–2001. [CrossRef]
54. Sarvepalli, K.; Nath, U. CIN-TCP transcription factors: Transiting cell proliferation in plants. *Iubmb Life* **2018**, *70*, 718–731. [CrossRef] [PubMed]
55. Nicolas, M.; Cubas, P. TCP factors: New kids on the signaling block. *Curr. Opin. Plant Biol.* **2016**, *33*, 33–41. [CrossRef] [PubMed]
56. Das Gupta, M.; Aggarwal, P.; Nath, U. CINCINNATA in *Antirrhinum majus* directly modulates genes involved in cytokinin and auxin signaling. *New Phytol.* **2014**, *204*, 901–912. [CrossRef]
57. Schommer, C.; Debernardi, J.M.; Bresso, E.G.; Rodriguez, R.E.; Palatnik, J.F. Repression of cell proliferation by miR319-regulated TCP4. *Mol. Plant* **2014**, *7*, 1533–1544. [CrossRef]
58. Koyama, T.; Mitsuda, N.; Seki, M.; Shinozaki, K.; Ohme-Takagi, M. TCP transcription factors regulate the activities of ASYMMETRIC LEAVES1 and miR164, as well as the auxin response, during differentiation of leaves in Arabidopsis. *Plant Cell* **2010**, *22*, 3574–3588. [CrossRef]
59. Koyama, T.; Sato, F.; Ohme-Takagi, M. Roles of miR319 and TCP transcription factors in leaf development. *Plant Physiol.* **2017**, *175*, 874–885. [CrossRef]
60. Du, J.; Hu, S.; Yu, Q.; Wang, C.; Yang, Y.; Sun, H.; Yang, Y.; Sun, X. Genome-wide identification and characterization of BrrTCP transcription factorsin *Brassica rapa* ssp rapa. *Front. Plant Sci.* **2017**, *8*, 1588. [CrossRef]
61. Seki, K.; Komatsu, K.; Tanaka, K.; Hiraga, M.; Kajiya-Kanegae, H.; Matsumura, H.; Uno, Y. A CIN-like TCP transcription factor (LsTCP4) having retrotransposon insertion associates with a shift from Salinas type to Empire type in crisphead lettuce (*Lactuca sativa* L.). *Hortic. Res.* **2020**, *7*, 15. [CrossRef] [PubMed]
62. Testone, G.; Baldoni, E.; Iannelli, M.A.; Nicolodi, C.; Di Giacomo, E.; Pietrini, F.; Mele, G.; Giannino, D.; Frugis, G. Transcription factor networks in leaves of *Cichorium endivia*: New insights into the relationship between photosynthesis and leaf development. *Plants-Basel* **2019**, *8*, 531. [CrossRef]
63. Ori, N.; Cohen, A.R.; Etzioni, A.; Brand, A.; Yanai, O.; Shleizer, S.; Menda, N.; Amsellem, Z.; Efroni, I.; Pekker, I.; et al. Regulation of LANCEOLATE by miR319 is required for compound-leaf development in tomato. *Nat. Genet.* **2007**, *39*, 787–791. [CrossRef] [PubMed]
64. Shleizer-Burko, S.; Burko, Y.; Ben-Herzel, O.; Ori, N. Dynamic growth program regulated by LANCEOLATE enables flexible leaf patterning. *Development* **2011**, *138*, 695–704. [CrossRef] [PubMed]
65. Burko, Y.; Shleizer-Burko, S.; Yanai, O.; Shwartz, I.; Zelnik, I.D.; Jacob-Hirsch, J.; Kela, I.; Eshed-Williams, L.; Ori, N. A Role for APETALA1/FRUITFULL transcription factors in tomato leaf development. *Plant Cell* **2013**, *25*, 2070–2083. [CrossRef] [PubMed]
66. Tanaka, Y.; Yamamura, T.; Oshima, Y.; Mitsuda, N.; Koyama, T.; Ohme-Takagi, M.; Terakawa, T. Creating ruffled flower petals in Cyclamen persicum by expression of the chimeric cyclamen TCP repressor. *Plant Biotech.* **2011**, *28*, 141–147. [CrossRef]
67. Mao, Y.; Wu, F.; Yu, X.; Bai, J.; Zhong, W.; He, Y. microRNA319a-targeted *Brassica rapa* ssp *pekinensis* TCP genes modulate head shape in Chinese cabbage by differential cell division arrest in leaf regions. *Plant Physiol.* **2014**, *164*, 710–720. [CrossRef]
68. Efroni, I.; Blum, E.; Goldshmidt, A.; Eshed, Y. A protracted and dynamic maturation schedule underlies arabidopsis leaf development. *Plant Cell* **2008**, *20*, 2293–2306. [CrossRef]

69. Huang, T.; Irish, V.F. Temporal control of plant organ growth by TCP transcription factors. *Curr. Biol.* **2015**, *25*, 1765–1770. [CrossRef]
70. Nag, A.; King, S.; Jack, T. MiR319a targeting of TCP4 is critical for petal growth and development in Arabidopsis. *Proc. Natl. Acad. Sci. USA* **2009**, *106*, 22534–22539. [CrossRef]
71. Li, J.; Wang, Y.; Zhang, Y.; Wang, W.; Irish, V.F.; Huang, T. RABBIT EARS regulates the transcription of TCP4 during petal development in Arabidopsis. *J. Exp. Bot.* **2016**, *67*, 6473–6480. [CrossRef] [PubMed]
72. Koyama, T.; Furutani, M.; Tasaka, M.; Ohme-Takagi, M. TCP transcription factors control the morphology of shoot lateral organs via negative regulation of the expression of boundary-specific genes in Arabidopsis. *Plant Cell* **2007**, *19*, 473–484. [CrossRef] [PubMed]
73. Ono, M.; Hiyama, S.; Higuchi, Y.; Kamada, H.; Nitasaka, E.; Koyama, T.; Mitsuda, N.; Ohme-Takagi, M.; Sage-Ono, K. Morphological changes in *Ipomoea nil* using chimeric repressors of Arabidopsis TCP3 and TCP5. *Plant Biotech.* **2012**, *29*, 457–463. [CrossRef]
74. Narumi, T.; Aida, R.; Koyama, T.; Yamaguchi, H.; Sasaki, K.; Shikata, M.; Nakayama, M.; Ohme-Takagi, M.; Ohtsubo, N. Arabidopsis chimeric TCP3 repressor produces novel floral traits in *Torenia fournieri* and *Chrysanthemum morifolium*. *Plant Biotech.* **2011**, *28*, 131–140. [CrossRef]
75. Koyama, T.; Ohme-Takagi, M.; Sato, F. Generation of serrated and wavy petals by inhibition of the activity of TCP transcription factors in *Arabidopsis thaliana*. *Plant Signal. Behav.* **2011**, *6*, 697–699. [CrossRef] [PubMed]
76. Vadde, B.V.L.; Challa, K.R.; Sunkara, P.; Hegde, A.S.; Nath, U. The TCP4 transcription factor directly activates *TRICHOMELESS1* and *2* and suppresses trichome initiation. *Plant Physiol.* **2019**, *181*, 1587–1599. [CrossRef] [PubMed]
77. Fan, D.; Ran, L.; Hu, J.; Ye, X.; Xu, D.; Li, J.; Su, H.; Wang, X.; Ren, S.; Luo, K. MiR319a/TCP module and DELLA protein regulate trichome initiation synergistically and improve insect defenses in Populus tomentosa. *New Phytol.* **2020**. Available online: https://pubmed.ncbi.nlm.nih.gov/32270484/ (accessed on 8 April 2020).
78. Cao, J.F.; Zhao, B.; Huang, C.C.; Chen, Z.W.; Zhao, T.; Liu, H.R.; Hu, G.J.; Shangguan, X.X.; Shan, C.M.; Wang, L.J.Y.; et al. The miR319-targeted GhTCP4 promotes the transition from cell elongation to wall thickening in cotton fiber. *Mol. Plant* **2020**. Available online: https://www.sciencedirect.com/science/article/pii/S1674205220301416 (accessed on 16 May 2020).
79. Challa, K.R.; Aggarwal, P.; Nath, U. Activation of YUCCA5 by the transcription factor TCP4 integrates developmental and environmental signals to promote hypocotyl elongation in Arabidopsis. *Plant Cell* **2016**, *28*, 2117–2130. [CrossRef]
80. Liu, J.; Cheng, X.; Liu, P.; Li, D.; Chen, T.; Gu, X.; Sun, J. MicroRNA319-regulated TCPs interact with FBHs and PFT1 to activate CO transcription and control flowering time in Arabidopsis. *PLoS Genet.* **2017**, *13*, e1006833. [CrossRef]
81. Silva, G.F.F.; Silva, E.M.; Correa, J.P.O.; Vicente, M.H.; Jiang, N.; Notini, M.M.; Junior, A.C.; De Jesus, F.A.; Castilho, P.; Carrera, E.; et al. Tomato floral induction and flower development are orchestrated by the interplay between gibberellin and two unrelated microRNA-controlled modules. *New Phytol.* **2019**, *221*, 1328–1344. [CrossRef]
82. Zhang, C.; Ding, Z.; Wu, K.; Yang, L.; Li, Y.; Yang, Z.; Shi, S.; Liu, X.; Zhao, S.; Yang, Z.; et al. Suppression of jasmonic acid-mediated defense by viral-inducible microRNA319 facilitates virus infection in rice. *Mol. Plant* **2016**, *9*, 1302–1314. [CrossRef]
83. Chen, M.; Chory, J.; Fankhauser, C. Light signal transduction in higher plants. *Annu. Rev. Genet.* **2004**, *38*, 87–117. [CrossRef] [PubMed]
84. Feng, S.; Martinez, C.; Gusmaroli, G.; Wang, Y.; Zhou, J.; Wang, F.; Chen, L.; Yu, L.; Iglesias-Pedraz, J.M.; Kircher, S.; et al. Coordinated regulation of *Arabidopsis thaliana* development by light and gibberellins. *Nature* **2008**, *451*, 475–479. [CrossRef] [PubMed]
85. Song, Y.H.; Shim, J.S.; Kinmonth-Schultz, H.A.; Imaizumi, T. Photoperiodic flowering: Time measurement mechanisms in leaves. *Annu. Rev. Plant Biol.* **2015**, *66*, 441–464. [CrossRef] [PubMed]
86. Dong, J.; Sun, N.; Yang, J.; Deng, Z.; Lan, J.; Qin, G.; He, H.; Deng, X.W.; Irish, V.F.; Chen, H.; et al. The transcription factors TCP4 and PIF3 antagonistically regulate organ-specific light induction of *SAUR* genes to modulate cotyledon opening during de-etiolation in Arabidopsis. *Plant Cell* **2019**, *31*, 1155–1170. [CrossRef]
87. von Arnim, A.; Deng, X.W. Light control of seedling development. *Annu. Rev. Plant Physiol. Mol. Biol.* **1996**, *47*, 215–243. [CrossRef]

88. Kami, C.; Lorrain, S.; Hornitschek, P.; Fankhauser, C. Light-regulated plant growth and development. *Curr. Top. Dev. Biol.* **2010**, *91*, 29–66.
89. Leivar, P.; Monte, E.; Oka, Y.; Liu, T.; Carle, C.; Castillon, A.; Huq, E.; Quail, P.H. Multiple phytochrome-interacting bHLH transcription factors repress premature seedling photomorphogenesis in darkness. *Curr. Biol.* **2008**, *18*, 1815–1823. [CrossRef]
90. Shi, H.; Lyu, M.; Luo, Y.; Liu, S.; Li, Y.; He, H.; Wei, N.; Deng, X.W.; Zhong, S. Genome-wide regulation of light-controlled seedling morphogenesis by three families of transcription factors. *Proc. Natl. Acad. Sci. USA* **2018**, *115*, 6482–6487. [CrossRef]
91. Mathews, S. Phytochrome-mediated development in land plants: Red light sensing evolves to meet the challenges of changing light environments. *Mol. Ecol.* **2006**, *15*, 3483–3503. [CrossRef]
92. Zhou, Y.; Zhang, D.; An, J.; Yin, H.; Fang, S.; Chu, J.; Zhao, Y.; Li, J. TCP transcription factors regulate shade avoidance via directly mediating the expression of both PHYTOCHROME INTERACTING FACTORs and auxin biosynthetic genes. *Plant Physiol.* **2018**, *176*, 1850–1861. [CrossRef]
93. Quint, M.; Delker, C.; Franklin, K.A.; Wigge, P.A.; Halliday, K.J.; van Zanten, M. Molecular and genetic control of plant thermomorphogenesis. *Nat. Plants* **2016**, *2*, 15190. [CrossRef]
94. Wigge, P.A. Ambient temperature signalling in plants. *Curr. Opin. Plant Biol.* **2013**, *16*, 661–666. [CrossRef] [PubMed]
95. Zhou, Y.; Xun, Q.; Zhang, D.; Lv, M.; Ou, Y.; Li, J. TCP transcription factors associate with PHYTOCHROME INTERACTING FACTOR 4 and CRYPTOCHROME 1 to regulate thermomorphogenesis in *Arabidopsis thaliana*. *Iscience* **2019**, *15*, 600–610. [CrossRef] [PubMed]
96. Franklin, K.A.; Lee, S.H.; Patel, D.; Kumar, S.V.; Spartz, A.K.; Gu, C.; Ye, S.Q.; Yu, P.; Breen, G.; Cohen, J.D.; et al. PHYTOCHROME-INTERACTING FACTOR 4 (PIF4) regulates auxin biosynthesis at high temperature. *Proc. Natl. Acad. Sci. USA* **2011**, *108*, 20231–20235. [CrossRef] [PubMed]
97. Sun, J.Q.; Qi, L.L.; Li, Y.N.; Chu, J.F.; Li, C.Y. PIF4-mediated activation of YUCCA8 expression integrates temperature into the auxin pathway in regulating Arabidopsis hypocotyl growth. *PLoS Genet.* **2012**, *8*, e1002594. [CrossRef] [PubMed]
98. Hogenhout, S.A.; Oshima, K.; Ammar, E.D.; Kakizawa, S.; Kingdom, H.N.; Namba, S. Phytoplasmas: Bacteria that manipulate plants and insects. *Mol. Plant Pathol.* **2008**, *9*, 403–423. [CrossRef]
99. Reichel, M.; Millar, A.A. Specificity of plant microRNA target MIMICs: Cross-targeting of miR159 and miR319. *J. Plant Physiol.* **2015**, *180*, 45–48. [CrossRef]
100. Llave, C.; Xie, Z.X.; Kasschau, K.D.; Carrington, J.C. Cleavage of Scarecrow-like mRNA targets directed by a class of Arabidopsis miRNA. *Science* **2002**, *297*, 2053–2056. [CrossRef]
101. Palatnik, J.F.; Wollmann, H.; Schommer, C.; Schwab, R.; Boisbouvier, J.; Rodriguez, R.; Warthmann, N.; Allen, E.; Dezulian, T.; Huson, D.; et al. Sequence and expression differences underlie functional specialization of Arabidopsis microRNAs miR159 and miR319. *Dev. Cell* **2007**, *13*, 115–125. [CrossRef]
102. Sun, X.D.; Wang, C.D.; Xiang, N.; Li, X.; Yang, S.H.; Du, J.C.; Yang, Y.P.; Yang, Y.Q. Activation of secondary cell wall biosynthesis by miR319-targeted TCP4 transcription factor. *Plant Biotechnol. J.* **2017**, *15*, 1284–1294. [CrossRef]
103. Vadde, B.V.L.; Challa, K.R.; Nath, U. The TCP4 transcription factor regulates trichome cell differentiation by directly activating *GLABROUS INFLORESCENCE STEMS* in *Arabidopsis thaliana*. *Plant J.* **2018**, *93*, 259–269. [CrossRef]
104. Wang, H.; Mao, Y.; Yang, J.; He, Y. TCP24 modulates secondary cell wall thickening and anther endothecium development. *Front. Plant Sci.* **2015**, *6*, 436. [CrossRef] [PubMed]
105. Li, Y.; Li, C.Q.; Ding, G.H.; Jin, Y.X. Evolution of MIR159/319 microRNA genes and their post-transcriptional regulatory link to siRNA pathways. *Bmc Evol. Biol.* **2011**, *11*, 122. [CrossRef]
106. Tsuzuki, M.; Nishihama, R.; Ishizaki, K.; Kurihara, Y.; Matsui, M.; Bowman, J.L.; Kohchi, T.; Hamada, T.; Watanabe, Y. Profiling and characterization of small rnas in the liverwort, *Marchantia polymorpha*, belonging to the first diverged land plants. *Plant Cell Physiol.* **2016**, *57*, 359–372. [CrossRef] [PubMed]
107. Flores-Sandoval, E.; Dierschke, T.; Fisher, T.J.; Bowman, J.L. Efficient and inducible use of artificial microRNAs in *Marchantia polymorpha*. *Plant Cell Physiol.* **2016**, *57*, 281–290. [CrossRef]
108. Axtell, M.J.; Snyder, J.A.; Bartel, D.P. Common functions for diverse small RNAs of land plants. *Plant Cell* **2007**, *19*, 1750–1769. [CrossRef] [PubMed]

109. Talmor-Neiman, M.; Stav, R.; Frank, W.; Voss, B.; Arazi, T. Novel micro-RNAs and intermediates of micro-RNA biogenesis from moss. *Plant J.* **2006**, *47*, 25–37. [CrossRef] [PubMed]
110. Warthmann, N.; Das, S.; Lanz, C.; Weigel, D. Comparative analysis of the MIR319a microRNA locus in Arabidopsis and related Brassicaceae. *Mol. Biol. Evol.* **2008**, *25*, 892–902. [CrossRef] [PubMed]
111. Clapier, C.R.; Cairns, B.R. The biology of chromatin remodeling complexes. *Annu. Rev. Biochem.* **2009**, *78*, 273–304. [CrossRef]
112. Ho, L.; Crabtree, G.R. Chromatin remodelling during development. *Nature* **2010**, *463*, 474–484. [CrossRef]
113. Tang, X.R.; Hou, A.F.; Babu, M.; Nguyen, V.; Hurtado, L.; Lu, Q.; Reyes, J.C.; Wang, A.M.; Keller, W.A.; Harada, J.J.; et al. The Arabidopsis BRAHMA chromatin-remodeling ATPase is involved in repression of seed maturation genes in leaves. *Plant Physiol.* **2008**, *147*, 1143–1157. [CrossRef] [PubMed]
114. Hurtado, L.; Farrona, S.; Reyes, J.C. The putative SWI/SNF complex subunit BRAHMA activates flower homeotic genes in *Arabidopsis thaliana*. *Plant Mol. Biol.* **2006**, *62*, 291–304. [CrossRef] [PubMed]
115. Efroni, I.; Han, S.-K.; Kim, H.J.; Wu, M.-F.; Steiner, E.; Birnbaum, K.D.; Hong, J.C.; Eshed, Y.; Wagner, D. Regulation of leaf maturation by chromatin-mediated modulation of cytokinin responses. *Dev. Cell* **2013**, *24*, 438–445. [CrossRef] [PubMed]
116. Tao, Q.; Guo, D.; Wei, B.; Zhang, F.; Pang, C.; Jiang, H.; Zhang, J.; Wei, T.; Gu, H.; Qu, L.-J.; et al. The TIE1 transcriptional repressor links TCP transcription factors with TOPLESS/TOPLESS-RELATED corepressors and modulates leaf development in Arabidopsis. *Plant Cell* **2013**, *25*, 421–437. [CrossRef] [PubMed]
117. Diao, Y.; Zhan, J.; Zhao, Y.; Liu, L.; Liu, P.; Wei, X.; Ding, Y.; Sajjad, M.; Hu, W.; Wang, P.; et al. GhTIE1 regulates branching through modulating the transcriptional activity of TCPs in cotton and Arabidopsis. *Front. Plant Sci.* **2019**, *10*, 1348. [CrossRef]
118. Wei, B.; Zhang, J.; Pang, C.; Yu, H.; Guo, D.; Jiang, H.; Ding, M.; Chen, Z.; Tao, Q.; Gu, H.; et al. The molecular mechanism of SPOROCYTELESS/NOZZLE in controlling Arabidopsis ovule development. *Cell Res.* **2015**, *25*, 121–134. [CrossRef]
119. Guilfoyle, T.; Hagen, G.; Ulmasov, T.; Murfett, J. How does auxin turn on genes? *Plant Physiol.* **1998**, *118*, 341–347. [CrossRef]
120. Kepinski, S.; Leyser, O. The Arabidopsis F-box protein TIR1 is an auxin receptor. *Nature* **2005**, *435*, 446–451. [CrossRef]
121. Chapman, E.J.; Estelle, M. Mechanism of auxin-regulated gene expression in plants. *Annu. Rev. Genet.* **2009**, *43*, 265–285. [CrossRef]
122. Zhang, J.; Wei, B.; Yuan, R.; Wang, J.; Ding, M.; Chen, Z.; Yu, H.; Qin, G. The Arabidopsis RING-Type E3 ligase TEAR1 controls leaf development by targeting the TIE1 transcriptional repressor for degradation. *Plant Cell* **2017**, *29*, 243–259. [CrossRef]
123. Reinhardt, B.; Hanggi, E.; Muller, S.; Bauch, M.; Wyrzykowska, J.; Kerstetter, R.; Poethig, S.; Fleming, A.J. Restoration of DWF4 expression to the leaf margin of a *dwf4* mutant is sufficient to restore leaf shape but not size: The role of the margin in leaf development. *Plant J.* **2007**, *52*, 1094–1104. [CrossRef] [PubMed]
124. Li, S.; Zachgo, S. TCP3 interacts with R2R3-MYB proteins, promotes flavonoid biosynthesis and negatively regulates the auxin response in *Arabidopsis thaliana*. *Plant J.* **2013**, *76*, 901–913. [CrossRef] [PubMed]
125. Rameau, C.; Bertheloot, J.; Leduc, N.; Andrieu, B.; Foucher, F.; Sakr, S. Multiple pathways regulate shoot branching. *Front. Plant Sci.* **2014**, *5*, 741. [CrossRef] [PubMed]
126. Braun, N.; de Saint Germain, A.; Pillot, J.-P.; Boutet-Mercey, S.; Dalmais, M.; Antoniadi, I.; Li, X.; Maia-Grondard, A.; Le Signor, C.; Bouteiller, N.; et al. The Pea TCP transcription factor PsBRC1 acts downstream of strigolactones to control shoot branching. *Plant Physiol.* **2012**, *158*, 225–238. [CrossRef] [PubMed]
127. Qin, G.J.; Gu, H.Y.; Zhao, Y.D.; Ma, Z.Q.; Shi, G.L.; Yang, Y.; Pichersky, E.; Chen, H.D.; Liu, M.H.; Chen, Z.L.; et al. An indole-3-acetic acid carboxyl methyltransferase regulates Arabidopsis leaf development. *Plant Cell* **2005**, *17*, 2693–2704. [CrossRef]

International Journal of *Molecular Sciences*

Article

Class I KNOX Is Related to Determinacy during the Leaf Development of the Fern *Mickelia scandens* (Dryopteridaceae)

Rafael Cruz [1,2,*], Gladys F. A. Melo-de-Pinna [2], Alejandra Vasco [3], Jefferson Prado [1,4] and Barbara A. Ambrose [5]

1 Instituto de Botânica, Av. Miguel Estéfano 3687, São Paulo (SP) CEP 04301-902, Brazil; jprado.01@uol.com.br
2 Instituto de Biociências, Universidade de São Paulo, Rua do Matão 277, São Paulo (SP) CEP 05422-971, Brazil; gfmpinna@usp.br
3 Botanical Research Institute of Texas, 1700 University Drive, Fort Worth, TX 76107-3400, USA; avascog@gmail.com
4 UNESP, IBILCE, Depto. de Zoologia e Botânica, Rua Cristóvão Colombo, 2265, São José do Rio Preto (SP) CEP 15054-000, Brazil
5 The New York Botanical Garden, 2900 Southern Blvd, Bronx, NY 10458-5126, USA; bambrose@nybg.org
* Correspondence: rafaeldscruz@gmail.com

Received: 2 May 2020; Accepted: 12 June 2020; Published: 16 June 2020

Abstract: Unlike seed plants, ferns leaves are considered to be structures with delayed determinacy, with a leaf apical meristem similar to the shoot apical meristems. To better understand the meristematic organization during leaf development and determinacy control, we analyzed the cell divisions and expression of *Class I KNOX* genes in *Mickelia scandens*, a fern that produces larger leaves with more pinnae in its climbing form than in its terrestrial form. We performed anatomical, in situ hybridization, and qRT-PCR experiments with *histone H4* (cell division marker) and *Class I KNOX* genes. We found that *Class I KNOX* genes are expressed in shoot apical meristems, leaf apical meristems, and pinnae primordia. During early development, cell divisions occur in the most distal regions of the analyzed structures, including pinnae, and are not restricted to apical cells. Fern leaves and pinnae bear apical meristems that may partially act as indeterminate shoots, supporting the hypothesis of homology between shoots and leaves. *Class I KNOX* expression is correlated with indeterminacy in the apex and leaf of ferns, suggesting a conserved function for these genes in euphyllophytes with compound leaves.

Keywords: apical meristems; *Class I KNOX* genes; compound leaves; determinacy; Dryopteridaceae; ferns; leaf development; pinna development; shoot development

1. Introduction

Vascular plant organs are classically defined based on their position; on their tissue organization (symmetry axes and vascular tissue); and on the presence, position, and activity of their meristems [1,2]. With these criteria, leaves are lateral determinate organs generally with an abaxial-adaxial asymmetry, and these features seem to generally apply well to leaves in seed plants. On the other hand, shoots are characterized by indeterminacy and are marked by the expression of *Class I KNOTTED-LIKE HOMEOBOX* (*KNOX*) genes in the shoot apical meristem (SAM) [3]. This class of genes belongs to the superclass three amino acid loop extension (*TALE*) of homeodomain proteins [4,5]. The downregulation of *Class I KNOX* is one of the first indications of the development of a determinate leaf primordium in seed plants [6,7]. Plants with defective *Class I KNOX* genes may be unable to maintain the SAM, as indicated by the mutants *shoot meristemless* (STM) in *Arabidopsis thaliana* (L.) Heynh. [8,9] and by mutants carrying malfunctioning alleles such as *knotted1* in maize that have defective branching and lateral organ formation [6,10]. In most vascular plants, when *Class I KNOX* genes are naturally expressed in the leaf primordium, the resulting morphology usually will be a compound leaf, as demonstrated by Bharathan et al. [7] in an extensive exploration of different groups (including many angiosperms, the cycad *Zamia floridana* A. DC., and the fern *Anogramma chaerophylla* (Desv.) link). This rule seems to have few known exceptions, except for legumes [11]. By analyzing *Cardamine hirsuta* L., a crucifer related to *Arabidopsis* with dissected leaves, Hay and Tsiantis [12] concluded that the expression of the STM homolog in the leaf primordium delays differentiation pathways, allowing leaflet initiation, while *Arabidopsis thaliana* produces simple leaves due to the exclusion of *Class I KNOX* expression from the leaf primordium. *Class I KNOX* genes have been found to be expressed and have a function in the meristematic regions of various organs in seed plants, and as such have been related to indeterminacy [10,13,14]. Thus, compound leaves can be interpreted as structures with a delayed determinacy during their development, and Class *I KNOX* facilitates leaflet formation [7,15,16].

Fern leaves are different from most seed plant leaves. For example, unlike seed plants, many fern leaves have a leaf apical meristem (LAM). In ferns, the LAM is responsible for a transient indeterminacy during leaf development, usually producing lateral pinnae during a longer period than the regular compound leaf of a seed plant. The LAM and SAM structure of ferns is also remarkably unique, in that they both have a distinctive prominent apical cell and a peripheral zone that together compose the entire LAM or SAM [17–22]. Some ferns in the orders Marattiales and Ophioglossales do not have only a single apical cell but instead have a group of apical initial cells [18,23]. However, the marattioid fern *Angiopteris lygodiifolia* Rosenst. has only a single initial apical cell in its shoot apex, as detected in a more detailed analysis [24]. Another key difference between fern and seed plants' leaf developments is that fern leaves mainly develop from the LAM and a marginal meristem (MM) composed of marginal and submarginal initials [21,22].

The expression analyses of two *Class I KNOX* genes in *Elaphoglossum peltatum* (Sw.) Urb. f. *peltatum* (Dryopteridaceae) characterize it as having a multicellular SAM with an apical initial and actively dividing surrounding cells [20], supporting previous work that proposed zonation for a multicellular meristem based on anatomical evidence [17,18]. *Class I KNOX* transcripts were also detected in leaf primordia and in the multicellular apex of the ferns *Anogramma chaerophyla* [7] and *Ceratopteris richardii* Brongn. [25]. Proteins coded by this class of genes were detected in the same regions in *Osmunda regalis* L. [26]. Few details are available about the expression in the pinnae primordia or in the LAM, but the expression reported of *Class I KNOX* in the leaf primordium may be the cause of the delayed determinacy of fern leaves [26].

Meristems seem to be the key character to understand the evolution and development of fern leaves (fronds). Fern leaves resemble the indeterminate shoot by having an apical meristem, producing lateral organs and having a transient or even persistent indeterminacy (as in the genera *Lygodium* Sw., *Nephrolepis* Schott, *Salpichlaena* J. Sm. and *Jamesonia* Hook. and Grev., as reviewed in Vasco et al., [27]). These features of the fern leaf do not fit the classical morphological concept of leaves as they do for seed plants. According to these classical concepts, leaves are a distinct set of features (e.g., determinacy and flattened morphology) that perfectly exclude stem features (e.g., indeterminacy and cylindrical morphology) [28–30]. There is evidence that *Class I KNOX* genes are directly associated with indeterminacy and are required to make compound leaves in many cases, representing a partial homology with the shoot [15]. *Class I KNOX* genes are also an important marker of meristematic activity in fern shoots [3].

Studies of Class *I KNOX* outside of spermatophytes are still necessary to better understand their role in the development of leaves, particularly in ferns. For instance, transgenic *Arabidopsis* plants overexpressing *Ceratopteris richardii Class I KNOX* genes have lobed leaves [25]. *Ceratopteris richardii Class I KNOX* genes in *Arabidopsis* mutants only partially restore their functions, even with high levels of transgene transcripts detected in complementation assays [3]. Because Class *I KNOX* proteins act together with the other class of *TALE* proteins *BELL* to target the nucleus, it is possible that *Ceratopteris* orthologs cannot interact with different *BELL* proteins in *Arabidopsis* [3]. In angiosperms, *ARP* genes, related to the development of lateral organs, are well-known to maintain the KNOX-off state in leaves [16], but these two classes of genes seem to co-occur in meristems and leaf primordia in the fern *Osmunda regalis* [26], suggesting that other regulatory mechanisms may be present. These data reinforce the importance of new experiments about *Class I KNOX* genes in ferns to better understand their potential function and role in shaping the fern body plan.

In order to gather more information concerning leaf and apical meristem organization, we studied the expression of *Class I KNOX* genes in *Mickelia scandens* (Raddi) R. C. Moran, Labiak and Sundue (Dryopteridaceae), a leptosporangiate fern endemic to the Brazilian Atlantic Rainforest. *Mickelia scandens* has pinnate leaves that have distinct forms during its life cycle as a hemiepiphyte. It bears small leaves and thin rhizomes in its terrestrial form and longer leaves with more pinnae in the thicker rhizomes of its climbing form. This difference between terrestrial and climbing leaves is an important feature of the genus [31–33]. A similar morphology was described for *Mickelia guianensis* (Aubl.) R.C. Moran, Labiak and Sundue based on specimens from the French Guiana [31,32]. *Mickelia guianensis* terrestrial leaves are three times narrower and with less pinnae than climbing leaves [31,32]. These differences are well documented for several *Mickelia* species [33]. This abrupt and substantial change in form is a feature that characterizes this plant as a species with a heteroblastic development, in contraposition to homoblastic species that show only small and gradual changes during their development [34]. *Mickelia* R. C. Moran, Labiak and Sundue is also the sister group of the genus *Elaphoglossum* Schott, a genus with mostly simple-leaved species, whose plants have been the targets of evolutionary and developmental studies [20,35,36], making *Mickelia scandens* a promising model to understand leaf development in ferns with compound leaves successfully applying molecular techniques. Assuming that the differences in the size and number of pinnae between the different forms of this plant represent a differential degree of determinacy, we aim to better understand if the association of *Class I KNOX* expression with determinacy is true for ferns as it is for seed plants, since these groups are separated by c. 327 Myr of evolution [37]. Our hypothesis is that the *Class I KNOX* expression is stronger and longer in developing leaves of the climbing form and that is possibly the form that presents more delayed determinacy when compared with the terrestrial form. We also aim to better understand the meristematic structure of the developing leaves by studying cell division patterns.

2. Results

During field collections, the clear dimorphism between the terrestrial and climbing forms of leaves was obvious (Figure 1). Although we also noticed a slight variation in size inside each of these categories, the terrestrial forms always have smaller leaves and shorter pinnae (Figure 1a–e) when compared with the climbing forms (Figure 1f–j). We also made a single observation in the field where one leaf from the climbing form had one anomalous pinna containing a basiscopic pinnule (Figure 1k).

We isolated three different *KNOX* gene homologs from *Mickelia scandens* (Supplementary Figure S1, Supplementary Table S1) using degenerate PCR. Phylogenetic analyses indicate that two of them, *MsC1KNOX1* and *MsC1KNOX2*, are closely related to known fern *Class I KNOX* genes (Supplementary Figure S2), while *MsC2KNOX1* is a *Class II KNOX* gene. We also cloned one gene that codes for *Histone H4*, *MsH4*, that was used as a positive control and cell division marker.

To determine if there are differences in the patterns of expression of *Class I KNOX* genes in *Mickelia scandens* terrestrial and climbing leaf types, we assessed their expression by *in situ* hybridization (ISH). The ISH experiments show that *MsC1KNOX1* and *MsC1KNOX2* have similar temporal and spatial expression patterns during shoot, leaf, and pinnae development (Figures 2–4). The *MsC1KNOX1* expression appeared stronger than *MsC1KNOX2* (Figures 2–4), although ISH experiments are not quantitative. In the shoot apical meristem, *MsC1KNOX1* and *MsC1KNOX2* are expressed in the apical cell and the derivative cells in the peripheral zone in most of the experiments (Figure 2a–e). The expression of *MsC1KNOX1* and *MsC1KNOX2* is also detected in the procambial cells, which are continuous with the SAM (Figure 2a,c). *MsC1KNOX1* and *MsC1KNOX2* expression were not detected in the boundaries between the leaf primordium and the shoot apex (Figure 2b). We also detected *Class I*

KNOX expression in the LAM and MM of developing leaves (Figure 2d). Occasionally, our experiments show a clear expression of *Class I KNOX* genes in the SAM peripheral zone, but no expression in the apical cell and in some of the surrounding prismatic cells (Figure 2d). This cell is bigger than derivative cells that gradually reduce in size in the SAM (Figure 2e). To compare the patterns of cell division in the developing shoot apex, we assessed the expression of *Histone H4* (*MsH4*), which has a slightly similar pattern when compared with *Class I KNOX*, being expressed in the shoot apex and in the developing vascular system (Figure 2f).

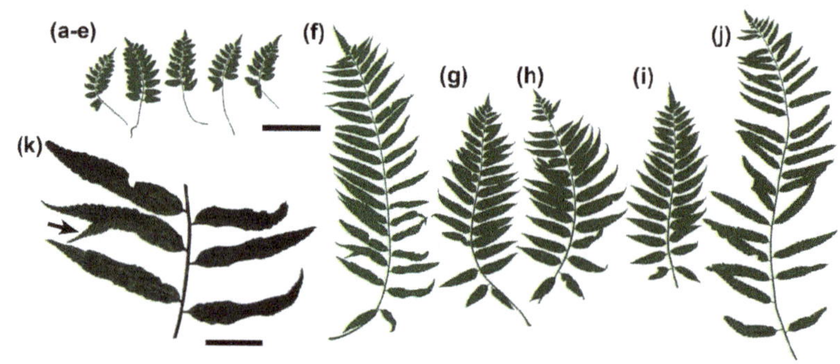

Figure 1. Leaves in *Mickelia scandens*. (**a**–**e**) Samples of the terrestrial form and (**f**–**j**) from the climbing form. The silhouettes are scans from actual leaves at the same scale. (**k**) Detail of the leaf of Figure 1f, with a pinna bearing a basiscopic anomalous pinnule (arrow). Bars: (**a**–**j**) 10 cm; (**k**) 4 cm.

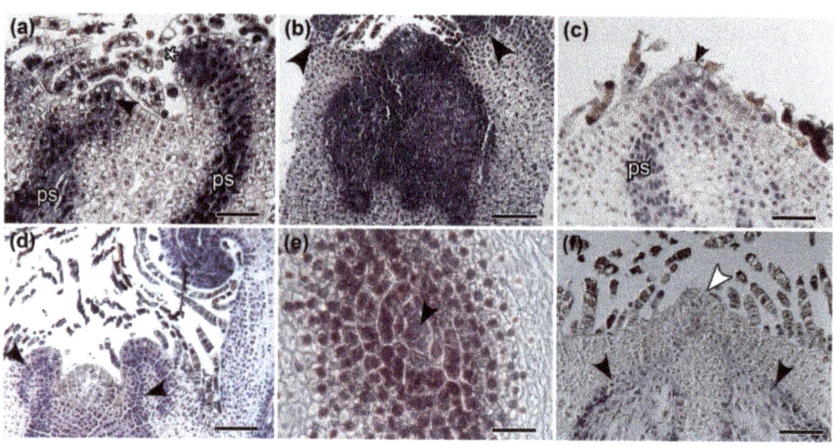

Figure 2. In situ hybridization of *Class I KNOX* genes and *MsH4* in shoot apices of terrestrial (Tf) and climbing (Cf) forms of *Mickelia scandens*. (**a**) Longitudinal section of Tf. Expression of *MsC1KNOX1* in the shoot apical cell (arrowhead), leaf apical cell (*), derivative cells, and procambial strands (ps). (**b**) Cf, longitudinal section. *MsC1KNOX1* expression in a peripheral zone and procambium. Leaf primordia position pointed out by arrowheads. (**c**) Longitudinal section of Tf. *MsC1KNOX2* expression in the shoot apical cell (arrowhead), derivative cells, and procambial strands (ps). (**d**) Longitudinal section of Cf. In this apex, *MsC1KNOX2* expression is not detected in the shoot apical cell. There are procambial strands (arrowheads) connecting the SAM and the leaf primordia. (**e**) Transverse section of Tf. *MsC1KNOX2* expression in a large apical cell (arrowhead) and in the surrounding derivative cells. (**f**) Longitudinal section of Cf. Scattered *MsH4* expression in the SAM (white arrowhead) and in the developing vascular system (black arrowheads). Bars: (**a**) 100, (**b**) 200, (**c**) 75, (**d**) 200, (**e**) 50, (**f**) 125 μm.

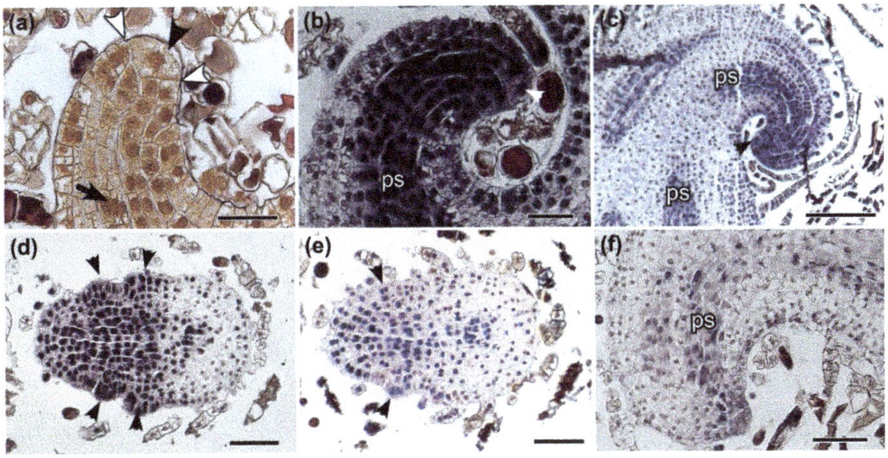

Figure 3. Investigations of leaf development in terrestrial (Tf) and climbing (Cf) forms in *Mickelia scandens* by anatomical and expression analyses. Longitudinal sagittal sections in (**a**–**c**,**f**); paradermal sections in (**d**,**e**). (**a**) Anatomical section of leaf apex bearing apical cell with a distal lenticular face (black arrowhead) and two dividing proximal cutting faces (their limits pointed out by white arrowheads). Inner derivative cells form the procambium (arrow). In situ hybridization (**b**–**f**). (**b**) *MsC1KNOX1* (Tf) and (**c**) *MsC1KNOX2* (Cf) are expressed in the leaf apical cell (arrowhead), derivatives cells, and in procambial strands (ps). (**d**) *MsC1KNOX1* (Tf) and (**e**) *MsC1KNOX2* (Tf) are expressed in young pinnae primordia (arrowheads). (**f**) *MsH4* expression indicates cell division in multiple cells in the apical region and procambial strands (ps). Tf. Bars: (**a**,**b**) 50, (c) 200, (**d**–**f**) 100 μm.

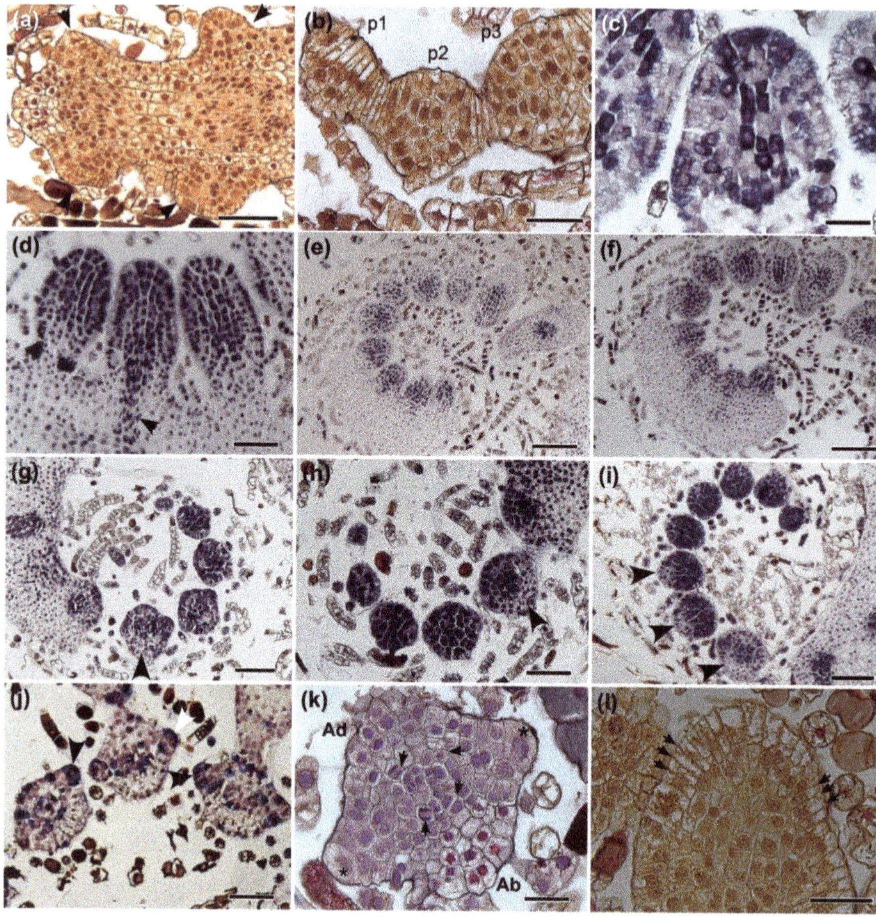

Figure 4. Late leaf development in terrestrial (Tf) and climbing (Cf) forms of *Mickelia scandens* by anatomical and expression analyses. Anatomical sections (**a**,**b**,**k**,**l**). (**a**) Pinnae primordia (arrowheads) emerge from the margins of the Tf leaf. (**b**) Transverse section of the youngest pinna (p1) shows grouped cells on its apex with evident periclinal divisions in Tf. The base has divisions in multiple planes, visible in older primordia (p2 and p3). In situ hybridization (**c**–**j**). (**c**) *MsH4* expression indicates cells division in multiple adjacent cells at the apex of the pinna primordium and in the central axis, where the vasculature will develop in the Tf. (**d**) As the pinnae primordium increases in size, developing vascular traces express *Class I KNOX* genes, exemplified by *MsC1KNOX1* in Cf. (**e**) *MsC1KNOX1* in Cf and (**f**) *MsC1KNOX2* in Cf are expressed in the entire young pinnae primordia. (**g**) *MsH4* in Tf, (**h**) t*MsC1KNOX1* in Tf and (**i**) *MsC1KNOX2* in Cf are all expressed throughout the entire pinnae primordia, expression is gradually reduced in the abaxial side of older pinnae (arrowheads). (**j**) Cell divisions detected by *MsH4* expression in marginal cells, some indicated by arrows in Tf. (**k**) Anatomical transverse section of the pinna primordium showing marginal cells (*) with outer lenticular faces of the wall and submarginal initials, between adaxial (Ad) and abaxial (Ab) sides, and radial divisions in the center (arrows) in Cf. (**l**) Anatomical paradermal section of the pinna primordium showing rows of marginal cells, with anticlinal cutting faces, some indicated with arrows in Cf. Bars: (**a**,**d**,**h**,**j**) 100, (**b**,**c**,**k**,**l**) 50, (**e**,**f**) 200, (**g**,**i**) 150 μm.

We found that during leaf development, both leaf types have similar *Class I KNOX* expression patterns in the LAM (apical initial and its derivatives, the peripheral cells). In general, the leaf primordium has a distinct apical cell with a lenticular distal face of the wall and two cutting faces (Figure 3a). The derivative cells undergo divisions, and inner cells are responsible for the establishment of the procambium (Figure 3a). In the leaf apex, we detected the expression of *MsC1KNOX1* and *MsC1KNOX2* in the LAM during leaf development (Figure 3b,c). *MsC1KNOX1* and *MsC1KNOX2* expression was also detected in the apical cell of the SAM and in the procambium (Figure 3b,c). *MsC1KNOX1* and *MsC1KNOX2* expression was also detected in the margin of developing leaves (Figure 2c) and in a punctate pattern along the margin as the pinnae primordia arise (Figure 3d,e). We detected a scattered *MsH4* expression in the leaf apical initial, immediate derivative cells, and in some procambial cells of the developing leaf primordium, confirming that a multicellular apical region of the leaf is undergoing active cell division (Figure 3f).

Pinnae primordia emerge laterally on the developing leaf (Figure 4a). The pinna primordium has in its apex grouped cells instead of a distinct single apical cell like the one that occurs in the LAM (Figure 4b). The central region of the pinna, where the vasculature of the costa will develop, also shows an evident expression of *MsH4*, as well as the grouped apical cells (Figure 4c). The expression of *MsC1KNOX1* and *MsC1KNOX2* is detected in all the regions of the pinnae primordia at the beginning of their development (Figure 4d–f) but is gradually reduced in the abaxial region in the older developing primordia (Figure 4g,h). The expression of *MsH4* also decreases in the abaxial region of the older pinnae primordia, indicating an earlier cessation of cell division in the abaxial region compared to the adaxial region of the pinnae (Figure 4i). In the older pinnae primordia, *MsH4* expression is concentrated in marginal cells (Figure 4j). In an anatomical analysis of the pinna primordium, it is possible to detect cell divisions in the central vascular system and a denser cytoplasm in the adaxial side when compared to the more vacuolized abaxial side, also revealing a late development of the adaxial region (Figure 4k). The marginal cells of the pinna primordium are pyramidal, with an outer lenticular face, and are bigger than other cells, making them remarkably similar to leaf apical cells in transverse sections, with two cutting faces responsible for abaxial and adaxial divisions (Figure 4k). Marginal cells are organized in longitudinal rows with other cutting faces that may play some role in the proximodistal growth (Figure 4l). However, based on our experiments, it is likely that the divisions responsible for the growth in length of the pinnae take place mainly in the apical region (acropetal growth), and marginal cells act later in development, being more responsible for cell divisions that will contribute to the lamina development.

Our initial hypothesis was that *Class I KNOX* expression was stronger and longer in the developing leaves of the climbing form (that are bigger and bear more pinnae, supposedly indeterminate for a longer time), when compared with the developing leaves of the terrestrial form. Because we did not find differences in the *Class I KNOX* expression patterns in any tissues between the two forms during our ISH analysis, we measured the relative expression of *Class I KNOX* through qRT-PCR experiments (Figure S3, Supplementary Table S2). Since our study species was collected in the wild, we had limited material and also pooled individuals for each sample. Our preliminary analyses suggest that the relative expression amongst the four different samples (developing leaves of terrestrial form, shoot apices of terrestrial form, developing leaves of climbing form, and shoot apices of climbing form) is significantly different for each gene (ANOVA; *MsC1KNOX1* $p = 0.0037$ and *MsC1KNOX2* $p = 0.0278$; Supplementary Tables S3 and S4). A Tukey test showed that the relative expression of *MsC1KNOX1* is significantly different between the sample containing developing leaves in the terrestrial form compared to all the other analyzed samples (Supplementary Table S3). However, the Tukey test showed that the relative expression of *MsC1KNOX2* was only significantly different between the sample containing the developing leaves of the terrestrial and climbing forms (Supplementary Table S4).

3. Discussion

Four main conclusions can be made from our results presented here: (i) there is a multicellular structure at the tip of developing shoots, leaves, and pinnae expressing *Class I KNOX* and bearing dividing cells (based on *MsH4* expression) that may include or not (for pinnae apices) a prominent apical cell; (ii) despite the differences in the overall size and morphology of the leaves of climbing and terrestrial forms, their pattern of development is similar, possibly differing only in how long the determination is delayed; (iii) a reduction in the size and number of pinnae—interpreted by us as the result of earlier determination—is possibly correlated to a reduction in *Class I KNOX* expression; (iv) fern leaves have two types of meristems, LAM and MM, in part specified by Class I KNOX that are integral for leaf development.

The anatomical structure of shoots, leaves, and pinnae apices is very similar. In addition, all of these meristems express *Class I KNOX*. The main differences between them are the presence of single apical cells (absent in the pinnae apices) and their number of dividing planes (three in the SAM apical cell and two in the LAM apical cell). The absence of a distinct apical cell in the apices of pinnae primordia cannot be interpreted as an absence of meristems. Although more studies concerning other genes and leaf morphologies are necessary to expand this conclusion to other fern groups, our data indicates the presence of a transient apical meristem in the pinnae of *Mickelia scandens* without the presence of a prominent apical cell. The widespread reference to a unicellular meristem for ferns by some authors may be the result of many textbooks that describe in detail the single apical cell and its cutting faces, while lacking further information about the other meristematic cells in this group (e.g., [22,38–40]), contributing to the propagation of this concept. A well-documented work that strongly defended the idea of a single-celled meristem was based on the observation of apical cell division planes in more than 50 genera of ferns [21]. However, several authors [17,18,41,42] proposed cytohistological zonation schemes for a multicellular structure, based mainly on the fact that the apical initial cell rarely divides. Based on the *Class I KNOX* expression data in *Elaphoglossum peltatum* f. *peltatum* and reviewing these previous studies, a recent study proposed a simplified zonation for the shoot apical meristem of ferns: a single apical cell that rarely divides and may not express *Class I KNOX* genes in some apices, and a peripheral zone with rapidly dividing cells [20]. It is important to highlight that even with data supporting a reduced mitotic activity for the single apical cells, they can still divide in the SAM and LAM and are likely the ultimate source of all cells similar to the quiescent center of seed plant meristems. The significance of the occasional absence of *Class I KNOX* expression in the apical cells is not clear; however, these shoots were still active, as *Class I KNOX* genes were detected in the peripheral zone of the SAM.

Our preliminary qRT-PCR experiments are the first ones to show that, in ferns, lower *Class I KNOX* expression is possibly correlated to a more determinate structure. These data support the conclusion that, based on *Class I KNOX* expression patterns, complex leaves should be interpreted as partially indeterminate structures [7]. However, more adequate conditions are needed to confirm this possibility, such as biological replicates and the use of plants grown in very controlled conditions, as well as exploring the expression of Class *I KNOX* genes in other ferns with different leaf forms. Further studies exploring the phenotype of fern mutants for Class *I KNOX* are also needed to test our hypothesis when they become available. In the *Mickelia scandens* developing leaf, *Class I KNOX* genes are expressed throughout the apical region, encompassing the apical cell to the cells of the first pinnae primordia and reinforcing the presence of a multicellular apical meristem in the leaf, similar to the shoot apex. The expression of *Class I KNOX* in the pinnae primordium, even as a terminal unit, suggests some degree of indeterminacy. This is reinforced by the anomalous leaf (Figure 1) that resembles other species of the genus, *Mickelia furcata* R.C. Moran, Labiak and Sundue, a plant with bipinnate leaves at the basal pinnae [33]. Possibly, a plant overexpressing *Class I KNOX* will show a similar phenotype. Additionally, cell division patterns at the pinna apex together with *Class I KNOX* expression suggest a meristematic activity in this region, even though those pinnae apices do not have an evident single apical cell or smaller derivative cells (Figure 4c,d). After the initial acropetal growth, the abaxial differentiation occurs preceding the adaxial differentiation, a phenomenon well-known for flowering plants [43]. In ferns, this may be responsible for the typical coiling of young fern leaves (known as a fiddlehead or crozier), thus protecting these meristematic apices of the leaf and pinnae.

Fern apical meristems should be interpreted as a complex and highly organized interconnected network of cells with indeterminate fates, specialized zones (apical cells vs. peripheral cells), and the capacity for producing new organs (leaves or pinnae). Interestingly, many studies have interpreted fern leaves as reiterative and fractal systems, in which the shoot apices generate structures that can repeat some degree of their own shoot development [19,44,45]. In this sense, as already have been stated by some authors [26,46,47], fern leaves and their segments could be interpreted evolutionarily and ontogenetically as reduced shoots, and the presence of similar characteristics detected by us in *Mickelia scandens* (i.e., *Class I KNOX* expression during initial development and cell divisions concentrated in apical and surrounding cells) gives support for this interpretation. The presence of such features in developing leaves is strong evidence that Agnes Arber's Partial Shoot Theory [28,29] is correct. Arber said that "the leaf is a partial-shoot, arising laterally from a parent whole-shoot", based mainly on the presence of lateral structures arising from axial elements in the leaf, as well as in shoots. According to her, the shoot has a gradient of determination between stems and leaves, and compound leaves present the same gradient. Her theory should be strongly discussed now that new molecular evidence, as our results and other studies discussed here, is available. Our results point to multicellular meristematic structures in the shoot, leaf, and pinna apices, also reinforcing her idea of "identity-in-parallel", in which structures may be put in a relation of the part to the whole, but is also equivalent as a whole [28]. The pinna is part of the shoot, but ultimately is equivalent to a whole shoot, carrying the potential of producing new lateral structures. The observed anomalous pinna with a lateral segment (Figure 1k) could be evidence of this potential.

The future of fern studies is promising, as new sequences are available in transcriptome projects like oneKP [48] and the first fern genomes are already available for *Salvinia cucullata* Roxb. and *Azolla filiculoides* Lam. [49,50]. The discovery of fern genes related to apical meristems and their regulation will certainly increase our understanding and can even detail better the zonation and functions of different cell niches. New developmental studies with multiple approaches, uniting these modern molecular analyses with classical anatomical data for developmental studies in ferns will certainly help us to better understand the evolution of all leaves.

4. Materials and Methods

4.1. Plant Material

Shoot apices (usually containing small leaf primordia covered by scales) and developing leaves of the terrestrial and climbing forms of *Mickelia scandens* sporophytes were collected from specimens that occur in a dense population in Fontes do Ipiranga State Park (São Paulo, Brazil). A voucher specimen is deposited in the SP Herbarium (Prado and Cruz 2332). Part of the material was stored in RNAlater® for RNA extraction and some were fixed in formalin-acetic acid-ethanol 50% (FAA) for in situ hybridization (ISH) experiments and anatomy.

4.2. RNA extraction and cDNA Synthesis

The total RNA of the shoot apices and developing leaves was extracted with QIAGEN RNeasy mini kit (Qiagen, Hilden, Germany). A cDNA synthesis was performed using Superscript III (Invitrogen, Carlsbad, CA, USA), following the manufacturers' protocols for these procedures (except for qRT-PCR). The cDNA for qRT-PCR was obtained with SuperScript IV VILO Master Mix (Invitrogen, Carlsbad, CA, USA).

4.3. Genes Isolation and Phylogenetic Analyses

Degenerate primers were used for KNOX genes (F: 5′ -CCBGARCTBGACMABTTYATGG-3′, R: 5′-CCAGTGSCKYTTCCKYTGRTTDATRAACC-3′), based on a previous study [20] for H4 genes (F:

5′-ATGTCWGGMMGRGGWAAGGGAGG-3′, R: 5′-CCRAADCCRTARAGVGTHCKKCC-3′) designed for this study to be used as a cell division marker, as used in previous studies [51,52]. Fragments were cloned in Invitrogen™ pCR™ 2.1–TOPO™ 3.9 kb plasmids and sequenced with M13 primers. The sequences (Supplementary Table S1) were then analyzed by the NCBI Conserved Domain Search tool [53] to detect the presence of KNOX and H4 domains. Unlike Class I, the phylogenetically distinct Class II KNOX that is also a target of these primers is related to tissue differentiation and not to cell proliferation in land plants [3]. In order to identify our cloned *KNOX* fragments, the sequences were aligned with other known *KNOX* genes (sequences referenced in two previous studies) [3,20] with Geneious version 10.1.2 [54]. Phylogenetic relationships were inferred from the nucleotide data using maximum likelihood (ML) analyses. ML searches for the best tree and bootstrap were performed simultaneously with 300 replicates with RaxML version 8 [55], partitioned by codon position with GTR + Γ + I model as recommended in a PartitionFinder2 analysis [56].

4.4. Anatomy and In Situ Hybridization (ISH) Experiments

Fixed material was embedded in paraplast (Fisher) and sectioned on a rotary microtome. For histological analyses, sections were stained with Safranin O 1% in ethanol, Crystal Violet 1% aqueous and Orange G 1% in clove oil [57]. For the ISH experiments, we followed the procedures previously described [58,59] using specific probes for *Class I KNOX* and *H4* generated with specific primers designed for them (Supplementary Table S5). The similarity between probes is 57% in *Class I KNOX* genes (Supplementary Figure S1).

4.5. Quantitative Real-Time PCR

In order to quantify the expression of *Class I KNOX* genes in the shoot apices and developing leaves of terrestrial and climbing forms, we assessed the transcript abundance by a qRT-PCR analysis using a 7500 Real-Time PCR system (Applied Biosystems® by Life Technologies, NY, USA). A β-actin specific sequence was accessed with PCR reactions with primers (F: 5′-GATGGATCCTCCAATCCAGACACTGTA-3′ and R: 5′-GTATTGTGTTGGACTCTGGTGATGGTGT-3′) and was used as a housekeeping gene. The PCR reactions were performed with 5 μl of cDNA; 12.5 μl of SYBR Green Master Mix (Applied Biosystems); 10 pmol/μl concentration of primers (specifically designed for qRT-PCR analysis, Supplementary Table S5); and the following cycling conditions: 95 °C for 10 min, 44 cycles of 95 °C for 15 s, 55 °C for 30 s and 72 °C for 1 min. All the reactions were performed in three technical replicates, each one analyzing the expression in four different samples: the developing leaves of the terrestrial form, the shoot apices of the terrestrial form, the developing leaves of the climbing form, and the shoot apices of the climbing form. Each sample was extracted from a single pool of material containing at least five different individuals randomly collected from the population. The expression was calculated using the ΔC_T (difference between threshold cycles) method [60], and the statistical significance was determined with ΔC_T values by using a one-way ANOVA test followed by Tukey's pairwise comparison ($p < 0.05$). This preliminary analysis can detect only that these four samples have different relative gene expression, potentially underestimating some of the differences due to biological variations between different individuals within the population. Raw data and calculations are available (Supplementary Tables S2–S4).

Supplementary Materials: Supplementary Materials can be found at http://www.mdpi.com/1422-0067/21/12/4295/s1. Figure S1, Alignment of partial KNOX sequences with conserved domains and probe binding sites; Figure S2, Phylogenetic tree showing relationships between KNOX genes; Figure S3, Relative expression of Class I KNOX genes in different pools; Table S1, GenBank accession numbers of obtained sequences; Table S2, Ct values of qRT-PCR and relative expression calculation; Table S3, ANOVA and Tukey's pairwise comparison of MsC1KNOX1 expression in different tissues; Table S4, ANOVA and Tukey's pairwise comparison of MsC1KNOX2 expression in different tissues; Table S5, Primer sequences.

Author Contributions: All the authors conceived the study, prepared the manuscript, and contributed to the discussion. R.C., G.F.A.M.-d.-P., and J.P. collected the material and performed the anatomical analysis. R.C. and B.A.A. performed and analyzed ISH and qRT-PCR experiments. R.C. and A.V. performed the phylogenetic analysis and designed the probes for ISH experiments. All authors have read and agreed to the published version of the manuscript.

Funding: This work was supported by the São Paulo Research Foundation (FAPESP) [grant numbers 2015/15158-4, 2015/15920-3, and 2013/26191-7 to R.C. & G.F.A.M.P.]; National Science Foundation [DEB-1020443 to B.A.A.]; and by the National Council for Scientific and Technological Development (CNPq) [grant numbers 301157/2010-3 to J.P., and 300810/2016-4 and 303962/2019-4 to G.F.A.M.P.].

Acknowledgments: We thank Tynisha Smalls and Cecilia Zumajo for their precious technical assistance during the development of the experiments at NYBG. We also thank the three anonymous reviewers and the editor of this article for their valuable suggestions.

Conflicts of Interest: The authors declare no conflict of interest. The funders had no role in the design of the study; in the collection, analysis, or interpretation of data; in the writing of the manuscript; or in the decision to publish the results.

Abbreviations

Class I KNOX	*Class I KNOTTED-LIKE HOMEOBOX*
ISH	In-situ hybridization
LAM	Leaf apical meristem
MM	Marginal meristem
qRT-PCR	Real-time quantitative reverse transcription polymerase chain reaction
SAM	Shoot apical meristem

References

1. Kaplan, D.R. Fundamental concepts of leaf morphology and morphogenesis: A contribution to the interpratation of developmental mutants. *Int. J. Plant Sci.* **2001**, *162*, 465–474. [CrossRef]
2. Dengler, N.G.; Tsukaya, H. Leaf Morphogenesis in Dicotyledons: Current Issues. *Int. J. Plant Sci.* **2001**, *162*, 459–464. [CrossRef]
3. Frangedakis, E.; Saint-Marcoux, D.; Moody, L.A.; Rabbinowitsch, E.; Langdale, J.A. Nonreciprocal complementation of KNOX gene function in land plants. *New Phytol.* **2017**, *216*, 591–604. [CrossRef]
4. Bürglin, T.R. Analysis of TALE superclass homeobox genes (MEIS, PBC, KNOX, Iroquois, TGIF) reveals a novel domain conserved between plants and animals. *Nucleic Acids Res.* **1997**, *25*, 4173–4180. [CrossRef] [PubMed]
5. Hamant, O.; Pautot, V. Plant development: A TALE story. *C. R. Biol.* **2010**, *333*, 371–381. [CrossRef] [PubMed]
6. Smith, L.G.; Greene, B.; Veit, B.; Hake, S. A dominant mutation in the maize homeobox gene, Knotted-1, causes its ectopic expression in leaf cells with altered fates. *Development* **1992**, *116*, 21–30. [PubMed]
7. Bharathan, G.; Goliber, T.E.; Moore, C.; Kessler, S.; Pham, T.; Sinha, N.R. Homologies in leaf form inferred from KNOX1 gene expression during development. *Science* **2002**, *296*, 1858–1860. [CrossRef]
8. Barton, M.K.; Poethig, R.S. Formation of the shoot apical meristem in Arabidopsis thaliana: An analysis of development in the wild type and in the shoot meristemless mutant. *Development* **1993**, *119*, 823–831.
9. Endrizzi, K.; Moussian, B.; Haecker, A.; Levin, J.Z.; Laux, T. The *SHOOT MERISTEMLESS* gene is required for maintenance of undifferentiated cells in Arabidopsis shoot and acts at a different regulatory level than the meristem genes *WUSCHEL* and *ZWILLE*. *Plant J.* **1996**, *10*, 967–979. [CrossRef] [PubMed]
10. Kerstetter, R.A.; Laudencia-Chingcuanco, D.; Smith, L.G.; Hake, S. Loss-of-function mutations in the maize homeobox gene, knotted1, are defective in shoot meristem maintenance. *Development* **1997**, *124*, 3045–3054. [PubMed]
11. Champagne, C.E.M.; Goliber, T.E.; Wojciechowski, M.F.; Mei, R.W.; Townsley, B.T.; Wang, K.; Paz, M.M.; Geeta, R.; Sinha, N.R. Compound Leaf Development and Evolution in the Legumes. *Plant Cell* **2007**, *19*, 3369–3378. [CrossRef] [PubMed]
12. Hay, A.; Tsiantis, M. The genetic basis for differences in leaf form between Arabidopsis thaliana and its wild relative Cardamine hirsuta. *Nat. Genet.* **2006**, *38*, 942–947. [CrossRef] [PubMed]
13. Venglat, S.P.; Dumonceaux, T.; Rozwadowski, K.; Parnell, L.; Babic, V.; Keller, W.; Martienssen, R.; Selvaraj, G.; Datla, R. The homeobox gene *BREVIPEDICELLUS* is a key regulator of inflorescence architecture in *Arabidopsis*. *Proc. Natl. Acad. Sci. USA* **2002**, *99*, 4730–4735. [CrossRef]
14. Pham, T.; Sinha, N. Role of KNOX genes in shoot development of Welwitschia mirabilis. *Int. J. Plant Sci.* **2003**, *164*, 333–343. [CrossRef]

15. Champagne, C.; Sinha, N. Compound leaves: Equal to the sum of their parts? *Development* **2004**, *131*, 4401–4412. [CrossRef] [PubMed]
16. Hay, A.; Tsiantis, M. KNOX genes: Versatile regulators of plant development and diversity. *Development* **2010**, *137*, 3153–3165. [CrossRef]
17. Stevenson, D.W. The Cytohistological and Cytohistochemical Zonation of the Shoot Apex of Botrychium multifidum. *Am. J. Bot.* **1976**, *63*, 852–856. [CrossRef]
18. Stevenson, D.W. Observations on Shoot Apices of Eusporangiate Ferns. *Kew Bull.* **1978**, *33*, 279. [CrossRef]
19. White, R.A.; Turner, M.D. Anatomy and development of the fern sporophyte. *Bot. Rev.* **1995**, *61*, 281–305. [CrossRef]
20. Ambrose, B.A.; Vasco, A. Bringing the multicellular fern meristem into focus. *New Phytol.* **2016**, *210*, 790–793. [CrossRef]
21. Bierhorst, D.W. On the stem apex, leaf initiation and early leaf ontogeny in filicalean ferns. *Am. J. Bot.* **1977**, *64*, 125–152. [CrossRef]
22. Wardlaw, C.W. Reflections on the unity of the embryonitc tissues in ferns. *Phytomorphology* **1958**, *8*, 323–327.
23. Bower, F.O. The comparative examination of the meristems of Ferns, as a Phylogenetic Study. *Ann. Bot.* **1889**, *3*, 305–392. [CrossRef]
24. Imaichi, R. Surface-viewed shoot apex of *Angiopteris lygodiifolia* Ros. (Marattiaceae). *Bot. Mag. Tokyo* **1986**, *99*, 309–317. [CrossRef]
25. Sano, R.; Juárez, C.M.; Hass, B.; Sakakibara, K.; Ito, M.; Banks, J.A.; Hasebe, M. KNOX homeobox genes potentially have similar function in both diploid unicellular and multicellular meristems, but not in haploid meristems. *Evol. Dev.* **2005**, *7*, 69–78. [CrossRef] [PubMed]
26. Harrison, C.J.; Coriey, S.B.; Moylan, E.C.; Alexander, D.L.; Scotland, R.W.; Langdale, J.A. Independent recruitment of a conserved developmental mechanism during leaf evolution. *Nature* **2005**, *434*, 509–514. [CrossRef] [PubMed]
27. Vasco, A.; Moran, R.C.; Ambrose, B.A. The evolution, morphology, and development of fern leaves. *Front. Plant Sci.* **2013**, *4*, 345. [CrossRef]
28. Arber, A. *The Natural Philosophy of Plant Form*; Cambridge University Press: Cambridge, UK, 1950.
29. Rutishauser, R.; Isler, B. Developmental genetics and morphological evolution of flowering plants, especially bladderworts (Utricularia): Fuzzy Arberian Morphology complements Classical Morphology. *Ann. Bot.* **2001**, *88*, 1173–1202. [CrossRef]
30. Rutishauser, R.; Grob, V.; Pfeifer, E. *Plants Are Used to Having Identity Crises*; Cambridge University Press: Cambridge, UK, 2008; pp. 194–213. [CrossRef]
31. Hebant-Mauri, R.; Gay, H. Morphogenesis and its relation to architecture in the dimorphic clonal fern *Lomagramma guianensis* (Aublet) Ching (Dryopteridaceae). *Bot. J. Linn. Soc.* **1993**, *112*, 257–276. [CrossRef]
32. Gay, H. The architecture of a dimorphic clonal fern, *Lomagramma guianensis* (Aublet) Ching (Dryopteridaceae). *Bot. J. Linn. Soc.* **1993**, *111*, 343–358. [CrossRef]
33. Moran, R.C.; Labiak, P.H.; Sundue, M. Synopsis of *Mickelia*, a newly recognized genus of bolbitidoid ferns (Dryopteridaceae). *Brittonia* **2010**, *62*, 337–356. [CrossRef]
34. Zotz, G.; Wilhelm, K.; Becker, A. Heteroblasty—A Review. *Bot. Rev.* **2011**, *77*, 109–151. [CrossRef]
35. Moran, R.C.; Labiak, P.H.; Sundue, M. Phylogeny and Character Evolution of the Bolbitidoid Ferns (Dryopteridaceae). *Int. J. Plant Sci.* **2010**, *171*, 547–559. [CrossRef]
36. Vasco, A.; Lóriga, J.; Rouhan, G.; Ambrose, B.A.; Moran, R.C. Divided leaves in the genus *Elaphoglossum* (Dryopteridaceae): A phylogeny of *Elaphoglossum* section *Squamipedia*. *Syst. Bot.* **2015**, *40*, 46–55. [CrossRef]
37. Smith, S.A.; Beaulieu, J.M.; Donoghue, M.J. An uncorrelated relaxed-clock analysis suggests an earlier origin for flowering plants. *Proc. Natl. Acad. Sci. USA* **2010**, *107*, 5897–5902. [CrossRef] [PubMed]
38. Esau, K. *Plant Anatomy*, 1st ed.; John Wiley & Sons: Hoboken, NJ, USA, 1953.
39. Fahn, A. *Plant Anatomy*, 3rd ed.; Pergamon Press: Cambridge, UK, 1982.
40. Evert, R.F. *Esau's Plant Anatomy—Meristems, Cells, and Tissues of the Plant Body—Their Structure, Function, and Development*, 3rd ed.; John Wiley & Sons: Hoboken, NJ, USA, 2006; ISBN 9780471738435.
41. Ogura, Y. *Comparative Anatomy of Vegetative Organs of the Pteridophytes*; Borntraeger: Berlin, Germany, 1972; ISBN 9783443140069.
42. McAlpin, B.; White, R. Shoot organization in the Filicales: The promeristem. *Am. J. Bot.* **1974**, *61*, 562–579. [CrossRef]

43. Beck, C.B. *An Introduction to Plant Structure and Development*, 2nd ed.; Cambridge University Press: Cambridge, UK, 2010; ISBN 9780521518055.
44. Barnsley, M.F. *Fractals Everywhere*; Morgan Kaufmann: Burlington, MA, USA, 2000; ISBN 9780120790692.
45. Sanders, H.L.; Darrah, P.R.; Langdale, J.A. Sector analysis and predictive modelling reveal iterative shoot-like development in fern fronds. *Development* **2011**, *138*, 2925–2934. [CrossRef]
46. Plackett, A.R.G.; Di Stilio, V.S.; Langdale, J.A. Ferns: The missing link in shoot evolution and development. *Front. Plant Sci.* **2015**, *6*, 972. [CrossRef]
47. Harrison, C.J.; Morris, J.L. The origin and early evolution of vascular plant shoots and leaves. *Philos. Trans. R. Soc. B Biol. Sci.* **2018**, *373*, 20160496. [CrossRef]
48. Matasci, N.; Hung, L.; Yan, Z.; Carpenter, E.J.; Wickett, N.J.; Mirarab, S.; Nguyen, N.; Warnow, T.; Ayyampalayam, S.; Barker, M.; et al. Data access for the 1,000 Plants (1KP) project. *Gigascience* **2014**, *3*, 17. [CrossRef]
49. Banks, J.A. Fern genomes finally here. *Nat. Plants* **2018**, *4*, 404–405. [CrossRef] [PubMed]
50. Li, F.W.; Brouwer, P.; Carretero-Paulet, L.; Cheng, S.; De Vries, J.; Delaux, P.M.; Eily, A.; Koppers, N.; Kuo, L.Y.; Li, Z.; et al. Fern genomes elucidate land plant evolution and cyanobacterial symbioses. *Nat. Plants* **2018**, *4*, 460–472. [CrossRef] [PubMed]
51. Brandstädter, J.; Rossbach, C.; Theres, K. The pattern of histone H_4 expression in the tomato shoot apex changes during development. *Planta* **1994**, *192*, 69–74. [CrossRef]
52. Groot, E.P.; Sinha, N.; Gleissberg, S. Expression patterns of STM-like KNOX and Histone H_4 genes in shoot development of the dissected-leaved basal eudicot plants Chelidonium majus and Eschscholzia californica (Papaveraceae). *Plant Mol. Biol.* **2005**, *58*, 317–331. [CrossRef]
53. Lu, S.; Wang, J.; Chitsaz, F.; Derbyshire, M.K.; Geer, R.C.; Gonzales, N.R.; Gwadz, M.; Hurwitz, D.I.; Marchler, G.H.; Song, J.S.; et al. CDD/SPARCLE: The conserved domain database in 2020. *Nucleic Acids Res.* **2020**, *48*, D265–D268. [CrossRef]
54. Kearse, M.; Moir, R.; Wilson, A.; Stones-Havas, S.; Cheung, M.; Sturrock, S.; Buxton, S.; Cooper, A.; Markowitz, S.; Duran, C.; et al. Geneious Basic: An integrated and extendable desktop software platform for the organization and analysis of sequence data. *Bioinformatics* **2012**, *28*, 1647–1649. [CrossRef]
55. Stamatakis, A. RAxML version 8: A tool for phylogenetic analysis and post-analysis of large phylogenies. *Bioinformatics* **2014**, *30*, 1312–1313. [CrossRef]
56. Lanfear, R.; Calcott, B.; Ho, S.Y.W.; Guindon, S. PartitionFinder: Combined selection of partitioning schemes and substitution models for phylogenetic analyses. *Mol. Biol. Evol.* **2012**, *29*, 1695–1701. [CrossRef]
57. Johansen, D. *Plant Microtechnique*; McGraw-Hill Book Company Inc.: New York, NY, USA, 1940.
58. Ambrose, B.A.; Lerner, D.R.; Ciceri, P.; Padilla, C.M.; Yanofsky, M.F.; Schmidt, R.J. Molecular and genetic analyses of the silky1 gene reveal conservation in floral organ specification between eudicots and monocots. *Mol. Cell* **2000**, *5*, 569–579. [CrossRef]
59. Vasco, A.; Smalls, T.L.; Graham, S.W.; Cooper, E.D.; Wong, G.K.S.; Stevenson, D.W.; Moran, R.C.; Ambrose, B.A. Challenging the paradigms of leaf evolution: Class III HD-Zips in ferns and lycophytes. *New Phytol.* **2016**, *212*, 745–758. [CrossRef]
60. Cantero, A.; Barthakur, S.; Bushart, T.J.; Chou, S.; Morgan, R.O.; Fernandez, M.P.; Clark, G.B.; Roux, S.J. Expression profiling of the Arabidopsis annexin gene family during germination, de-etiolation and abiotic stress. *Plant Physiol. Biochem.* **2006**, *44*, 13–24. [CrossRef] [PubMed]

International Journal of
Molecular Sciences

MDPI

Article

Simple and Divided Leaves in Ferns: Exploring the Genetic Basis for Leaf Morphology Differences in the Genus *Elaphoglossum* (Dryopteridaceae)

Alejandra Vasco [1,*] and Barbara A. Ambrose [2,*]

1 Botanical Research Institute of Texas, 1700 University Drive, Fort Worth, TX 76107-3400, USA
2 The New York Botanical Garden, 2900 Southern Blvd, Bronx, NY 10458-5126, USA
* Correspondence: avascog@gmail.com (A.V.); bambrose@nybg.org (B.A.A.);
Tel.: +1-817-546-1840 (A.V.); +1-718-817-8185 (B.A.A.)

Received: 1 May 2020; Accepted: 18 July 2020; Published: 22 July 2020

check for updates

Abstract: Despite the implications leaves have for life, their origin and development remain debated. Analyses across ferns and seed plants are fundamental to address the conservation or independent origins of megaphyllous leaf developmental mechanisms. *Class I KNOX* expression studies have been used to understand leaf development and, in ferns, have only been conducted in species with divided leaves. We performed expression analyses of the *Class I KNOX* and *Histone H4* genes throughout the development of leaf primordia in two simple-leaved and one divided-leaved fern taxa. We found *Class I KNOX* are expressed (1) throughout young and early developing leaves of simple and divided-leaved ferns, (2) later into leaf development of divided-leaved species compared to simple-leaved species, and (3) at the leaf primordium apex and margins. *H4* expression is similar in young leaf primordia of simple and divided leaves. Persistent *Class I KNOX* expression at the margins of divided leaf primordia compared with simple leaf primordia indicates that temporal and spatial patterns of *Class I KNOX* expression correlate with different fern leaf morphologies. However, our results also indicate that *Class I KNOX* expression alone is not sufficient to promote divided leaf development in ferns. *Class I KNOX* patterns of expression in fern leaves support the conservation of an independently recruited developmental mechanism for leaf dissection in megaphylls, the shoot-like nature of fern leaves compared with seed plant leaves, and the critical role marginal meristems play in fern leaf development.

Keywords: *Class I KNOX*; Dryopteridaceae; *Elaphoglossum*; ferns; fronds; leaf diversity; leaf evolution and development; megaphyll

1. Introduction

Leaves are the dominant organ in most extant vascular plants and their evolutionary origin, likely in the early Devonian, fundamentally changed not only life on earth, but also the basic Bauplan of vascular plants [1,2]. Despite this profound importance, the number of times leaves have evolved in vascular plants is still debated, and it is mainly within Euphyllophytes (ferns and seed plants) that the number of times leaves have evolved is still not settled. In Euphyllophytes, leaves have been hypothesized to have evolved from one up to nine times [3–6]. Particularly in the ferns, there is currently no consensus on whether the leaves of major lineages such as the Equisetaceae (horsetails), Psilotaceae (whisk ferns), Ophioglossaceae, Marattiaceae, and the leptosporangiate ferns are homologous [7,8].

The leaves of the Euphyllophytes are also called megaphylls and are characterized by an enormous morphological diversity; they can be simple, lobed, pedate, digitate, or divided (also termed compound or dissected). The genetic developmental basis of this enormous diversity has been mainly studied in model angiosperms, but it is largely unknown in ferns. Comparative analyses across Euphyllophyte lineages are essential to gain the comparative data needed to resolve the long-standing questions of leaf evolution and development. Comparative approaches to understand the genetic pathway affecting megaphyll shape outside of angiosperms have mainly focused on the *Class I KNOTTED1-like HOMEOBOX* (*Class I KNOX*) genes [9–11].

In angiosperms, genetic studies have explored the basis for differences in leaf division in the species with simple leaves: *Arabidopsis thaliana*, *Zea mays*, and *Antirrhinum majus*, compared with species with divided leaves: *Cardamine hirsuta*, *Lycopersicon esculentum*, *Pisum sativum*, and *Medicago truncatula* [12–14]. In angiosperms, Class I KNOX proteins are generally necessary for meristem maintenance and are expressed throughout the vegetative and floral shoot apical meristems (SAMs) and down-regulated in leaf primordia and floral organs [14–21]. In angiosperms with simple leaves, Class I KNOX proteins are expressed in the SAM, and down-regulated in incipient leaf primordia and throughout primordium development [14,17,20]. In many angiosperms with divided leaves, Class I KNOX proteins are also expressed in the SAM and down-regulated in incipient leaf primordia; however, they are expressed later in young leaf primordia and in sites of leaflet initiation [9,22]. Class I KNOX expression and function in leaves have been shown to underlie divided leaf morphology in angiosperms [9,22,23]. In angiosperms with divided leaves, over-expression of Class I KNOX produces mature leaves that are highly divided [24–26]. Meanwhile, in plant species with simple leaves, an overexpression of *Class I KNOX* genes does not result in divided leaves, but leaves with lobes or crenulated margins [17,24,27,28]. Therefore, *Class I KNOX* genes are required, but not sufficient, to produce divided leaves in angiosperms [9,29].

In gymnosperms, *Class I KNOX* expression has been reported for *Picea abies* (simple leaves) [30], *Zamia floridans* (divided leaves) [9], and *Welwitschia mirabilis* (simple leaves) [31]. These studies have shown that *Class I KNOX* are expressed in the SAM and down-regulated in incipient leaf primordia of simple and divided gymnosperms leaves, and up-regulated in divided leaved species. These patterns of expression are similar to those found in angiosperms, providing support for the homology of seed plant leaves.

For ferns, the expression profiles of *Class I KNOX* genes have been studied only in the species with divided leaves: *Osmunda regalis* [10], *Anogramma chaerophylla* [9], *Ceratopteris richardii* [32], and *Elaphoglossum peltatum* f. *peltatum* [33]. These studies showed that *Class I KNOX* genes are expressed in the fern's SAM, in young leaf primordia, and in the margins of old leaf primordia, similar to seed plants; but unlike in most seed plants, *Class I KNOX* genes were found not to be down-regulated in incipient leaf primordia of ferns with divided leaves [9,32]. This lack of down-regulation has been interpreted either as leaves of ferns and seed plants having evolved independently [9], or as a reflection of the delayed determinacy (i.e., persistent meristematic activity) exhibited by fern leaves [10]. However, *Class I KNOX* expression in the margins of old leaf primordia in fern species with divided leaves suggests that the same network for divided leaf development might be conserved in ferns and seed plants [9,10,32]. Central to resolving this debate are ferns with simple leaves, whose *Class I KNOX* expression patterns have not been studied before. The expression of *Class I KNOX* in simple leaves in ferns will help to better understand if the differences in expression found between ferns and angiosperms are linked to leaf morphology or if they can explain the different evolutionary origins of fern and seed plants leaves. Such a comparative approach is fundamental to address questions about the conservation or independent origins of megaphyllous leaf developmental mechanisms in plants.

Among leptosporangiate ferns, *Elaphoglossum* is one of the most diverse genera of ferns and its nearly 600 species are characterized by simple entire leaves [34]. There are only six species of *Elaphoglossum* that have divided leaves and four of them belong to a monophyletic group of 20 species, *Elaphoglossum* section *Squamipedia* [35–37]. The species with divided leaves belonging to section *Squamipedia* are *E. colombianum* (Maxon) Mickel, *E. moorei* (E. Britton) Christ, *E. peltatum* (Sw.) Urban, and *E. tripartitum* (Hook. & Grev.) Mickel (Figure 1a). Phylogenetic molecular studies have shown that, within section *Squamipedia*, the four species with divided leaves are not monophyletic and instead have had independent evolutionary origins from ancestors with simple, entire leaves [36] (Figure 1a). This suggests that the four species with divided leaves in section *Squamipedia* represent four independent reversions or new acquisitions of the divided condition [36,38], providing a fascinating system to study the evolution and development of leaf division in ferns within a robust phylogenetic framework.

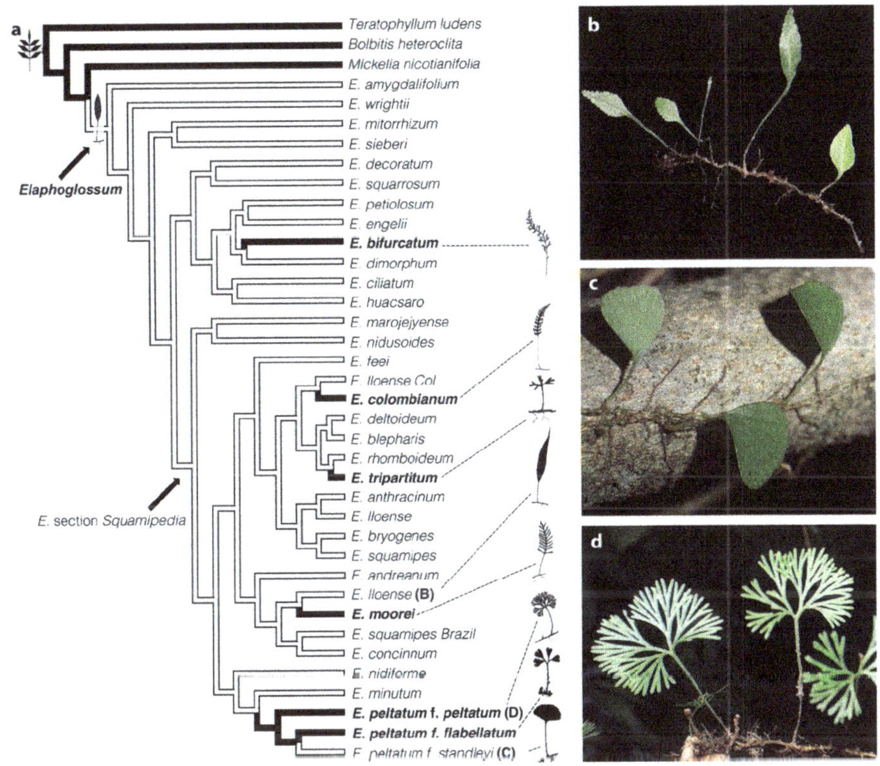

Figure 1. Leaf morphological variation in the genus *Elaphoglossum*. (**a**) Phylogeny of *Elaphoglossum* with leaf dissection optimized onto the tree (characters were optimized under a parsimony criterium with Mesquite V. 3.5). Black branches = divided leaves, white branches = simple leaves (modified from [36]). Divided-leaf taxa are in bold and displayed as shadow diagrams (not to scale), and the letter after species indicates species included in the expression studies. (b–d) The three closely related taxa studied of *Elaphoglossum* section *Squamipedia*. (**b**) *Elaphoglossum lloense* (simple leaves). (**c**) *Elaphoglossum peltatum* f. *standleyi* (simple leaves). (**d**) *Elaphoglossum peltatum* f. *peltatum* (divided leaves).

To better understand the genetic and developmental basis underlying fern leaf morphological diversity and to compare this with what is known for ferns and seed plants, we isolated *Class I KNOX* orthologs from ferns, investigated their evolution, and studied their expression in three taxa with different leaf morphologies belonging to *Elaphoglossum* section *Squamipedia* (Figure 1). We also used the expression of *Histone H4* to better understand the leaf development of fern species with simple and divided leaves. *H4* genes have been previously used to assay cell-cycle activity in lateral organs and, as such, they can be used as a cellular division marker [22,39].

The selected three taxa were as follows: *Elaphoglossum lloense* (Hook.) T. Moore with simple, entire leaves, typical of most species within the genus (Figure 1b); *Elaphoglossum peltatum* f. *standleyi* (Maxon) Mickel with simple leaves that are circular to lunate (Figure 1c); and *E. peltatum* f. *peltatum*, with divided leaves cleaved medially into two halves, where the two halves are divided subdichotomously up to seven times (Figure 1d). *Elaphoglossum peltatum* and its forms (two of them included in this study, forma *standleyi* and forma *peltatum*) have perplexed taxonomists for years [36,40–44]. Several authors have considered these forms as different species and not as merely phenotypic variants, because they appear quite distinct from each other and may even grow intermixed and maintain their distinctness. However, recent studies considered them as different forms of the same species, because examination of herbarium specimens reveals many intermediates among all the forms and because phylogenetic studies recover all three forms as part of the same clade, but not reciprocally monophyletic [35,36].

By studying these three closely related fern taxa with different leaf morphologies, two with simple leaves, we wanted to better understand if the leaf developmental genes and their expression are conserved among ferns and angiosperms with different leaf morphologies, and if differences in the patterns of expression of *Class I KNOX* genes correlate with different fern leaf morphologies.

2. Results

2.1. Evolutionary History of Class I KNOX Transcription Factors in Ferns

To gain a more detailed evolutionary history of *Class I KNOX* in ferns and to discover the putative *Elaphoglossum Class I KNOX* gene copies for our expression studies, we isolated putative homologs from selected species spanning the phylogeny of ferns, and all three orders of lycophytes by PCR and database mining (Appendix A). We identified 13 new sequences by PCR. The final analyzed matrix included 53 sequences, of which 22 belonged to ferns. The aligned matrix had 732 nucleotide and 244 amino acid characters and included the four domains encoded by *KNOX* genes (KNOX1, KNOX2, ELK, and TALE-HD). The final data set is deposited in figshare (10.6084/m9.figshare.12576581). Analyses of the nucleotide and amino acid sequences yielded congruent tree topologies. The phylogenetic relationships found are presented as majority-rule consensus trees, including branch lengths and posterior probability values for nodes (Figure 2).

Lycophyte sequences are not recovered as monophyletic, but in four different clades successively sister to euphyllophytes. Each lineage of Lycophytes, Selaginellales, Lycopodiales, and Isoetales, has at least two copies of *Class I KNOX*. Fern sequences form a monophyletic group sister to all seed plant *Class I KNOX* genes. During the evolution of ferns, at least two major duplication events are inferred (Figure 2, arrows), thus ferns have at least three copies of *Class I KNOX* genes. One copy (Copy 3 in Figure 2), which is sister to the other two, consists exclusively of sequences from the heterosporous ferns (order Salviniales). The other two copies of ferns *Class I KNOX* are sister to each other and include all the major lineages of ferns. The two *Class I KNOX* copies previously reported for the fern *Ceratopteris richardii* (*CrKNOX1* and *CrKNOX2*; Sano et al., 2005) belong to Fern *Class I KNOX* Copy 1 (Figure 2).

For two of our study species, *Elaphoglossum peltatum* f. *peltatum* (divided leaves) and *E. peltatum* f. *standleyi* (simple leaves), our mining for *KNOX* genes using degenerate primers recovered Copy1 and Copy 2 of ferns *Class I KNOX* (*EppC1KNOX1*, *EppC1KNOX2*, *EpsC1KNOX1*, and *EpsC1KNOX2*, Figure 2); for *E. lloense* (simple leaves), we only recovered Copy 2 (*ElC1KNOX2*, Figure 2). We found that all these *Elaphoglossum Class I KNOX* genes are recovered in a clade sister to all the well-known *Class I KNOX* angiosperm genes (Figure 2). Comparison of both *Class I KNOX* copies does not indicate that they are differentially spliced (Figure S1).

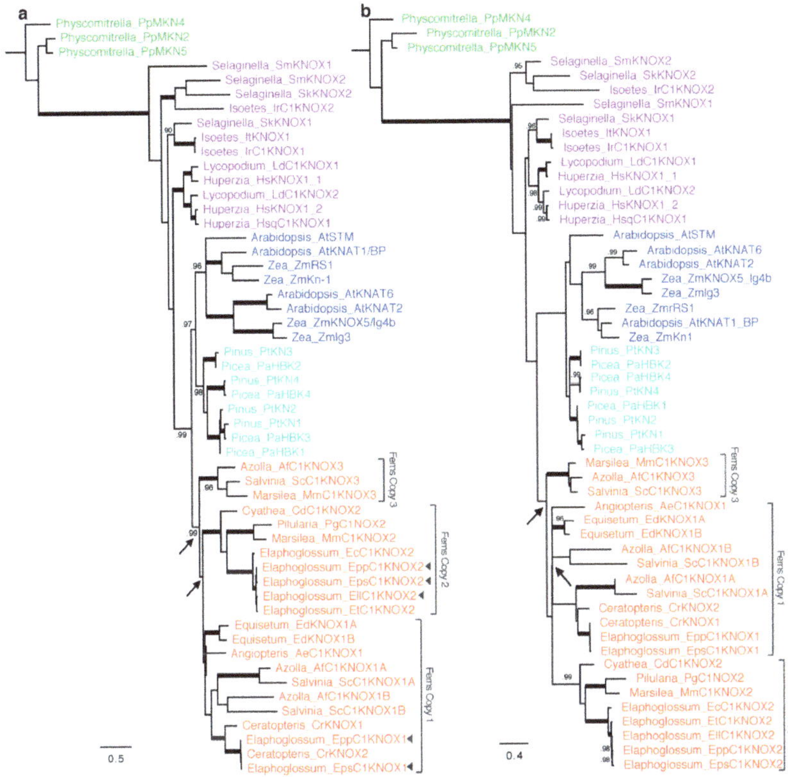

Figure 2. Phylogeny and evolution of *Class I KNOX* genes in ferns. (**a**,**b**) Phylograms inferred with (**a**) nucleotides and (**b**) amino acids presented as majority-rule consensus trees recovered in Bayesian inference (BI) analysis, including branch lengths and posterior probability (PP) values for nodes. Thick branches indicate PP = 1. PP values below 0.90 are not displayed. Colors of clade names correspond to the sources of the genes: green, bryophytes; purple, lycophytes; orange, ferns; light blue, gymnosperms; dark blue, angiosperms. Species abbreviations are listed in Appendix A. Arrowheads in (**a**), genes used for in situ hybridization analyses; arrows, inferred duplications within ferns.

2.2. Development of Simple and Divided Leaves of *Elaphoglossum*

One characteristic typical of most fern leaves is their coiled young emerging leaves. Those have been referred to as crosiers or fiddleheads. Just as the whole leaf is coiled in bud, so too are its subdivisions, the pinnae and/or pinnules. Presumably, the function of coiling is to protect the soft meristematic parts concealed within the fiddlehead. Fiddleheads are highly distinctive of ferns because they are absent from lycophytes and nearly all seed plants [7].

The three *Elaphoglossum* species studied here have long creeping stems (Figure 1b–d). Leaves are distichous (two vertical columns on opposite sides of the stem), alternate, and distant. Leaf primordia develop and grow slowly compared with the stem elongation rate, which is why in our studied species there is a relatively long distance among the visible developing leaves. In general, when plants are in the field growing with sufficient space and humidity, leaves only start to uncoil after the sixth visible leaf (Vasco, pers. Obs.). Delayed leaf expansion seems to be a characteristic of ferns with long creeping stems (Vasco, pers. Obs). Many studies of leaf development label leaves using plastochron numbers. Using a similar terminology for this study was not possible, mainly because of the delayed leaf expansion as described. Here, we defined five developmental stages based on leaf primordium morphology following our observations of histological sections and previously published fern leaf morphological and anatomical analyses [45–49] (Figure 3).

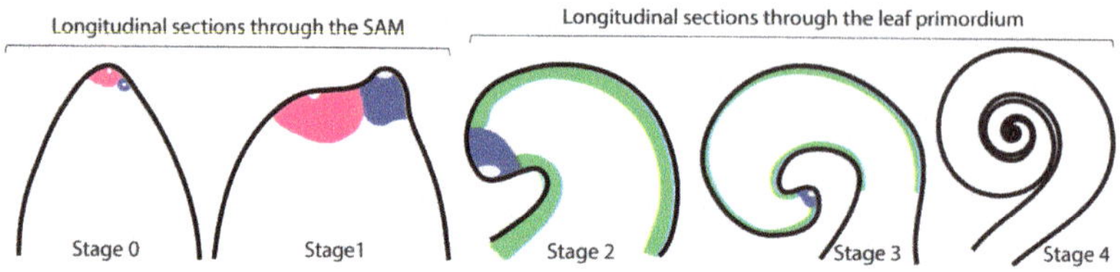

Figure 3. Stages of leaf development in the studied ferns and the meristems involved (pink = SAM, blue = LAM, green = MM). **Stage 0** (leaf initiation) Leaf initiation begins with the enlargement of an epidermal cell close to the shoot apical meristem. **Stage 1** (early leaf development): the leaf primordium is a protrusion that is more or less circular in outline; it has a prominent leaf apical initial (LAI). The LAI cuts off two files of cells that will become the marginal meristem (MM). Anatomically, the leaf apical meristem (LAM) resembles the shoot apical meristem (SAM). **Stage 2** (middle leaf development): the apex of the leaf primordium is clearly curved with an apparent LAI and MM. Basipetal procambium development is apparent. **Stage 3** (late leaf development): the apex of the leaf primordium apex is extremely curved towards the shoot apex owing to more cell divisions on the abaxial side. **Stage 4** (late crosier formation; pinna emergence): the crosier is apparent and the LAI is the same size as the rest of the cells and no longer dominant; in divided leaves, acropetal development of pinnae is apparent (not shown). Different stages not to scale. White regions in the SAM and LAM indicate leaf apical initial/s.

All leaves from the three *Elaphoglossum* species studied arise as lateral organs from the flank of the SAM. Leaf initiation (Stage 0) is detected by the enlargement of a superficial cell on the flank of the SAM. The morphology of leaf primordia of species with simple and divided leaves appears similar from Stages 0–2; all primordia are simple, and no outgrowths are detectable in both simple and divided leaved species (Figures 4 and 5). All leaf primordia are characterized by the presence of an enlarged cell at the apex—the leaf apical initial (LAI), surrounded by small cells forming a wedge shape around it, together comprising the leaf apical meristem (LAM) (LAI clearly seen in Figure 4b,c,f,g,j and Figure 5c,g,k,l). In Stage 3 of leaf development, the apex of the leaf primordium apex is extremely curved towards the shoot apex (Figure 4c,l). In Stage 4 of development, in species with divided leaves, subdivisions are detectable in the apical portion of the leaf primordium, but the primordium and its pinnae are still coiled (Figure 5n).

2.3. Patterns of Cell Division in Simple and Divided Leaves of *Elaphoglossum*

Generally, the first approach to study leaf development is to look at leaf primordia at different stages of development under a scanning electron microscope (SEM). In our studied species, this approach was not feasible owing to the presence of scales, which develop early, are large, and are copious around the SAM and leaf primordia (Figure S2). Instead, to better understand the patterns of cell division in developing simple and divided leaves of the three closely related species of *Elaphoglossum* with diverse leaf morphologies, we used in situ hybridization analyses of the *H4* genes. Using degenerate primers, we recovered one copy of *H4* for each of the three studied species (*EllH4*, *EppH4*, *EpsH4*, Appendix B).

We found that *H4* is expressed in punctate patterns during Stage 1 (Figure 4e,j) and Stage 2 (Figure 4b,f,g,k). In these developmental stages, the *H4* expression pattern is similar in species with simple and divided leaves, suggesting that cell divisions in young leaf primordia are similar, occurring randomly throughout the primordium and infrequently in the LAI, but more frequently in the cells surrounding the LAI. In primordia of species with simple and divided leaves, *H4* expression in Stage 3 is detected in the apical region behind the LAI and in the procambium (Figure 4c,l). However, at Stage 3, in the species with simple leaves, *H4* expression along the margins is continuous (Figure 4c), while H4 expression in divided leaves is discontinuous in the abaxial side, being detected in discrete regions of the leaf margins (Figure 4l). Although *H4* expression is clearly different at Stage 3 between species with simple and divided leaves, morphologically, these primordia are indistinguishable (compare Figure 4c with Figure 4l). Transverse sections of Stage 4 leaf primordia of simple and divided developing leaves show little expression of *H4* in the petiole and random expression in the lamina (Figure 4h,m).

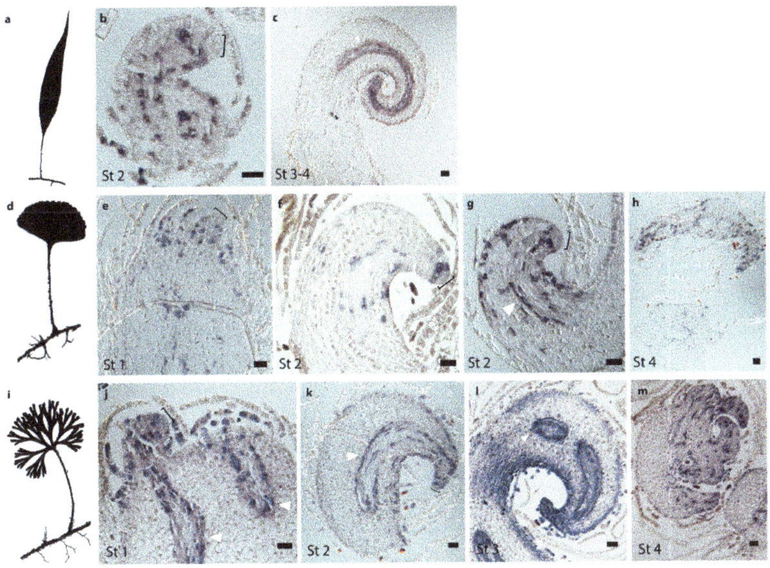

Figure 4. Cell division patterns as indicated by expression of *Histone H4* genes during leaf development in species of the fern genus *Elaphoglossum* with simple and divided leaves. (**a–c**) *Elaphoglossum lloense* (simple leaves). (**b**,**c**) Expression patterns of *ElH4*, longitudinal sections through the leaf primordium. (**b**) Expression in cells surrounding the leaf apical initial (LAI) and random cells throughout the primordium. (**c**) Expression in the apical region behind the LAI, in the procambium, and in the apical region of the margins. (**d–h**) *Elaphoglossum peltatum* f. *standleyi* (simple leaves). (**e–h**) Expression patterns of *EpsH4*. (**e–g**) Longitudinal sections through the shoot apical meristem (SAM) and/or leaf primordium. (**e**) Expression in random cells throughout the primordium that do not include the LAI. (**f**) Expression in the apical region right behind the LAI and in random cells throughout the primordium. (**g**) Expression in the apical region behind the LAI, in the procambium, and in cells of the margins. (**h**) Transverse section of old developing leaf, little expression in petiole and random expression throughout the lamina. (**i–m**) *Elaphoglossum peltatum* f. *peltatum* (divided leaves). (**j–m**) Expression patterns of *EppH4*. (**j–l**) Longitudinal sections through the SAM and/or leaf primordium. (**j**) Expression in the apical region including the LAI and in random cells throughout the primordium. (**k**) Expression in the apical region behind the LAI, in the procambium, and in cells of the margins. (**l**) Expression in the apical region, the procambium, and the margins; expression is discrete on the adaxial margin. (**m**) Transverse section of old developing leaf, little expression in petiole and random expression throughout the lamina. St, leaf developmental Stages following Figure 3; asterisk, SAM; brackets, LAI; white arrowhead, procambium. Bars = 40 um.

2.4. Class I KNOX Gene Expression Patterns in Simple and Divided Leaves of the Fern Genus Elaphoglossum

To better understand the molecular genetic basis underlying fern leaf morphological diversity and to compare our data with what is known for other ferns and seed plants, we used in situ hybridization to determine whether changes in gene expression correlate with changes in leaf morphology in the species with simple and divided leaves of our study group in the genus *Elaphoglossum* (Figure 1). We compared and analyzed the expression profiles of two of the fern copies of the meristem maintenance *Class I KNOX* genes (orthologous to all the well-known *Class I KNOX* angiosperm genes) specific to *Elaphoglossum lloense* (simple leaves, only *EllC1KNOX2* copy), *E. peltatum* f. *peltatum* (divided leaves,

EppC1KNOX1, *EppC1KNOX2*), and *E. peltatum* f. *standleyi* (simple leaves, *EpsC1KNOX1*, *EpsC1KNOX2*) (Figure 2 arrow heads).

We found that, in our studied species of *Elaphoglossum*, patterns of expression of both *Class I KNOX* copies are similar to each other throughout leaf development (compare patterns of expression in Figure 5 with Figure S3). *Class I KNOX* genes are expressed throughout the entire apical dome of the shoot meristem and the procambium regardless of leaf morphology (Figure 5b,f,k,l). Interestingly, we found evidence that indicates *Class I KNOX* are downregulated in incipient leaf primordia in at least one of the species with simple leaves (Figure 5b). Because fern roots develop in the stem just beneath the leaf primordium [50], we also detected *Class I KNOX* expression at developing roots, likely at the root apical meristem (RAM) (Figure S3).

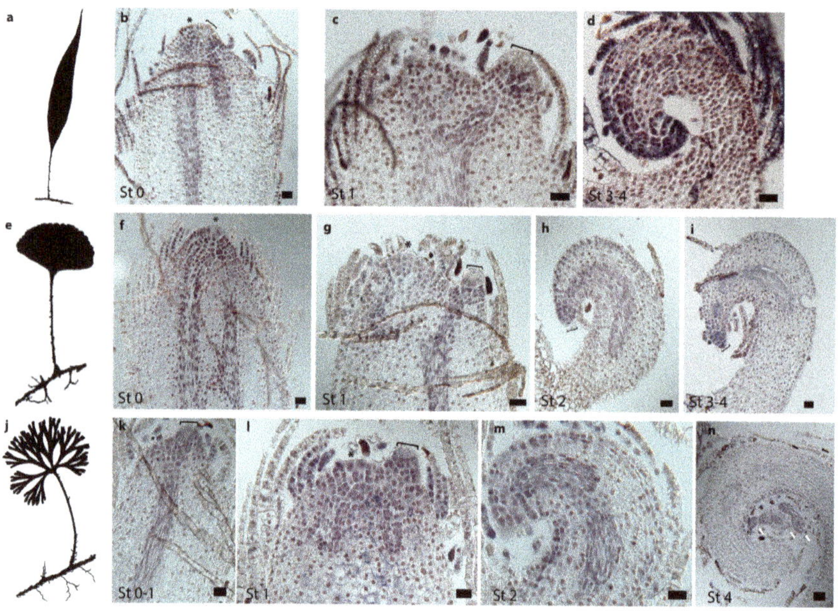

Figure 5. Expression patterns of *Class I KNOX* genes during leaf development in species of the fern genus *Elaphoglossum* with simple and divided leaves, longitudinal sections through the shoot apical meristem (SAM) and/or leaf primordia. (**a–d**) *Elaphoglossum lloense* (simple leaves). (**b–d**) Expression patterns of *EllC1KNOX2*. (**b**) Expression throughout the entire apical dome of the SAM and procambium, expression lacking from the incipient leaf primordium. (**c**) Expression throughout the entire young leaf primordium, lacking from the LAI. (**d**) Expression in the apical region, procambium, and the margins distally particularly abaxially. (**e–i**) *Elaphoglossum peltatum* f. *standleyi* (simple leaves). (**f–i**). Expression patterns in *E. peltatum* f. *standleyi* of (**f–h**) *EpsC1KNOX2* and (**i**) *EpsC1KNOX1*. (**f**) Expression throughout the entire apical dome of the SAM and procambium. (**g**) Expression in the SAM, procambium, and throughout the entire young leaf primordium, lacking from the LAI. (**h**) Expression in the apical region, procambium, and the margins, lacking from the LAI. (**i**) Expression in the apical region including the LAI and in the procambium; expression is absent from the margins. (**j–n**) *Elaphoglossum peltatum* f. *peltatum* (divided leaves). (**k–n**) Expression patterns of *EppC1KNOX2*. (**k**) Expression throughout the entire apical dome of the SAM and procambium, including the incipient leaf primordium. (**l**) Expression throughout the entire apical dome of the SAM, procambium, and throughout the entire young leaf primordium, lacking from the LAI. (**m**) Expression in the apical region, procambium, and margins; expression is discrete on the adaxial margin. (**n**) Expression in the apical region where pinnae are developing (white arrows), procambium, and adaxial margin. St, leaf developmental stages following Figure 3; asterisks, SAM; brackets, LAI. Bars = 40 um, (except n = 80 um).

In *E. lloense* and *E. peltatum* f. *standleyi*, the species with simple leaves, *Class I KNOX* are expressed throughout the entire young leaf primordium (not including the LAI) at Stage 1 (Figure 5c,g). This expression is maintained throughout Stage 2 in the leaf apical region (including the LAI), procambium, and in the margins (Figure 5h). Later in development, in Stages 3 and 4, *Class I KNOX* expression is restricted to the leaf primordium apical region, procambium, and it starts disappearing or it is absent from the margins (Figure 5d,i).

In *E. peltatum* f. *peltatum*, the species with divided leaves, *Class I KNOX* are expressed throughout the entire young leaf primordium (not including the LAI) at Stage 1 (Figure 5k,l). This expression is maintained throughout Stage 2 in the leaf apical region (including the LAI), the procambium, and in a discontinuous pattern in the margins (Figure 5m). Later in development in Stage 4, when divisions are evident at the apical region in older leaf primordia of the species with divided leaves, *Class I KNOX* expression is detected at the apical region and developing divisions, in the procambium, and in the margins of the leaf primordium adaxially (Figure 5n).

3. Discussion

3.1. Evolutionary History of *Class I KNOX* Transcription Factors in Ferns

Our results showed that, within vascular plants, *Class I KNOX* lycophyte sequences are recovered in four different clades, successively sister to euphyllophytes, and not reciprocally monophyletic (Figure 2). The non-monophyly of lycophyte sequences might suggest ancient duplication events of the only inherited *Class I KNOX* gene in the ancestor of all vascular plants, or it might be the result of high rates of evolution combined with limited sequence data (see [51,52] for similar results in different gene phylogenies). Further analyses of additional lycophyte genomes are necessary to better understand the evolutionary history of *Class I KNOX* in lycophytes and their relationship with those of other vascular plants.

Our phylogenetic hypothesis recovered fern sequences as a monophyletic group sister to *Class I KNOX* genes of seed plants, which suggests that all fern sequences are putative orthologs to the one known *Class I KNOX* seed plant lineage. Although additional expression analyses using RNAseq techniques might reveal additional *Class I KNOX* copies in certain fern groups, we found ferns have at least three copies of *Class I KNOX* genes (Figure 2). The ferns *Class I KNOX* Copy3, which is recovered sister to the other two, was found exclusively in sequences of the water fern order Salviniales (sensu [53]). Two of those ferns, *Azolla filiculoides* and *Salvinia cucullata*, correspond to the family Salviniaceae and are the only fern species whose genomes are currently sequenced and publicly available [54]. The other sequence corresponds to *Marsilea minuta* in the family Marsiliaceae and was revealed during our mining for *KNOX* genes using degenerate primers. Salviniaceae and Marsiliaceae, which are sister families, not only predominantly grow in water, but also are the only ferns that are heterosporic [53]. Although further genome sequencing may reveal additional fern taxa that have the *Class I KNOX* Copy3, it is also possible that this copy may be restricted to heterosporic ferns and play a role in heterospory.

The other two copies of fern *Class I KNOX* found, Copy1 and Copy2, are sister to each other and show phylogenetic relationships largely consistent with recently published fern species phylogenies [53,55,56], suggesting the two copies diversified during the evolution of ferns (Figure 2). Obtaining representative fern genomes and conducting further comparative analyses of the evolutionary history of different gene families will be important to determine if the duplication that led to Copies 1 and 2 of *Class I KNOX* genes in ferns was the result of the whole-genome duplication predating the core Leptosporangiate ferns inferred by Li et al. [54] or of another mechanism of gene duplication.

3.2. *Class I KNOX* Genes Are Expressed in Shoot and Leaf Fern Meristems

Developmentally, both seed plant and fern leaves (megaphylls) arise as lateral organs from the flank of an indeterminate SAM in a distinct phyllotaxy, have adaxial/abaxial identities, and are determinate organs. Fern leaves, however, differ from seed plant leaves in several aspects. Morphological and anatomical studies have shown that, in general, development of the fern leaf is from the leaf apical meristem (LAM) and the marginal meristem (MM). The LAM is composed by a leaf apical initial (LAI) and its derivatives, the LAI is an enlarged cell located at the tip of the fern leaf primordium [57–59]. The MM is located at the periphery of developing leaf primordia and is composed of marginal and submarginal initials and has been argued to be organized similar to a SAM [49,57,60–63]. In ferns, the MM makes a major contribution to lamina formation, and remains active until the general morphology of the leaf is established and the location of all procambium has been determined [47,62,64].

Class I KNOX proteins in angiosperms have been shown to be generally necessary for meristem maintenance [14–21]. We found *Class I KNOX* expression in the SAM, LAM, and RAM of ferns with simple and divided leaves (Figure 5 and Figure S3). We also found *Class I KNOX* expression in the margins in the early development of simple and divided leaved *Elaphoglossum* species (Figure 5). The *Class I KNOX* expression in the margins of fern leaf primordia reflects the persistent meristematic activity of the MM and the interpretation of the fern leaf margin as a region of sustained meristematic activity [47,49,63].

3.3. *Class I KNOX Gene Expression in Fern Leaves Recapitulates Shoot Expression*

Compared with angiosperms, fern leaves have longer meristematic activity and maturation toward the apex [65,66]. This has been explained by the presence in fern leaf primordia of the LAM (LAI and derivatives) [57–59]. Angiosperm leaves do not have apical initials and, contrary to ferns, leaf growth is not limited to the apex and margins; instead, it can be diffuse with meristematic activity throughout the developing leaf, with some angiosperms having an intercalary meristem and plate meristem that give rise to most cells of the lamina [45].

Anatomical and experimental studies have demonstrated that fern leaf primordia have shoot-like characteristics, transitioning later to determinate fate when compared with angiosperms [65–72]. The persistent *Class I KNOX* expression we found at the LAM (LAI and surrounding cells) of developing leaves in the three species of *Elaphoglossum* supports these anatomical and experimental studies and agrees with previous findings of other comparative genetic studies [9,73,74]. Studies of angiosperm species with divided leaves, such as tomato, have considered divided leaves more shoot like, and this was reflected by persistent *Class I KNOX* expression in the leaves [25].

Class I KNOX genes are expressed in the SAM of ferns [9,32,33]. Our previous study, concentrated on *Class I KNOX* expression in the shoot apical fern meristem, found expression throughout the shoot apical dome (apical initial and surrounding cells) and reported that, in 40% of the experiments, expression of *Class I KNOX* was not detected in the shoot apical initial [33]. We found that this pattern of expression is recapitulated in the LAM of both simple and divided leaves of ferns, with expression detected at the apical portion of the leaf primordium throughout development but captured in the LAI intermittently (Figure 5). The recapitulated *Class I KNOX* expression in the SAM and in the LAM of leaf primordia of both simple and divided leaves in ferns, suggest that a similar developmental mechanism is present during development in fern shoots and fern leaves, giving further genetic and molecular support for the shoot-like nature (persistent meristematic activity) of fern leaves compared with seed plants leaves. Our findings support the partial shoot theory of leaf evolution proposed by Arber [75,76], who considered the shoot to be the fundamental organ of the plant, and that all leaves were partial shoots because their indeterminate growth and radial symmetry are repressed.

3.4. *Development of Simple and Divided Leaves in Ferns*

Anatomical and morphological studies of fern leaf development have shown that primary fern leaf primordium development is owing to the growth and divisions of the LAM and the MM [57,60,61,63]. The patterns of *H4* expression we found in simple and divided leaves support these findings and suggest that, regardless of final morphology, cell divisions in early developing leaf primordia are similar in ferns (Figure 4).

The expression patterns of *Class I KNOXs* we found at Stages 0–2 of leaf development are also similar in the three species of *Elaphoglossum*, suggesting that *Class I KNOX* expression is necessary for leaf development, but that early *Class I KNOX* expression cannot explain the morphological differences between simple and divided leaves in ferns (Figure 5). Only later in leaf development (Stages 3 and 4) does *Class I KNOX* expression and cellular division patterns (*H4* expression) change between simple and divided leaves (Figures 4 and 5). Notably, at Stages 3 and 4, the expression of *Class I KNOX* persisted at the margins of leaf primordia of species with divided leaves (even after divisions develop) compared with species with simple leaves (Figure 5i,n). This persistent expression in the species with divided leaves compared with simple leaves indicates that temporal and spatial patterns of expression of *Class I KNOX* genes correlate with different fern leaf morphologies. This suggests that, although Class I KNOX alone is not sufficient to promote divided leaf development in ferns, *Class I KNOX* genes might contribute to the morphological variation between simple and divided leaves in ferns, as has been shown for angiosperms [22].

Previous anatomical and ontogenetic studies have suggested that, in leaf primordia of ferns with divided leaves, the LAM (LAI and derivatives) remains active and divisions occurred when the MM becomes interrupted in a regular manner, and some regions lose their meristematic potential [47,57,60,61,63,64,73]. The *Class I KNOX* expression in the margins of leaf primordia of ferns with simple and divided leaves highlights the critical role marginal meristems play in leaf development in ferns. Interestingly, both of our studied species with simple leaves have similar *Class I KNOX* expression patterns, even though the simple leaved *E. peltatum* f. *standleyi* is only a different form of the same species as the divided leaved *E. peltatum* f. *peltatum* [35].

Notably, we detected down-regulation of *Class I KNOX* in incipient leaf primordia in one of the *Elaphoglossum* species with simple leaves (Figure 5b), but not in the divided leaved species (Figure 5k and Figure S3c). Although a lack of downregulation at Stage 0 in fern species with divided leaves has also been reported before [9], it could be that this precise developmental stage is difficult to capture in fern leaf development or that there is a difference in downregulation of *Class I KNOX* between species of ferns with simple and divided leaves.

3.5. A Conserved Mechanism of Leaf Dissection in Megaphylls

Class I KNOX expression and function in leaf primordia of angiosperms has been shown to underlie leaf morphology differences in angiosperms [9,22,77]. In angiosperms with simple leaves, *Class I KNOX* are only expressed in the SAM, and down-regulated in incipient leaf primordia and in mature leaves [14,17,20]. Whereas, in most angiosperms with divided leaves, *Class I KNOX* are expressed in the SAM, down-regulated in incipient leaf primordia, expressed throughout the young leaf primordia, and expressed in sites of leaflet initiation in older leaf primordia [9,22]. A notable exception is the divided leaved species of tomato, where down-regulation of *Class I KNOX* in P0 has not been found [24,25].

We found a fundamental difference in leaf development in the *Class I KNOX* expression patterns between ferns and seed plants with simple leaves. In ferns with simple leaves, contrary to what has been reported for most angiosperms, *Class I KNOX* are expressed early in leaf development and maintained in the apical region of the developing leaf (Figure 5). On the other hand, similar to angiosperms with divided leaves, we found that, in ferns with divided leaves, *Class I KNOX* are expressed throughout the young leaf primordium and that this expression persists at the margins of leaf primordia (Figure 5). Moreover, in ferns with divided leaves, *Class I KNOX* expression persists in the apical region of the leaf primordium until very late in development. It has been shown that *Class I KNOX* expression in leaves was independently recruited to control divided leaf development in multiple seed plant lineages [9]. Our *Class I KNOX* expression data in a fern with divided leaves suggest that this genetic mechanism might have also been independently recruited in ferns to control divided leaf development.

In angiosperms, there are several proteins and hormones known to act in the Class I KNOX pathway that affect leaf shape, including the proteins belonging to the NO APICAL MERISTEM/CUP-SHAPED COTYLEDON (NAM/CUC) and ASYMMETRIC LEAVES/ROUGH SHEATH2/PHANTASTICA (ARP) families of transcription factors, that are known to be redeployed to make leaflets in a divided leaf [29,78–81]. The complex patterns of *Class I KNOX* expression we found in all our studied fern species, and the persistent *Class I KNOX* expression in leaf margins and divisions of species with divided leaves compared with the species with simple leaves, could also be mediated by auxin maxima that are generated by PIN1, an auxin efflux transporter [22,29,82,83], as well as changes in protein partners such as members of the *NAM/CUC* family of transcription factors that maintain *Class I KNOX* expression in a positive feedback loop in the SAM and within divided leaves for leaflet formation [77,79,84]. Phylogenetic, expression, and functional studies of these genes in all the major lineages of vascular plants will be important to fully understand to what extent the developmental genetic network underlying megaphyll morphological diversity is conserved in Euphyllophytes (ferns and seed plants).

3.6. Class I KNOX Genes and Megaphyll Evolution

The homology of megaphylls is still highly debated, and even within the ferns, it is not clear if leaves are homologous [5,7]. Previous comparative studies have come to different conclusions about the conservation of a leaf developmental network between ferns and seed plants [9,10,52,85,86]. Conservation in a leaf developmental program across ferns and seed plants was suggested by comparative expression studies of two leaf developmental genes, *Class I KNOX* and *Class III HD-Zip* [10,52]. The *Class I KNOX* downregulation in leaf primordia of fern species with simple leaves we reported here, along with the other similarities in *Class I KNOX* expression between angiosperms and ferns (SAM and margins of leaf primordia in species with divided leaves, Figure 5), supports the hypothesis of a conservation in a leaf developmental program across ferns and seed plants, suggesting an independent co-option of a common ancestral mechanism for leaf development.

Overexpression and complementation studies in angiosperms suggest that *Class I KNOX* homologs from ferns can provide some of the same functions as endogenous angiosperm genes [10,32]. However, a recent comparative study across ferns showed differential expression of another leaf transcription factor, *Class III HD-Zip* in the SAM of ferns, where expression was not detected in *Equisetum* and *Osmunda*, but was detected in leptopsorangiate ferns [52]. Additional expression studies in diverse fern species as well as knockouts will be necessary to better understand what these genes do in their native context, to further test hypotheses of leaf evolution, and to better understand the differences in connection with the leaf developmental network across ferns.

4. Materials and Methods

4.1. Sampling for the Phylogenetic Analyses of Class I KNOX Genes

To gain a more detailed evolutionary history of *Class I KNOX* in ferns and to discover the putative *Elaphoglossum Class I KNOX* gene copies for our expression studies, we obtained representative species across the fern and lycophyte phylogeny from publicly available databases and by cloning. We included *Class I KNOX* genes previously published from the lycophytes *Selaginella krausiana* [10], *Huperzia selago* and *Isoetes tegetiformans* [87], and *Lycopodium deuterodensum* [88]; from the ferns *Ceratopteris richardii* [32], *Elaphoglossum peltatum* f. *peltatum* [33], and *Equisetum diffusum* [88]. We got these sequences from GenBank, the 1KP plant transcriptome project (http://www.onekp.com, accessed May 2018) databases, or directly from the published papers. BLAST similarity searches (Altschul et al., 1990) in the lycophyte *Selaginella moellendorffii* genome available in Phytozome (https://phytozome.jgi.doe.gov, last accessed May 2018) were used to identify *Class I KNOX* copies in *S. moellendorffii*. BLAST searches were also conducted in the fern genomes of *Azolla filiculoides* and *Salvinia cucullata*, available in Fernbase (https://www.fernbase.org/, last accessed May 2019). Further lycophyte and fern sequences were obtained using degenerate primers previously published [33]. For the phylogenetic analyses, published sequences from GenBank for the other lineages of vascular plants (gymnosperms and angiosperms) were also included. *Class I KNOX Physcomitrium patens* sequences available at GenBank were used as outgroups and to root the trees. A list of all sampled species, provenance, and accession numbers is provided in Appendix A (these will be provided during review).

4.2. Sequence Analysis, Alignment, and Phylogenetic Analysis

New sequence contigs were assembled using Geneious V. 11 (Biomatters Ltd., New Zealand). Sequences were compiled and cleaned to keep just the open reading frame. Nucleotide sequences were aligned using the online version of MAFFT v.7 [89]. The alignment was refined by hand, using Mesquite V. 3.5 [90], considering protein domains and amino acid motifs that have been reported as conserved for *KNOX* genes. A matrix that included KNOX, the ELK, and the TALE-HD was used for phylogenetic analyses. Phylogenetic relationships were inferred from the nucleotide data using Bayesian inference (BI). Analyses were performed on CIPRES (http://www.phylo.org) [91]. The best partition scheme was found with PartitionFinder2 [92], for the nucleotides matrix 15 data blocks were defined by dividing the matrix into five regions (KNOX 1 (first KNOX domain), KNOX 1–KNOX 2 (region between KNOX 1 and KNOX 2), KNOX 2 (second KNOX domain), KNOX 2–HD (region between KNOX 2 and the HD), and HD (ELK and TALE-HD domains)), and by dividing each region by codon position. Analyses were performed with nine subsets as estimated by the corrected Akaike Information Criterion (AICc) implemented in PartitionFinder2 (Table S1; see Supplemental Data with this article). For the amino acids matrix, the JTT+I+G model was used as estimated by the corrected Akaike Information Criterion (AICc) implemented in PartitionFinder2 [93]. For both matrices, BI analyses were conducted using MrBAYES 3.2.6 [93]. Two independent runs of 10 million generations were completed, with four chains each (three heated, one cold), using a chain temperature of 0.2 and uniform priors. Trees and parameters were sampled every 1000th generation. Samples corresponding to the initial phase of the Markov chains (25%) were discarded as burn-in. The applicability of this burn-in value was determined by the inspection of the likelihood scores and effective sample sizes. Post-burn-in trees were combined to obtain a single majority rule consensus tree and the respective posterior probabilities (PPs) of nodes. Trees were depicted using FigTree v1.4.3 (http://tree.bio.ed.ac.uk/software/figtree/).

4.3. Taxonomic Sampling for Gene Expression Studies

To better understand the molecular genetic basis for the differences in leaf form in ferns, we studied gene expression patterns of *Histone H4* genes (used as a cell division marker) and of *Class I KNOX* genes in developing leaf primordia of two taxa with simple leaves (*E. lloense* and *E. peltatum* f. *standleyi*) and one taxon with divided leaves (*E. peltatum* f. *peltatum*) belonging to the fern genus *Elaphoglossum* (Figure 1b–d). Additionally, we compared our results of *Class I KNOX* expression to what is known from similar studies performed in seed plants and lycophytes, in order to better understand what these leaf developmental genes tell us about megaphyll leaf evolution.

For the expression analyses, the material of *E. lloense* was collected in the field in Ecuador (Vasco 865, *NY*), and both *E. peltatum* forms were sourced from specialist fern growers and kept in the Nolen glasshouses at the New York Botanical Garden (NYBG).

4.4. RNA and DNA Extraction and cDNA Synthesis

For RNA and DNA extraction, we preserved the material collected in the field in Ecuador in RNA*later* (Life Technologies, Carlsbad, CA, USA); for the living plants growing in the NYBG greenhouses, we preserved the material in liquid nitrogen. Total RNA was extracted from sporophyte shoot apices (including the SAM and young leaves), as previously described [94] with some modifications as follows. Approximately 5 g of tissue was ground to a fine powder in liquid nitrogen with a mortar and pestle. The powder was added to 25 mL of lysis buffer containing 0.1 M NaCl, 50 mM TrisHCl (pH 7.4), 50 mM EDTA (pH 8), 2% SDS, and proteinase K (200 ug/mL), and stirred at room temperature for 10–15 min. Cell debris was centrifuged at 10,000 rpm for 5 min, and the supernatant was extracted twice with an equal volume of phenol/chloroform/isoamylalcohol (50:48:2) and once with chloroform/isoamylalcohol (96:4), centrifuging each time at 10,000 rpm for 10 min. A volume of 0.1 of 3M NaOAc and 2.5 volumes Ethanol (ETOH) were added to the aqueous phase and centrifuged at 10,000 rpm for 5 min. The pellet was air dried from 5 min and resuspended in 700 uL DEPC-water. Then, 700 uL of LiCl (4M) was added and incubated overnight at 4 °C. The sample was centrifuged at 10,000 rpm for 10 min at 4 °C, the supernatant was discarded, and the pellet was resuspended in 200 uL DEPC water. Then, 20 uL of NaOAc (3M) and 500 uL ETOH were added and left at −20 °C for 30 min. The sample was centrifuged at 10,000 rpm for 10 min at 4 °C, and then washed with 70% ETOH made with DEPC water. Finally, the pellet was air dried and resuspended in 20 uL of DEPC water. Samples collected in RNA*later* were extracted with the same protocol, but the tissue was ground in the lysis buffer. cDNA was synthesized using Superscipt III (Invitrogen, Carlsbad, CA, USA) according to the manufacturer's instructions.

4.5. Primer Design and PCR

Histone H4 and *Class I KNOX* sequences of the three *Elaphoglossum* species were isolated by PCR with degenerate primers (01H4f5′ATGTCWGGMMGRGGWAAGGGAGG, 01H4r5′ CCRAAD CCRTARAGVGTHCKKCC, 01KNOXf5′ CCBGARCTBGACMABTTYATGG, and 02KN OXr5′ CCAGT GSCKYTTCCKYTGRTTDATRAACC) and by 5′ RACE (Clontech Laboratories Inc., Mountain View, CA, USA) according to the manufacturer's protocol. PCR reactions used cDNA as template and forward and reverse degenerate primers. PCR products were cleaned and cloned directly into the pCRII vector (TOPO TA cloning kit, Invitrogen, Carlsbad, CA, USA). A total of 20–30 colonies were grown in LB culture and plasmid DNA was isolated. Clones representing different banding patterns were sequenced by the Sanger method (Macrogen, USA) and BLAST was used to compare sequences in NCBI.

4.6. In Situ Hybridization

Tissues were fixed in the field for *E. lloense* or at NYBG for both *E. peltatum* forms in formaldehyde acetic acid for 2–4 h, and then dehydrated through a graded ethanol series to 100% ethanol. Tissue was embedded in Paraplast X-tra (Fisher brand) and sectioned on a microtome to 10 um. Sections were placed on ProbeOn Plus slides (Fisherbrand, Pittsburgh, PA). Gene-specific fragments for all the recovered *Elaphoglossum H4* and *Class I KNOX* copies (see results) were amplified using primers designed for this study (Figure S1 and Table S2). Digoxigenin labeled gene-specific probes were generated according to the manufacturer's instructions (Roche Applied Science, Indianapolis, IN, USA). Slides were left on a hot plate at 42 °C overnight. Treatment of cells and tissues prior to hybridization was performed as previously described [95]. Hybridizations, washes, blocking, antibody incubation, and detection were performed as in Torres et al. [96], except hybridization was performed overnight in 50% formamide humidified box at 55 °C. Sense probes were used as negative controls on pairs of slides and run in parallel with antisense probes. Sense probes gave no staining to illustrate that none of the tissue was sticky, as already indicated by different expression patterns exhibited by antisense *KNOX* and *H4*. Slides were examined and photographed on a Zeiss Axioskop microscope equipped with a Zeiss Axiocam digital camera.

Supplementary Materials: Supplementary Materials can be found at http://www.mdpi.com/1422-0067/21/15/5180/s1. Table S1. Best partition scheme and models for the aligned *Class I KNOX* matrix as estimated by the corrected Akaike Information Criterion (AICc) implemented in PartitionFinder2. Table S2. Forward and reverse primers designed for in-situ hybridizations. Figure S1. (a) Nucleotide and (b) amino acid alignment of the three copies of *Class I KNOX* genes recovered in ferns. In the nucleotide alignment, dark and white bars show the location of the forward and reverse primers respectively, designed for in-situ hybridizations. Figure S2. Scanning electron microscope images of shoot apices of *Elaphoglossum peltatum* f. *peltatum* showing massive presence of scales over the SAM and coiled young leaf primordia. (a) Shoot apex completely covered by scales. (b) Stage 3, leaf primordium completely covered by scales. (c) Stage 4 leaf primordium with coiled subdivisions (pinnae) completely covered by scales. (d) Late Stage 4 leaf primordium, only at this stage of development is lamina visible.

Star indicates the putative location of the SAM; L, leaf primordium; P, pinna; scales are highlighted with dotted lines. Figure S3. Additional expression of *Class I KNOX* genes during leaf development in species of the fern genus *Elaphoglossum* with simple and divided leaves. (a,b) *Elaphoglossum peltatum* f. *standleyi* (simple leaves), expression of *EpsC1KNOX1*, longitudinal sections through the SAM, and/or leaf primordia. (a) Expression throughout the entire apical dome of the SAM and procambium, including expression in the incipient leaf primordium (Stage 0). (b) Expression throughout the entire young leaf primordium including the leaf apical initial (LAI); expression in the root apical meristem (RAM). (c–g) *Elaphoglossum peltatum* f. *peltatum* (divided leaves). (c–f) Expression patterns of *EppC1KNOX1*, longitudinal sections through the SAM, and/or leaf primordia. (c) Expression throughout the entire apical dome of the SAM and procambium, including expression in the incipient leaf primordium (Stage 0). (d) Expression throughout the entire young leaf primordium including the LAI. (e) Expression throughout the entire apical dome of the SAM and procambium; expression in the root apical meristem (RAM). (f) Expression throughout the entire apical dome of the SAM and procambium, including expression in the incipient leaf primordium; expression in the apical region of an older leaf primordium (right). (e) Expression of *EppC2KNOX1* in the leaf primordium, procambium and root primordium. Black arrowheads, leaf primordia; brackets, LAI; stars, SAM; white arrows = root primordium; white arrowheads, procambium; bars = 40 um.

Author Contributions: Conceptualization, A.V. and B.A.A.; Methodology, A.V. and B.A.A.; Investigation, A.V. and B.A.A.; Data Curation, A.V.; Writing—Original Draft Preparation, A.V.; Writing—Review & Editing, A.V. and B.A.A.; Visualization, A.V.; Funding Acquisition, A.V. and B.A.A. All authors have read and agreed to the published version of the manuscript.

Funding: This research was funded by National Science Foundation DEB-1020443 to B.A.A. and from Programa UNAM-DGAPA-PAPIIT IA201416 to A.V.

Acknowledgments: Catalina Quintana, Renato Valencia, Hugo Navarrete, and the herbarium personnel of the Catholic University of Quito (QCA) were enormously helpful during our field work in Ecuador, where we collected *E. lloense* and *E. tripartitum*. We thank the staff of the Nolen glasshouses at the New York Botanical Garden for taking care of some of the species used for isolating *Class I KNOX* and *H4* genes and for in situs, and to Charles Alford (www.rareferns.com) for donating plants of E. *peltatum* f. *standleyi*. Tynisha Smalls assisted with the molecular work at NY. Advice and discussions with Robbin Moran were instrumental for this project. Insights from Dennis Stevenson greatly helped with interpreting some of the results. Thanks also to the anonymous reviewers and the associated editor for their time and suggestions.

Conflicts of Interest: The authors declare no conflict of interest.

Abbreviations

C1KNOX	*Class I KNOTTED1-like HOMEOBOX*
LAI	Leaf Apical Initial
LAM	Leaf Apical Meristem
MM	Marginal Meristem
RAM	Root Apical Meristem
SAM	Shoot Apical Meristem

Appendix A

Class I KNOX sequences used in this study. The information is presented in the following order: **Lineage**, ***species***, name of the sequence in the tree of Figure 2, database, accession number. The first instance of a lineage and species is given in bold. Sequences MT680030- MT680042 are new sequences generated for this study.

Bryophytes: ***Physcomitrium patens*** (Hedw.) Bruch & Schimp., Physcomitrella_PpMKN2, NCBI, XM_001758540. Physcomitrella_PpMKN4, NCBI, XM_001781425. Physcomitrella_PpMKN5, NCBI, XM_001778213. **Lycophytes:** ***Huperzia selago*** (L.) Bernh. ex Schrank & Mart., Huperzia_HsKNOX1_1, NCBI, KX761181. Huperzia_HsKNOX1_2, NCBI, KX761182. ***Huperzia squarrosa*** (G. Forst.) Trevis., Huperzia_HsqC1KNOX1, NCBI, MT680030. ***Isoetes riparia*** Engelm. ex A. Braun, Isoetes_IrC1KNOX1, NCBI, MT680031. Isoetes_IrC1KNOX2, NCBI, MT680032. ***Isoetes tegetiformans*** Rury, Isoetes_ItKNOX1, 1KP, PKOX_2098898. ***Lycopodium deuterodensum*** Herter, Lycopodium_LdC1KNOX1, 1KP, PQTO-2010435. Lycopodium_LdC1KNOX2, 1KP, PQTO-2081329. ***Selaginella kraussiana*** (Kunze) A. Braun, Selaginella_SkKNOX1, NCBI, AY667449. Selaginella_SkKNOX2, NCBI, AY667450. ***Selaginella moellendorffii*** Hieron., Selaginella_SmKNOX1, NCBI, XM_002988279. Selaginella_SmKNOX2, NCBI, XM_002977393. **Gymnosperms,** ***Picea abies*** (L.) H. Karst., Picea_PaHBK1, NCBI, AF063248. Picea_PaHBK2, NCBI, AF483277. Picea_PaHBK3, NCBI, AF483278. Picea_PaHBK4, NCBI, DQ257981 & DQ258006. ***Pinus taeda*** L., Pinus_PtKN1, NCBI, AY680402. Pinus_PtKN2, NCBI, AY680403. Pinus_PtKN3, NCBI, AY680404. Pinus_PtKN4, NCBI, AY680387 & AY680398. **Angiosperms,** ***Arabidopsis thaliana*** (L.) Heynh., Arabidopsis_AtSTM, NCBI, NM_104916. Arabidopsis_AtKNAT1/BP, NCBI, U14174.1. Arabidopsis_AtKNAT2, NCBI, NM_105719. Arabidopsis_AtKNAT6, NCBI, NM_102187. ***Zea mays*** L., Zea_ZmKn-1, NCBI, X61308. Zea_ZmrRS1, NCBI, L44133. Zea_Zmlg3, NCBI, NM_001112038. Zea_ZmKNOX5/lg4b, NCBI, NM_001111615.2. **Ferns,** ***Angiopteris evecta*** (G. Forst.) Hoffm., Angiopteris_AeC1KNOX1, NCBI, MT680033. ***Azolla filiculoides***

Lam., Azolla_AfC1KNOX1A, FernBase, Azfi_s2491.g111832 and Azfi_s0350.g066570. Azolla_AfC1KNOX1B, FernBase, Azfi_s0006.g009595. Azolla_AfC1KNOX3, FernBase, Azfi_s2342.g110698 and Azfi_s0350.g066569. ***Ceratopteris richardii*** Brongn., Ceratopteris_CrKNOX1, NCBI, AB043954. Ceratopteris_CrKNOX2, NCBI, AB043956. ***Cyathea dregei*** Kunze, Cyathea_CdC1KNOX2, NCBI, MT680034. ***Elaphoglossum ciliatum*** (C. Presl) T. Moore, Elaphoglossum_EcC1KNOX2, NCBI, MT680035. ***Elaphoglossum lloense*** (Hook.) T. Moore, Elaphoglossum_EllC1KNOX2, NCBI, MT680036. ***Elaphoglossum peltatum f. peltatum*** (Sw.) Urb., Elaphoglossum_EppC1KNOX1, NCBI, KT382287. Elaphoglossum_EppC1KNOX2, NCBI, KT382288. ***Elaphoglossum peltatum f. standleyi*** (Maxon) Mickel, Elaphoglossum_EpsC1KNOX2, NCBI, MT680037. Elaphoglossum_EpsC1KNOX1, NCBI, MT680038. ***Elaphoglossum tripartitum*** (Hook. & Grev.) Mickel, Elaphoglossum_EtC1KNOX2, NCBI, MT680039. ***Equisetum diffusum*** D. Don, Equisetum _EdKNOX1A, 1kp, CAPN-2006400. Equisetum_EdKNOX1B, 1kp, CAPN-2006401. ***Marsilea minuta*** L., Marsilea_MmC1KNOX2, NCBI, MT680040. Marsilea_MmC1KNOX3, NCBI, MT680041. ***Pilularia globulifera*** L., Pilularia_PgC1NOX2, NCBI, MT680042. ***Salvinia cucullata*** Roxb., Salvinia_ScC1KNOX1A, FernBase, Sacu_v1.1_s0115.g020964. Salvinia_ScC1KNOX1B, FernBase, Sacu_v1.1_s0011.g005308. Salvinia_ScC1KNOX3, FernBase, Sacu_v1.1_s0091.g018785.

Appendix B

Histone H4 sequences recovered in this study. The information is presented in the following order: ***species***, name of the sequence, database, accession number. The first instance of a lineage and species is given in bold. All sequences are newly generated for this study.

Elaphoglossum lloense (Hook.) T. Moore, EllH4, NCBI, MT776685. ***Elaphoglossum peltatum* f. *peltatum*** (Sw.) Urb., EppH4, NCBI, MT776686. ***Elaphoglossum peltatum* f. *standleyi*** (Maxon) Mickel, EpsH4, NCBI, MT776687.

References

1. Gifford, E.M.; Foster, A.S. *Morphology and Evolution of Vascular Plants*, 3rd ed.; W.H. Freeman and Co.: New York, NY, USA, 1989; ISBN 0716719460.
2. Beerling, D.J. Leaf Evolution: Gases, Genes and Geochemistry. *Ann. Bot.* **2005**, *96*, 345–352. [CrossRef] [PubMed]
3. Boyce, C.K.; Knoll, A.H. Evolution of developmental potential and the multiple independent origins of leaves in Paleozoic vascular plants. *Paleobiology* **2002**, *28*, 70–100. [CrossRef]
4. Boyce, C.K. The evolution of plant development in a paleontological context. *Curr. Opin. Plant Biol.* **2010**, *13*, 102–107. [CrossRef] [PubMed]
5. Sanders, H.; Rothwell, G.W.; Wyatt, S.E. Key morphological alterations in the evolution of leaves. *Int. J. Plant Sci.* **2009**, *170*, 860–868. [CrossRef]
6. Tomescu, A.M.F. Megaphylls, microphylls and the evolution of leaf development. *Trends Plant Sci.* **2009**, *14*, 5–12. [CrossRef]
7. Vasco, A.; Moran, R.C.; Ambrose, B.A. The evolution, morphology, and development of fern leaves. *Front. Plant Sci.* **2013**, *4*, 345. [CrossRef]
8. Harrison, C.J.; Morris, J.L. The origin and early evolution of vascular plant shoots and leaves. *Philos. Trans. R. Soc. B Biol. Sci.* **2017**, *373*, 20160496. [CrossRef]
9. Bharathan, G.; Goliber, T.E.; Moore, C.; Kessler, S.; Pham, T.; Sinha, N.R. Homologies in leaf form inferred from KNOXI gene expression during development. *Science* **2002**, *296*, 1858–1860. [CrossRef]
10. Harrison, C.J.; Corley, S.B.; Moylan, E.C.; Alexander, D.L.; Scotland, R.W.; Langdale, J.A. Independent recruitment of a conserved developmental mechanism during leaf evolution. *Nature* **2005**, *434*, 509–514. [CrossRef]
11. Floyd, S.K.; Bowman, J.L. Gene expression patterns in seed plant shoot meristems and leaves: Homoplasy or homology? *J. Plant Res.* **2010**, *123*, 43–55. [CrossRef]
12. Bar, M.; Ori, N. Compound leaf development in model plant species. *Curr. Opin. Plant Biol.* **2015**, *23*, 61–69. [CrossRef]
13. Du, F.; Guan, C.; Jiao, Y. Molecular Mechanisms of Leaf Morphogenesis. *Mol. Plant* **2018**, *11*, 1117–1134. [CrossRef] [PubMed]
14. Vollbrecht, E.; Veit, B.; Sinha, N.R.; Hake, S. The developmental gene Knotted-1 is a member of a maize homeobox gene family. *Nature* **1991**, *350*, 241–243. [CrossRef] [PubMed]
15. Jackson, D.; Veit, B.; Hake, S. Expression of maize KNOTTED1 related homeobox genes in the shoot apical meristem predicts patterns of morphogenesis in the vegetative shoot. *Development* **1994**, *120*, 405–413.

16. Kerstetter, R.; Vollbrecht, E.; Lowe, B.; Veit, B.; Yamaguchi, J.; Hake, S. Sequence analysis and expression patterns divide the maize knotted1-like homeobox genes into two classes. *Plant Cell* **1994**, *6*, 1877–1887.
17. Lincoln, C.; Long, J.; Yamaguchi, J.; Serikawa, K.; Hake, S. A knotted1-like homeobox gene in Arabidopsis is expressed in the vegetative meristem and dramatically alters leaf morphology when overexpressed in transgenic plants. *Plant Cell* **1994**, *6*, 1859–1876.
18. Endrizzi, K.; Moussian, B.; Haecker, A.; Levin, J.Z.; Laux, T. The SHOOT MERISTEMLESS gene is required for maintenance of undifferentiated cells in Arabidopsis shoot and floral meristems and acts at a different regulatory level than the meristem genes WUSCHEL and ZWILLE. *Plant J.* **1996**, *10*, 967–979. [CrossRef]
19. Long, J.A.; Moan, E.I.; Medford, J.I.; Barton, M.K. A member of the KNOTTED class of homeodomain proteins encoded by the STM gene of Arabidopsis. *Nature* **1996**, *379*, 66–69. [CrossRef]
20. Nishimura, A.; Tamaoki, M.; Sato, Y.; Matsuoka, M. The expression of tobacco knotted1-type class 1 homeobox genes correspond to regions predicted by the cytohistological zonation model. *Plant J.* **1999**, *18*, 337–347. [CrossRef]
21. Reiser, L.; Sanchez-Baracaldo, P.; Hake, S.; Sánchez-Baracaldo, P. Knots in the family tree: Evolutionary relationships and functions of knox homeobox genes. *Plant Mol. Biol.* **2000**, *42*, 151–166. [CrossRef]
22. Hay, A.; Tsiantis, M. The genetic basis for differences in leaf form between Arabidopsis thaliana and its wild relative Cardamine hirsuta. *Nat. Genet.* **2006**, *38*, 942–947. [CrossRef]
23. Kimura, S.; Koenig, D.; Kang, J.; Yoong, F.Y.; Sinha, N.R. Natural variation in leaf morphology results from mutation of a novel KNOX gene. *Curr. Biol.* **2008**, 672–677. [CrossRef] [PubMed]
24. Hareven, D.; Gutfinger, T.; Parnis, A.; Eshed, Y.; Lifschitz, E. The making of a compound leaf: Genetic manipulation of leaf architecture in tomato. *Cell* **1996**, *84*, 735–744. [CrossRef]
25. Chen, J.J.; Janssen, B.J.; Williams, A.; Sinha, N.R. A gene fusion at a homeobox locus: Alterations in leaf shape and implications for morphological evolution. *Plant Cell* **1997**, *9*, 1289–1304. [PubMed]
26. Janssen, B.J.; Williams, A.; Chen, J.J.; Mathern, J.; Hake, S.; Sinha, N.R. Isolation and characterization of two knotted-like homeobox genes from tomato. *Plant Mol. Biol.* **1998**, *36*, 417–425. [CrossRef] [PubMed]
27. Sinha, N.R.; Williams, R.E.; Hake, S. Overexpression of the maize homeo box gene, KNOTTED-1, causes a switch from determinate to indeterminate cell fates. *Genes Dev.* **1993**, *7*, 787–795. [CrossRef] [PubMed]
28. Ori, N.; Eshed, Y.; Chuck, G.; Bowman, J.L.; Hake, S. Mechanisms that control knox gene expression in the Arabidopsis shoot. *Development* **2000**, *127*, 5523–5532.
29. Koenig, D.; Bayer, E.; Kang, J.; Kuhlemeier, C.; Sinha, N.R. Auxin patterns Solanum lycopersicum leaf morphogenesis. *Development* **2009**, *136*, 2997–3006. [CrossRef]
30. Sundås-Larsson, A.; Svenson, M.; Liao, H.; Engström, P. A homeobox gene with potential developmental control function in the meristem of the conifer Picea abies. *Proc. Natl. Acad. Sci. USA* **1998**, *95*, 15118–15122. [CrossRef]
31. Pham, T.; Sinha, N.R. Role of Knox Genes in Shoot Development of Welwitschia mirabilis. *Int. J. Plant Sci.* **2003**, *164*, 333–343. [CrossRef]
32. Sano, R.; Juárez, C.M.; Hass, B.; Sakakibara, K.; Ito, M.; Banks, J.A.; Hasebe, M.; Juarez, C.M. KNOX homeobox genes potentially have similar function in both diploid unicellular and multicellular meristems, but not in haploid meristems. *Evol. Dev.* **2005**, *7*, 69–78. [CrossRef] [PubMed]
33. Ambrose, B.A.; Vasco, A. Bringing the multicellular fern meristem into focus. *New Phytol.* **2016**, *210*, 790–793. [CrossRef]
34. Mickel, J.T.; Smith, A.R. The pteridophytes of Mexico. *Mem. N. Y. Bot. Gard.* **2004**, *88*, 1–1055.
35. Vasco, A.; Mickel, J.T.; Moran, R.C. Taxonomic revision of the neotropical species of Elaphoglossum Sect. Squamipedia (Dryopteridaceae). *Ann. Missouri Bot. Gard.* **2013**, *99*, 244–286. [CrossRef]
36. Vasco, A.; Lóriga, J.; Rouhan, G.; Ambrose, B.A.; Moran, R.C. Divided leaves in the genus Elaphoglossum (Dryopteridaceae): A phylogeny of Elaphoglossum section Squamipedia. *Syst. Bot.* **2015**, *40*, 46–55. [CrossRef]
37. Matos, F.B.; Vasco, A.; Moran, R.C. Elaphoglossum doanense and Elaphoglossum tonduzii: New members of Elaphoglossum Sect. Squamipedia (Dryopteridaceae) and their significance for inferring the evolution of rhizome habit and nest-forming leaves within the genus. *Int. J. Plant Sci.* **2018**, *179*, 296–313. [CrossRef]
38. Moran, R.C.; Labiak, P.H.; Sundue, M. Phylogeny and character evolution of the bolbitidoid ferns (Dryopteridaceae). *Int. J. Plant Sci.* **2010**, *171*, 547–559. [CrossRef]

39. Dinneny, J.R.; Yadegari, R.; Fischer, R.L.; Yanofsky, M.F.; Weigel, D. The role of JAGGED in shaping lateral organs. *Development* **2004**, *131*, 1101–1110. [CrossRef] [PubMed]
40. von Willdenow, C.L. *Species Plantarum. Editio quarta 5*; G.C. Nauk: Berlin, Germany, 1810.
41. Christ, H. Monographie des Genus Elaphoglossum. Neue Denkschriften der Allg. *Schweizerischen Gesellschaft für die gesammten Naturwissenschaften* **1899**, *36*, 1–159.
42. Morton, C.V. Notes on Elaphoglossum, III. The publication of Elaphoglossum and Rhipidopteris. *Am. Fern J.* **1955**, *45*, 11–14. [CrossRef]
43. Gómez, L.D. Contribuciones a la pteridología costarricense. VI. El género Peltapteris Link en Costa Rica. *Brenesia* **1975**, *6*, 25–31.
44. Mickel, J.T. Relationships of the dissected elaphoglossoid ferns. *Brittonia* **1980**, *32*, 109–117. [CrossRef]
45. Steeves, T.A.; Sussex, I.M. *Patterns in Plant Development*; Cambridge University Press: New York, NY, USA, 1989.
46. Wardlaw, C.W. Experimental and analytical studies of pteridophytes: XIV. Leaf formation and phyllotaxis in Dryopteris aristata Druce. *Ann. Bot.* **1949**, *13*, 163–198. [CrossRef]
47. Saha, B. Morphogenetic studies on the distribution and activities of leaf meristems in ferns. *Ann. Bot.* **1963**, *27*, 269–279. [CrossRef]
48. Zurakowski, K.A.; Gifford, E.M. Quantitative studies of pinnule development in the ferns Adaiantum raddiantum and Cheilanthes viridis. *Am. J. Bot.* **1988**, *75*, 1559–1570. [CrossRef]
49. Bierhorst, D.W. On the stem apex, leaf initiation and early leaf ontogeny in filicalean ferns. *Am. J. Bot.* **1977**, *64*, 125–152. [CrossRef]
50. Bower, F.O. *The Ferns (Filicales) Treated Comparatively with a View to Their Natural Classification*; Cambridge University Press: Cambridge, UK, 1923.
51. Prigge, M.J.; Otsuga, D.; Alonso, J.M.; Ecker, J.R.; Drews, G.N.; Clark, S.E. Class III homeodomain-leucine zipper gene family members have overlapping, antagonistic, and distinct roles in Arabidopsis development. *Plant Cell* **2005**, *17*, 61–76. [CrossRef]
52. Vasco, A.; Smalls, T.L.; Graham, S.W.; Cooper, E.D.; Wong, G.K.-S.; Stevenson, D.W.; Moran, R.C.; Ambrose, B.A. Challenging the paradigms of leaf evolution: Class III HD-Zips in ferns and lycophytes. *New Phytol.* **2016**, *212*, 745–758. [CrossRef]
53. PPG I A community-derived classification for extant lycophytes and ferns. *J. Syst. Evol.* **2016**, *54*, 560–603.
54. Li, F.-W.; Brouwer, P.; Carretero-Paulet, L.; Cheng, S.; de Vries, J.; Delaux, P.-M.; Eily, A.; Koppers, N.; Kuo, L.-Y.; Li, Z.; et al. Fern genomes elucidate land plant evolution and cyanobacterial symbioses. *Nat. Plants* **2018**, *4*, 460–472. [CrossRef]
55. Rothfels, C.J.; Li, F.-W.; Sigel, E.M.; Huiet, L.; Larsson, A.; Burge, D.O.; Ruhsam, M.; Deyholos, M.; Soltis, D.E.; Stewart, C.N.; et al. The evolutionary history of ferns inferred from 25 low-copy nuclear genes. *Am. J. Bot.* **2015**, *102*, 1089–1107. [CrossRef] [PubMed]
56. Leebens-Mack, J.H.; Barker, M.S.; Carpenter, E.J.; Deyholos, M.K.; Gitzendanner, M.A.; Graham, S.W.; Grosse, I.; Li, Z.; Melkonian, M.; Mirarab, S.; et al. One thousand plant transcriptomes and the phylogenomics of green plants. *Nature* **2019**, *574*, 679–685.
57. Wardlaw, C.W. Experimental studies of the sporophytes of ferns. *J. Linn. Soc. London, Bot.* **1963**, *8*, 385–400. [CrossRef]
58. White, R.; Turner, M. Anatomy and development of the fern sporophyte. *Bot. Rev.* **1995**, *61*, 281–305. [CrossRef]
59. Imaichi, R. Meristem organization and organ diversity. In *Biology and Evolution of Ferns and Lycophytes*; Ranker, T.A., Haufler, C.H., Eds.; Cambridge University Press: Cambridge, UK, 2008; pp. 75–103.
60. Bower, F.O. The comparative examination of the meristems of ferns, as a phylogenetic study. *Ann. Bot.* **1889**, *3*, 305–392. [CrossRef]
61. Pray, T.R. Ontogeny of the open dichotomous venation in the pinna of the fern Nephrolepis. *Am. J. Bot.* **1960**, *47*, 319–328. [CrossRef]
62. Pray, T.R. Ontogeny of the Closed Dichotomous Venation of Regnellidium. *Am. J. Bot.* **1962**, *49*, 464–472. [CrossRef]
63. Wardlaw, C.W. Reflections on the unity of the embryonic tissues in ferns. *Phytomorphology* **1958**, *8*, 323–327.
64. Gaudet, J.J. Morphology of Marsilea vestita. II. Morphology of the Adult Land and Submerged Leaves. *Am. J. Bot.* **1964**, *51*, 591–597. [CrossRef]

65. Steeves, T.A. A study of the developmental potentialities of excised leaf primordia in sterile culture. *Phytomorphology* **1961**, *11*, 346–359.
66. Steeves, T.A. Morphogenetic studies of fern leaves. *J. Linn. Soc. London, Bot.* **1963**, *58*, 401–415. [CrossRef]
67. Wardlaw, C.W. Experiments on organogenesis in ferns. *Growth* **1949**, *13*, 93–131.
68. Wardlaw, C.W. Phyllotaxis and organogenesis in ferns. *Nature* **1949**, *164*, 167–169. [CrossRef]
69. Cutter, E.G. Experimental induction of buds from fern leaf primordia. *Nature* **1954**, *173*, 440–441. [CrossRef]
70. Cutter, E.G. Experimental and analytical studies of pteridophytes: XXXIII. The experimental induction of buds from leaf primordia in Dryopteris aristata Druce. *Ann. Bot.* **1956**, *20*, 143–165. [CrossRef]
71. White, R. Experimental and developmental studies of the fern sporophyte. *Bot. Rev.* **1971**, *37*, 509–540. [CrossRef]
72. Steeves, T.A.; Hicks, G.; Steeves, M.; Retallack, B. Leaf determination in the Fern Osmunda cinnamomea—A reinvestigation. *Ann. Bot.* **1993**, *71*, 511–517. [CrossRef]
73. Sanders, H.L.; Darrah, P.R.; Langdale, J. a Sector analysis and predictive modelling reveal iterative shoot-like development in fern fronds. *Dev. Cambridge Engl.* **2011**, *138*, 2925–2934. [CrossRef]
74. Plackett, A.R.G.; Di Stilio, V.S.; Langdale, J.A. Ferns: The missing link in shoot evolution and development. *Front. Plant Sci.* **2015**, *6*, 972. [CrossRef]
75. Arber, A. The interpretation of leaf and root in the angiosperms. *Biol. Rev.* **1941**, *16*, 81–105. [CrossRef]
76. Arber, A. *The Natural Philosophy of Plant Form*; Cambridge University Press: Cambridge, UK, 1950.
77. Hay, A.; Tsiantis, M. KNOX genes: Versatile regulators of plant development and diversity. *Development* **2010**, *137*, 3153–3165. [CrossRef]
78. Barkoulas, M.; Hay, A.; Kougioumoutzi, E.; Tsiantis, M. A developmental framework for dissected leaf formation in the Arabidopsis relative Cardamine hirsuta. *Nat. Genet.* **2008**, *40*, 1136–1141. [CrossRef] [PubMed]
79. Blein, T.; Pulido, A.; Vialette-Guiraud, A.; Nikovics, K.; Morin, H.; Hay, A.; Johansen, I.E.; Tsiantis, M.; Laufs, P. A Conserved Molecular Framework for Compound Leaf Development. *Science* **2008**, *322*, 1835–1839. [CrossRef] [PubMed]
80. Peng, J.; Chen, R. Auxin efflux transporter MtPIN10 regulates compound leaf and flower development in Medicago truncatula. *Plant Signal. Behav.* **2011**, *6*, 1537–1544. [CrossRef] [PubMed]
81. Byrne, M.E. Making leaves. *Curr. Opin. Plant Biol.* **2012**, *15*, 24–30. [CrossRef] [PubMed]
82. Reinhardt, D.; Mandel, T.; Kuhlemeier, C. Auxin Regulates the Initiation and Radial Position of Plant Lateral Organs. *Plant Cell* **2000**, *12*, 507–518. [CrossRef] [PubMed]
83. Benková, E.; Michniewicz, M.; Sauer, M.; Teichmann, T.; Seifertová, D.; Jürgens, G.; Friml, J. Local, Efflux-Dependent Auxin Gradients as a Common Module for Plant Organ Formation. *Cell* **2003**, *115*, 591–602. [CrossRef]
84. Berger, Y.; Harpaz-Saad, S.; Brand, A.; Melnik, H.; Sirding, N.; Alvarez, J.P.; Zinder, M.; Samach, A.; Eshed, Y.; Ori, N. The NAC-domain transcription factor GOBLET specifies leaflet boundaries in compound tomato leaves. *Development* **2009**, *136*, 823–832. [CrossRef]
85. Floyd, S.K.; Bowman, J.L. Distinct developmental mechanisms reflect the independent origins of leaves in vascular plants. *Curr. Biol.* **2006**, *16*, 1911–1917. [CrossRef]
86. Prigge, M.J.; Clark, S.E. Evolution of the class III HD-Zip gene family in land plants. *Evol. Dev.* **2006**, *8*, 350–361. [CrossRef]
87. Evkaikina, A.I.; Berke, L.; Romanova, M.A.; Proux-Wéra, E.; Ivanova, A.N.; Rydin, C.; Pawlowski, K.; Voitsekhovskaja, O.V. The Huperzia selago shoot tip transcriptome sheds new light on the evolution of leaves. *Genome Biol. Evol.* **2017**, *9*, 2444–2460. [CrossRef] [PubMed]
88. Frangedakis, E.; Saint-Marcoux, D.; Moody, L.A.; Rabbinowitsch, E.; Langdale, J.A. Nonreciprocal complementation of KNOX gene function in land plants. *New Phytol.* **2017**, *216*, 591–604. [CrossRef]
89. Katoh, K.; Standley, D.M. MAFFT multiple sequence alignment software version 7: Improvements in performance and usability. *Mol. Biol. Evol.* **2013**, *30*, 772–780. [CrossRef] [PubMed]
90. Maddison, W.P.; Maddison, D.R. Mesquite: A modular System for Evolutionary Analysis. Version 3.61. Available online: http://mesquiteproject.org (accessed on 21 July 2020).
91. Miller, M.A.; Pfeiffer, W.; Schwartz, T. Creating the CIPRES Science Gateway for inference of large phylogenetic trees. In Proceedings of the Gateway Computing Environments Workshop (GCE), New Orleans, LA, USA, 14 November 2010; pp. 1–8.

92. Lanfear, R.; Frandsen, P.B.; Wright, A.M.; Senfeld, T.; Calcott, B. PartitionFinder 2: New methods for selecting partitioned models of evolution for molecular and morphological phylogenetic analyses. *Mol. Biol. Evol.* **2017**, *34*, 772–773. [CrossRef]
93. Ronquist, F.R.; Huelsenbeck, J.P. MrBayes: Bayesian inference of phylogeny. *Bioinformatics* **2003**, *19*, 1572–1574. [CrossRef] [PubMed]
94. Cone, K.C.; Burr, F.A.; Burr, B. Molecular analysis of the maize anthocyanin locus C1. *Proc. Natl. Acad. Sci. USA* **1986**, *83*, 9631–9635. [CrossRef]
95. Ambrose, B.A.; Lerner, D.R.; Ciceri, P.; Padilla, C.M.; Yanofsky, M.F.; Schmidt, R.J. Molecular and genetic analyses of the Silky1 gene reveal conservation in floral organ specification between eudicots and monocots. *Mol. Cell* **2000**, *5*, 569–579. [CrossRef]
96. Torres, M.-A.; Rigau, J.; Puigdomenech, P.; Stiefel, V. Specific distribution of mRNAs in maize growing pollen tubes observed by whole-mount in situ hybridization with non-radioactive probes. *Plant J.* **1995**, *8*, 317–321. [CrossRef]

International Journal of
Molecular Sciences

MDPI

Article

Transformation of *Riccia fluitans*, an Amphibious Liverwort Dynamically Responding to Environmental Changes

Felix Althoff and **Sabine Zachgo** *

Botany Department, School of Biology and Chemistry, Osnabrück University, 49076 Osnabrück, Germany; felix.althoff@uni-osnabrueck.de
* Correspondence: szachgo@uni-osnabrueck.de

check for updates

Received: 2 July 2020; Accepted: 27 July 2020; Published: 29 July 2020

Abstract: The colonization of land by streptophyte algae, ancestors of embryophyte plants, was a fundamental event in the history of life on earth. Bryophytes are early diversifying land plants that mark the transition from freshwater to terrestrial ecosystems. The amphibious liverwort *Riccia fluitans* can thrive in aquatic and terrestrial environments and thus represents an ideal organism to investigate this major transition. Therefore, we aimed to establish a transformation protocol for *R. fluitans* to make it amenable for genetic analyses. An *Agrobacterium* transformation procedure using *R. fluitans* callus tissue allows to generate stably transformed plants within 10 weeks. Furthermore, for comprehensive studies spanning all life stages, we demonstrate that the switch from vegetative to reproductive development can be induced by both flooding and poor nutrient availability. Interestingly, a single *R. fluitans* plant can consecutively adapt to different growth environments and forms distinctive and reversible features of the thallus, photosynthetically active tissue that is thus functionally similar to leaves of vascular plants. The morphological plasticity affecting vegetative growth, air pore formation, and rhizoid development realized by one genotype in response to two different environments makes *R. fluitans* ideal to study the adaptive molecular mechanisms enabling the colonialization of land by aquatic plants.

Keywords: *Riccia fluitans*; liverwort; adaptation; terrestrialization; transformation; sexual induction

1. Introduction

The conquest of land by plants around 470 million years ago (MYA) formed new ecosystems as well as carbon cycles. This foundational event enabled the evolution of three-dimensional land plants with adaptations to new environments such that land plants represent today 80% of the total biomass on earth [1,2]. Recent molecular studies demonstrated that the monophyletic land plant group evolved from streptophyte algae and that the Zygnematophyceae, conjugating green algae, are sister to land plants [3–5]. The pioneering land colonization behavior of streptophyte algae is still present in modern habitats. *Zygnema* is a member of the advanced Zygnematales group of the streptophyte lineage that already possesses many genes required for a life on land. These streptophyte algae are abundant in freshwater as well as in hydro-terrestrial habitats and adapted to water loss and increased light exposure [6–9]. The earliest diverging land plants, the bryophytes, formed novel features to cope with the challenges associated with a permanent life on land. Breathing structures for CO_2 uptake and balancing of water loss along with anchoring rooting structures mediating also nutrition and water uptake were key adaptations enabling to increase plant size and complexity during further land plant evolution [5,10]. The complex liverwort *Marchantia polymorpha* has been informative to understand the evolution of key adaptations such as rhizoid and air pore formation [11–13]. This dioecious plant forms for sexual reproduction gametangiophores, erected reproductive structures with antheridia and archegonia comprising the respective male and female gametes [14]. Appreciation of *M. polymorpha* for molecular analyses has been boosted by the availability of molecular tools, particularly different genome editing techniques for sporeling, gemmae, and thallus transformations [15–20]. *M. polymorpha* is a terrestrial species that is well adapted to growth on land. For investigation of the molecular mechanisms enabling the water-to-land transition, an ideal model would be a basal land plant species with the ability to thrive in freshwater as well as in terrestrial environments. This requires phenotypic plasticity, defined as changes in the phenotype realized by a single genotype in response to different environments [21].

Riccia fluitans is an amphibious liverwort that grows in water transition zones where it develops in dependence on water levels either a water form (WF) or a land form (LF) [22,23]. This widespread species, belonging to the subclass of Marchantiidae, quickly adapts to different environments by undergoing severe and traceable morphological changes that have intrigued scientists already since the early 20th century [24]. Further analyses investigated *R. fluitans* rhizoid and air pore formation in adaptation of the WF to the LF and recently molecular analyses started with sequencing of its chloroplast genome [22,25–27]. The phenotypic plasticity of *R. fluitans* enables dynamic adaptations of its morphogenesis to changing environments and reveals its potential to investigate molecular mechanisms contributing to the successful conquest of land.

Here, we established a transformation protocol for this amphibious liverwort facilitating functional genetic analyses and we demonstrate dynamic and reversible adaptations of *R. fluitans* plants to aquatic and terrestrial life styles. Abiotic stresses were investigated to induce a switch from vegetative to reproductive *R. fluitans* growth and two ecotypes were characterized that are available for future studies.

2. Results

2.1. Comparison of *R. fluitans* Water and Land Form Thalli

The amphibious liverwort *R. fluitans* typically occurs as an aquatic species forming free floating mats and also grows as a terrestrial plant on moist grounds in drying ponds or streams [23]. *R. fluitans* is a popular aquarium plant and an axenic aquatic culture was purchased from an aquarium plant supplier and named *R. fluitans* 001TC according to the item number. *R. fluitans* 001TC was cultivated as WF (Figure 1A) and WF thallus pieces were spread on solid medium to induce growth of LF thalli (Figure 1B).

Next, morphologies of the two *R. fluitans* 001TC forms were compared to identify distinct characteristic features. The WF develops slender, elongated thalli with a width of about 250–500 μm that branch only occasionally (Figure 1A,C) and grow towards the water surface (inset in Figure 1A). Ventral scales but no rhizoids are regularly formed on ventral WF thallus sides (Figure 1E,J). Cross sections reveal a continuous epidermal cell layer encompassing the WF thallus lacking air pores and air chambers for gas exchange. Cells containing chloroplasts are scattered throughout the epidermis and parenchymatous thallus (Figure 1G). LF thalli display different morphological features, where thallus width is increased to about 500–1000 μm and dichotomously branching occurs more frequently than observed in the WF (Figure 1B,D,F). Numerous rhizoids are formed on the ventral LF thallus side and, as in the WF, scales are regularly initiated (Figure 1F,L). Air pores in the LF epidermis and subjacent air chambers facilitate gas exchange (Figure 1H,K). The roundish or elongated pores are simple opening structures, lacking the complexity known from pores of other liverworts such as *M. polymorpha*.

To summarize, *R. fluitans* 001TC forms under LF and WF growth conditions distinctive morphological features affecting thallus width, branching frequency, air pore, and air chamber formation for gas exchange as well as rhizoid development for anchoring. Our data support observations from earlier *R. fluitans* studies with other ecotypes [24–26], emphasizing that similar morphological adaptations of different ecotypes to two distinct growth environments enable molecular studies to unravel the underlying regulatory mechanisms.

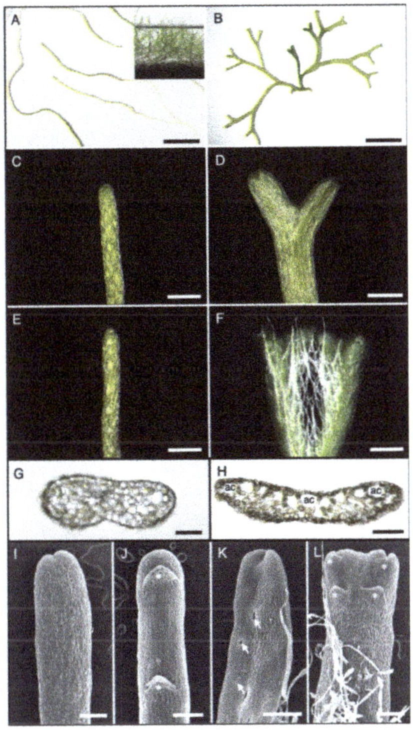

Figure 1. Land form (LF) and water form (WF) thallus morphologies of *Riccia fluitans* 001TC. (**A**) WF thalli were transferred to solid medium for image taking. Inset shows submersed growth of the WF. (**B**) LF thallus, grown on solid medium. (**C**) WF and (**D**) LF dorsal thallus side. (**E**) WF ventral thallus side. (**F**) LF ventral thallus side revealing thallus branching and rhizoid formation. (**G**) WF thallus cross section lacking air chambers. (**H**) LF cross section revealing air chambers (ac). (**I–L**) SEM pictures of thallus tips. (**I**) Dorsal WF thallus side lacking air pores. (**J**) Ventral WF side develops ventral scales (white asterisks) but no rhizoids. (**K**) Dorsal LF thallus side with air pores. Three air pores are indicated by white arrows. (**L**) Ventral LF thallus side forming rhizoids and ventral scales (white asterisks). Scale bars A,B: 5 mm; C–F: 500 μm; G–L: 200 μm.

2.2. *Induction of Reproductive Tissues*

The ability to induce the switch from vegetative to reproductive tissue formation is important for analysis of all *R. fluitans* life phases. It has been known for over 100 years that the *R. fluitans* WF easily propagates vegetatively via ruptured thallus pieces [24]. Sexual reproduction in bryophytes requires the formation of female and male gametangia, archegonia and antheridia, respectively. However, fertile *R. fluitans* producing archegonia as well as sporophytes have so far only been reported once and antheridia have never been observed [28]. Different approaches were tried to induce reproductive *R. fluitans* 001TC growth. At first, we tested the sexual induction treatment for the liverwort *M. polymorpha* [18] and grew both, the *R. fluitans* 001TC WF and LF, under long-day and also short-day conditions with far-red light supplementation. However, WF and LF plants continued vegetative growth even after more than 6 weeks of cultivation. Next, the impact of water availability was tested by transferring WF thalli to solid medium, where WF plants accordingly developed into the LF. After transfer of LF thalli to water, these floating plants were no longer anchored and developed WF features. Neither one of these transitions provided an appropriate stimulus to induce reproductive tissue formation.

Considering natural *R. fluitans* habitat conditions, flooding of LF thalli adhering on the ground can occur after strong rainfalls and could mediate inductive conditions. A LF culture was flooded with water to simulate raising water levels in ponds (Figure 2A,B) and their thalli did not drift to the water surface, as they were still anchored to the substrate by their rhizoids (Figure 2B).

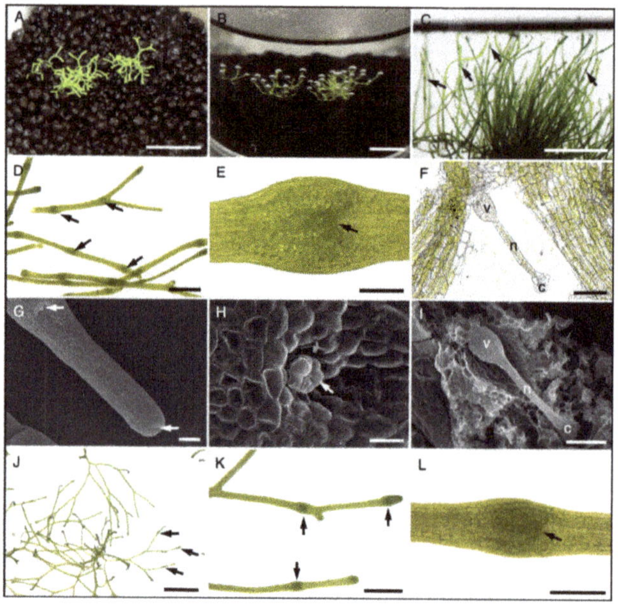

Figure 2. Reproductive tissue induction in *R. fluitans* 001TC by flooding and starvation. (**A**) Typical LF culture. (**B**) Culture shown in (A) 3 days after flooding with water, when thalli start to grow towards the water surface. (**C**) 14 days after flooding, the induced culture forms multiple archegonium cavities (arrows). (**D**) Close up of thalli with archegonium cavities (arrows). (**E**) Dorsal view of one archegonium cavity with entrance marked by an arrow. (**F**) Squash mount of an archegonium cavity to visualize the archegonium. Venter (v), neck (n), cover cells (c) of the archegonium. (**G–I**) SEM pictures. (**G**) Dorsal side of induced thallus with two archegonium cavities (arrows), the youngest is initiated close to the meristematic zone. (**H**) Archegonium cover cells (arrow) protruding the archegonium cavity. (**I**) Removal of dorsal thallus tissue uncovers the archegonium in the archegonium cavity. (**J–L**) *R. fluitans* 001TC forms archegonium cavities (arrows) 5 weeks after cultivation on solid 1/100 GB5 starvation medium. (**K**) Close up of archegonium cavities (arrows). (**L**) Archegonium cavity with entrance (arrow). Scale bars A,B,C,J: 1 cm; D, K: 2 mm; E,L: 500 μm; F,I: 100 μm; G: 200 μm; H: 50 μm.

Plants reacted to this stimulus after around 3 days by directing thallus growth towards the water surface (Figure 2B). Notably, after around 7 days, plants started to produce morphological changes, discernable as thallus swellings. After 14 days of submersed thallus growth, several swellings were regularly spaced every 5–10 mm along the thalli (Figure 2C,D). Closer inspection revealed that these thallus areas represent archegonium cavities, which contain a single archegonium with a length of about 300 μm (Figure 2E,F,I). *Riccia* archegonia are composed of a central egg cell enfolded by venter cells. Neck cells form a tubular structure ending with cover cells at the tip that are reaching out of the archegonium cavity (Figure 2F,H,I) [29]. Cover cells of the youngest archegonium are already visible close to the apical cells, indicating that archegonia formation is initiated in the meristematic region at the thallus tip (Figure 2G). Induction of female reproductive growth was reproduced several times independently, providing the robustness of the anchored LF-WF *R. fluitans* sexual induction system.

Selkirk [30] reported an induction of *R. fluitans* archegonia under nutrient starvation conditions. To test this condition, we transferred LF thalli to liquid and to solid starvation medium containing only 1/100 GB5. The plants on solid starvation medium developed elongated, yellowish thalli within a few days indicating nitrogen deficiency. After 5 weeks, first thallus swellings with archegonium cavity formation were observed (Figure 2J–L). Plants transferred to 1/100 GB5 liquid medium floated on the medium, likely mediated by air chambers of LF thalli. These plants remained in the vegetative phase, decreased growth, and ultimately died. Reproductive *R. fluitans* tissues have only very rarely been documented in the wild [28] and our analyses in the natural habitat of the Botanical Garden of Osnabrück support the assumption that *R. fluitans* reproduces mainly vegetatively.

Here, we report a reliable induction of female reproductive growth within 2–3 weeks by flooding LF thalli with water. Additionally, induction of *R. fluitans* archegonia formation is also possible by transferring LF thallus to a solid starvation medium, which then takes 5 weeks.

2.3. Analysis of Dynamic Growth Transitions in Individual R. fluitans Plants

We further investigated the adaptive morphological transformations occurring within individual plants during the switch from water to land under in vitro culture conditions. To mimic a shore situation with a water-to-land transition zone, growth containers with slanted agar surfaces were prepared and filled with liquid medium. Thalli from a WF culture were transferred to the water part of a slanted agar container. When growing WF thalli reached the imitated transition zone, their tips came into contact with agar. In response to the changed environment, thalli broadened and started to branch regularly, revealing typical LF features including rhizoid formation for anchorage on the solid medium (Figure 3A,B).

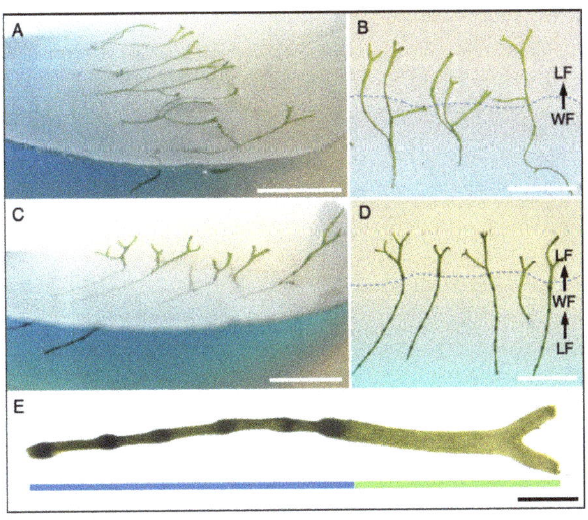

Figure 3. Phenotypic plasticity of *R. fluitans* 001TC plants. (**A**) WF thalli after growing for 13 days in a growth container with a slanted agar surface, liquid medium is stained with toluidine blue. WF thallus tips entered the solid phase and developed branching LF thalli. (**B**) Top view of (A) with a dashed line indicating the water/land border. (**C**) A LF culture was flooded to induce WF and archegonia formation and was then transferred into a slanted agar growth container and grown for 17 days. WF thallus tips entered the solid phase, stopped producing archegonium cavities and developed LF features, such as thallus branching. (**D**) Top view of (C). (**E**) Visualization of sequentially formed, distinct morphological adaptations within a single plant that has experienced LF–WF–LF transitions. The blue bar marks sexual WF thallus induced by flooding of the LF and the green bar indicates formation of the broadened, typical LF thallus with dichotomous branching. Scale bars A–D: 1 cm; E: 1 mm.

Next, we investigated the response of sexually induced thalli to cultivation on land by conducting double LF–WF–LF transition experiments. Archegonia formation was induced in LF thalli by water flooding and thalli harboring archegonium cavities were then transferred to the liquid part of slanted agar containers. When these thalli reached the solid phase, plants quickly adopted to the novel environment imitating the shore zone of a pond or stream. We then observed a broadening of the thallus, initiation of thallus branching, and thalli stopped producing further archegonium cavities (Figure 3C,D).

These data demonstrate that single *R. fluitans* plants are capable to respond consecutively to the LF–WF–LF transition conditions, which enables to observe dynamic vegetative and reproductive growth responses of one genotype to two different environments along the growth axis of a single plant (Figure 3E).

2.4. Establishment of a Transformation Protocol for *R. fluitans*

Spore, gemmae, and thallus transformation protocols have been established for the related liverwort species *M. polymorpha* [18–20]. Given the absence of gemmae production in *R. fluitans* and lack of a male line for spore generation, we aimed to establish a *R. fluitans* thallus transformation protocol.

At the start, four antibiotics frequently used for *M. polymorpha* selection were tested for *R. fluitans* 001TC. LF thallus was chopped and spread on medium containing hygromycin (2.5, 5, 10 μg/mL), G418 (5, 10, 20 μg/mL), chlorsulfuron (0.25, 0.5, 1 μM), and gentamycin (100, 200, 400 μg/mL) (Table S1). Observation of mortality rates after 1, 2, and 3 weeks identified 5–10 μg/mL hygromycin and 20 μg/mL G418 as suitable selections, whereas chlorsulfuron and gentamycin were not effective.

The binary pMpGWB-mCherry vector was used for establishing a transformation protocol. This pGWB2 derivate vector comprises the *hpt* gene mediating hygromycin resistance and the mCherry CDS, which was used as reporter fluorophore expressed under the control of the *M. polymorpha* Mp*EF1α* promoter [31,32]. At first, a method based on *M. polymorpha* transformation protocols [18] was tested using chopped *R. fluitans* 001TC LF thallus pieces, which failed to produce lines surviving the hygromycin selection. Several modifications such as injuring thallus pieces in a bead ruptor, enzymatic digestion of cell walls and testing different *Agrobacterium* strains (EHA105, LBA4404, GV3101) as well as co-cultivation on solid medium and in the dark, reported to increase *M. polymorpha* transformation efficiencies [33], also failed to generate hygromycin-resistant *R. fluitans* lines expressing mCherry.

To optimize transformation approaches for species that are difficult to transform, such as rice and maize, callus tissue has been successfully used [34,35]. Formation of callus tissue, comprising redifferentiated, pluripotent cells can be induced by auxin application [36]. Addition of 10 μM 1-naphthaleneacetic acid (NAA), a synthetic auxin, has been reported to affect growth of *M. polymorpha*, but was not sufficient to induce callus growth [37]. To determine a NAA concentration suitable to induce callus formation in *R. fluitans*, LF thalli were chopped into 1 mm pieces (Figure 4A) and spread on medium containing NAA concentrations of 50, 100, 200, and 400 μM. The 50 and 100 μM NAA caused only curling of the thalli and 400 μM NAA exerted a severe growth inhibiting effect causing early tissue death. Growth on 200 μM NAA was ideal, as it led to callus tissue formation of most tissue pieces and only a few pieces developed into small and curled thallus-like structures (Figure 4B).

Therefore, 250 mg chopped *R. fluitans* LF thallus pieces were plated on a 200 μM 1-naphthaleneacetic acid (NAA) plate. After 4 weeks of cultivation, all tissues were collected from the plate, chopped again to 1 mm pieces, and equally distributed into three Erlenmeyer flasks containing 25 mL liquid $\frac{1}{2}$ GB5 medium. The three cultures were each inoculated with 1 mL *Agrobacterium* cells harboring the pMpGWB-mCherry vector and further grown for 2 days. After removal of *Agrobacterium* cells, plant tissue from each Erlenmeyer flask was distributed on three selection plates (Figure 4C). Three independent experiments were carried out, each starting with 250 mg chopped thallus used for generating one callus plate per experiment. Concentration of 5 and 10 μg/mL hygromycin were used for selection and also 2.5 μg/mL hygromycin was tested as calli might be more sensitive to hygromycin than the investigated thallus tissue.

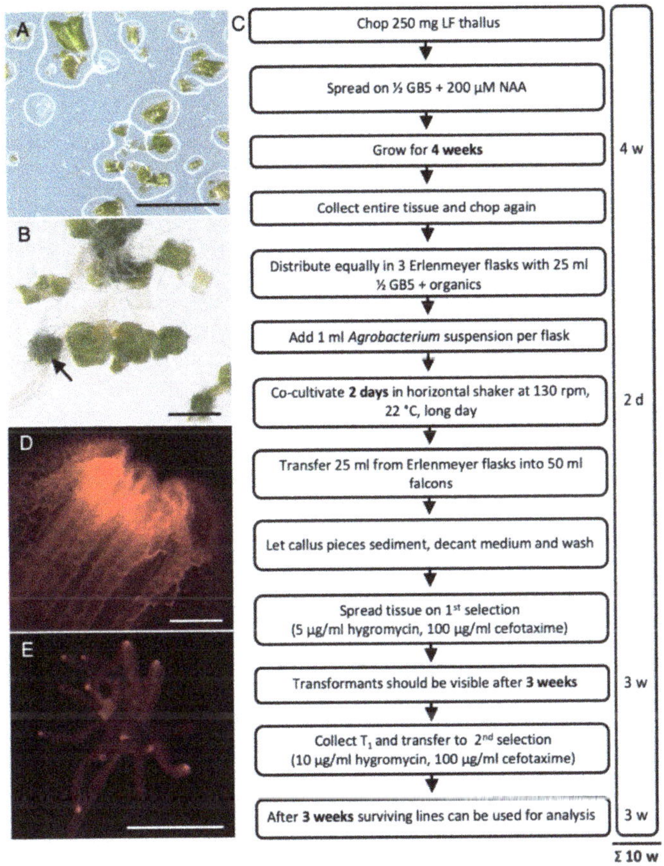

Figure 4. Scheme of the *R. fluitans* 001TC transformation procedure. (**A**) Chopped LF thallus pieces were plated on 200 µM NAA-containing medium. (**B**) Four weeks after plating NAA induced mainly callus (arrow) and also curled thallus tissue growth. (**C**) Overview of the transformation procedure. (**D**) mCherry expression in thallus tip region of a $_{pro}$Mp*EF1α::mCherry* T1 line. (**E**) mCherry signal detection in a transgenic line after three rounds of vegetative propagation. Scale bars A,B: 2 mm; D: 5 mm; E: 100 µm.

Selection with 2.5 µg/mL hygromycin resulted in growth of 144 surviving candidates that were transferred to a second selection step using 10 µg/mL hygromycin. Seven transformants survived the second selection, all of which expressed mCherry fluorescence (Table 1, Figure 4D). Using a 5 µg/mL hygromycin selection, the number of surviving lines was lower, and 39 lines could be transferred to plates with 10 µg/mL hygromycin. Of these, five lines survived and revealed mCherry fluorescence (Table 1). Only one plant survived the first selection on 10 µg/mL hygromycin, which also continued growth under the second 10 µg/mL hygromycin selection and showed mCherry signal (Table 1).

Table 1. *R. fluitans* 001TC transformation results. Three independent transformation experiments using the $_{pro}$Mp*EF1α::mCherry* construct were successively subjected to two hygromycin selection steps with indicated concentrations. Together, 13 T1 lines survived the selection procedure and expressed mCherry fluorescence.

Transformation Experiment (hyg µg/mL)	Plants Surviving 1st Selection	Plants Transferred to 2nd Selection (10 µg/mL hyg)	Plants Surviving 2nd Selection + Expressing mCherry
1 (2.5)	144	144	7
2 (5)	39	39	5
3 (10)	1	1	1

Strong mCherry expression was observed in cells of the thallus tips and further subcellular investigation detected mCherry fluorescence in the cytoplasm and nuclei of these meristematic cells (Figure 5A). The mCherry signals did not overlap with chlorophyll a autofluorescence (Figure 5B,C) and no signal could be detected in untransformed wild type controls (Figure 5D–F).

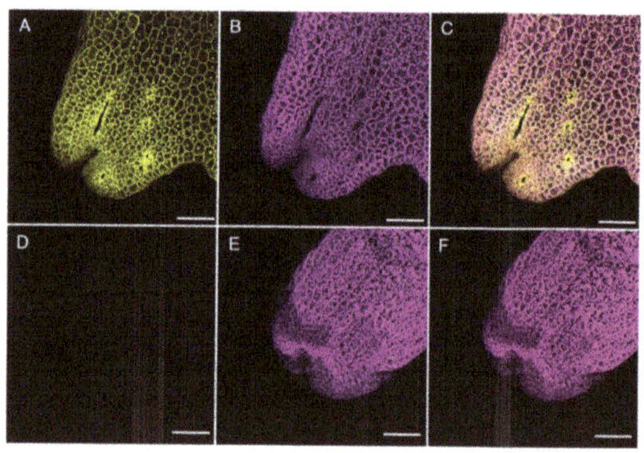

Figure 5. Transformed *R. fluitans* 001TC plant expressing mCherry. (**A–C**) $_{pro}$Mp*EF1α::mCherry* T1 line. (**D–F**) Wild type control. (**A**,**D**) mCherry fluorescence channel. (**B**,**E**) Chlorophyll a autofluorescence channel. (**C**) Merged channels of (A) and (B). (**F**) Merged channels of (D) and (E). Scale bars 200 µm.

The stability of the transformations was tested by chopping thalli from three lines ($_{pro}$Mp*EF1α::mCherry*$^{\#1}$, $_{pro}$Mp*EF1α::mCherry*$^{\#4}$, $_{pro}$Mp*EF1α::mCherry*$^{\#5}$) and regenerating growth on solid medium with 10 µg/mL hygromycin selection. After three consecutive repetitions of this vegetative propagation procedure, regenerated plants of the three lines were still expressing mCherry in meristematic regions of their regenerating thalli (Figure 4E). Additionally, thalli of the third propagation step from the three lines and from one wild type control line were chopped and again spread on solid medium without selection. After recovery of thallus growth for 4 weeks, 60 young plants per line were again transferred to selection medium containing 10 µg/mL hygromycin. All 60 plants of the three transgenic lines continued normal thallus LF growth, while the control wild type plants died (Figure S1). These data demonstrate that the hygromycin and mCherry expression remained stable throughout prolonged cultivation times and over four vegetatively propagated generations.

To summarize, we established a *R. fluitans* callus transformation protocol enabling to generate stably transformed T1 lines within 10 weeks. In three transformation experiments 13 independent lines were generated, which showed all fluorophore expression and revealed under selective conditions a wild type like growth.

2.5. Characterization of Two *R. fluitans* Ecotypes

High inter-species similarities exist within the *Riccia* genus. Furthermore, *R. fluitans* is difficult to be distinguished within the taxonomically challenging *R. fluitans* complex, which is further complicated by the two morphologically distinct *R. fluitans* LF and WF. These obstacles can lead to misidentifications and even *R. fluitans* LF and WF have been described as different species [22,38]. In addition to *R. fluitans* 001TC with an unknown origin, we established an axenic culture of an ecotype derived from the main pond of the Botanical Garden of the University of Osnabrück and named this ecotype *R. fluitans* BoGa. In this habitat, plants grow free floating in the water (Figure 6A lower side of white dashed line, Figure 6B) as well as terrestrial, close to the littoral zone (Figure 6A upper side of dashed line, Figure 6C).

Figure 6. *R. fluitans* BoGa ecotype. (**A**) *R. fluitans* BoGa in its natural habitat, a waterside of the main pond in the Botanical Garden of Osnabrück. White dashed line indicates the water to land border. (**B**) Floating WF thalli of *R. fluitans* BoGa. (**C**) LF forms dense mats close to the littoral zone. (**D**) Top view of a WF culture, where inset shows a side view of the floating culture. (**E**) LF grown in axenic culture. Thallus cross section with air chambers (ac) of the WF (**F**) and LF (**G**). Scale bars D: 1 cm; E: 5 mm; F, G: 100 μm.

Further analyses were conducted to exclude species misinterpretations. The *R. fluitans* LF can be mixed-up with *R. canaliculata*, forming as distinctive features a furrowed thallus and an apical scale overtopping the apex [39] not observed in BoGa and 001TC ecotypes (Figure 1J,L). The *R. fluitans* WF resembles *R. stenophylla*, which is monoecious and forms consistently sporophytes [23], which were not formed by both *R. fluitans* ecotypes. Moreover, the WF of *R. fluitans* and *R. rhenana* are morphologically highly similar. Univocal discrimination is possible by chromosome counting, as *R. fluitans* possesses eight and *R. rhenana* 16 chromosomes [23]. Chromosome analysis of *R. fluitans* BoGa and *R. fluitans* 001TC revealed that both ecotypes do not possess duplicated chromosome sets (Figure S2).

The *R. fluitans* BoGa WF floats just below the water surface, while the *R. fluitans* 001TC WF exhibits a more submersed growth (Figures 1A and 6D). Cross sections revealed that the *R. fluitans* BoGa WF develops air chambers, which likely mediate an impetus towards the water surface (Figure 6F), whereas the WF of the *R. fluitans* 001TC ecotype does not form air chambers, explaining a submersed growth (Figure 1G). LF comparisons showed that the *R. fluitans* BoGa LF forms slightly larger air chambers compared to *R. fluitans* 001TC (Figure 6G).

Taken together, morphological and chromosome analyses demonstrate a clear *R. fluitans* species affiliation for the two investigated different ecotypes, which can be used for future transformation approaches.

3. Discussion

The liverwort *M. polymorpha* is well adapted to a terrestrial life and has been established as a basal land plant model system [14,40,41]. Notably, liverworts also comprise the amphibious species *R. fluitans*, which lives in water and on land and is thus ideal to close the gap between aquatic and terrestrial plant model species. Here, we report the establishment of *R. fluitans* transformation and sexual induction protocols and characterized dynamic growth adaptations.

3.1. *Agrobacterium Mediated Transformation Makes* R. fluitans *Amenable for Molecular Analyses*

Given the absence of gemmae and spore formation in *R. fluitans*, various transformation approaches using thallus pieces were initially tested that were not successful. For optimization, we tried callus tissue, a cell mass of redifferentiated pluripotent cells known to mediate efficient and stable transformations in maize and rice [34,35,42,43]. Application of callus tissue enabled the successful transformation of *R. fluitans* and 13 transgenic lines were generated in three experiments. The transformation procedure starts with the induction of callus growth from LF thalli followed by *Agrobacterium* co-cultivation and two consecutive selection steps, together generating transgenic *R. fluitans* lines in 10 weeks.

We conclude from our tested conditions that a first selection with 5 μg/mL and a second selection with 10 μg/mL hygromycin is advisable to increase transformation results. It has been shown that the Mp*EF1α* promoter drives strongest reporter gene expression in meristematic zones of *M. polymorpha* [32], which was similarly observed in transformed *R. fluitan*s plants. The 13 transgenic *R. fluitans* lines showed strongest mCherry fluorophore signals in thallus tips. Three lines were further investigated and survived three successive vegetative propagation steps under selection and continuously expressed mCherry. After an additional cultivation for 4 weeks without selection followed by growth on selection plates, again all propagated lines survived and expressed mCherry.

Together, these alternating selection regimes over prolonged cultivation periods strongly support that continuous expression of the hygromycin resistance and fluorophore genes are the result of stable transformations.

3.2. *Abiotic Stress Stimuli Induce* R. fluitans *Reproductive Tissue Formation*

As sessile organisms, plants are exposed to various stresses from which they cannot escape and thus require developmental and metabolic adaptations. Starvation is an important stimulus to switch from vegetative to reproductive growth in chlorophytic algae as well as in angiosperms, benefiting from recombination events to adapt to novel environmental conditions [44,45]. A report that *R. fluitans* archegonia formation is induced under starvation conditions [30] prompted us to test our investigated ecotypes. *R. fluitans* 001TC and BoGa LF thalli that were cultivated on 1/100 GB5 medium developed archegonia after 5 weeks.

Additionally, we investigated whether flooding induces *R. fluitans* to enter the reproductive phase. Flooding is a major environmental change that is likely often experienced by the *R. fluitans* LF and requires rapid metabolic and morphological adaptations. We observed that thalli of flooded LF plants quickly developed the WF and initiated archegonia after only 2 weeks. Interestingly, under both stress conditions, starvation and flooding, induced *R. fluitans* thalli revealed a similar morphology and formed thin, elongated, less branched thalli with a brighter greenish coloration. The thallus stress-response phenotype was formed independently of aquatic or terrestrial growth and archegonia emerged on an induced WF and also on a starving LF. The two independent induction systems demonstrate that different abiotic stresses can cause an induction of reproductive tissues in *R. fluitans*.

Furthermore, we demonstrate for the first time that a single *R. fluitans* plant can switch from vegetative to reproductive growth and then back to vegetative growth, revealing a remarkable capability to adapt dynamically to different environments. Notably, Hellwege and colleagues showed that abscisic acid (ABA) levels are increasing in *R. fluitans* WF plants transforming into LF plants and that WF to LF transition can be induced by supplementation of ABA in an aquatic environment [26]. Charophyceae, freshwater green algae, possess homologs of ABA synthesis but lack the genes for the receptors [46], whereas liverworts already possess all components of the ABA signaling pathway [41]. Functional ABA signaling might therefore have exerted an important contribution to land plant adaptations.

Thus far, only one fertile *R. fluitans* population with archegonia and sporophytes has been reported, however without any male structure identification [28]. Donaghy [24] described an additional sporophyte formation in *R. fluitans* LF plants, but as he mentioned a furrowed thallus, the described plants could have been *R. canaliculata* and not *R. fluitans*. It is known that misdeterminations can occur in the *Riccia* genus that are caused by high inter-species similarities and also intra-species variations considering the different WF and LF morphologies of *R. fluitans* [22,38]. Hence, in addition to morphological analyses to discriminate the two investigated *R. fluitans* ecotypes from highly similar species, chromosome counting was carried out. Thereby, we demonstrate that *R. fluitans* 001TC and *R. fluitans* BoGa were correctly identified. Although antheridia formation in *R. fluitans* has not been reported, it is commonly accepted that *R. fluitans* is a dioecious species [23,39]. The *Riccia* genus contains dioecious and monoecious species [39] supporting that both possibilities could hold true. In our studies, we observed that *R. fluitans* 001TC forms female structures and also all isolated and induced *R. fluitans* BoGa plants only formed archegonia. This invariable gender occurrence suggests that each ecotype could have been formed by clonal reproduction of a single female individual. In our Botanical Garden, similar to what McGregor [47] observed, a thallus fragment might have been introduced by a water bird followed by further spreading throughout the pond. The two established sexual induction approaches will be helpful to further investigate whether amphibious *R. fluitans* forms female and male structures and whether this occurs on the same or on distinct plants.

3.3. *Adaptations to the Two R. fluitans Life Forms and Ecotype Differentiations*

In comparison to *M. polymorpha* that forms smooth and pegged rhizoids, *R. fluitans* only forms smooth rhizoids as pegged rhizoids were secondarily lost in this genus [48]. Air pores and air chambers are crucial structures for gas exchange and respiration. In *Riccia*, air pores are simple openings, while air pores in *Marchantia* are complex, forming a compound barrel shape [14]. Molecular data indicate that compound, complex air pores are ancestral in liverworts and that the simple pores formed by *Riccia* are derived [48]. Together, secondary losses of pegged rhizoids and complex air pores indicate that *Riccia* holds a derived and not basal phylogenetic position in the Marchantiidae sub-class [48,49].

Interestingly, most liverwort lineages have notably slow rates of evolution, including *Marchantia*, whereas *Riccia* reveals a faster evolution rate and is the most species-rich genus in complex liverworts [48,50]. The faster evolution rate might have supported the evolution of its phenotypic plasticity and ability to grow submersed, which is likely rather a secondary adaptation of *R. fluitans* given the exceptional amphibious lifestyle of *R. fluitans* within the *Riccia* genus. Additionally, it might also have supported further adaptations to different ecological niches. *R. fluitans* 001TC, with an unknown origin, might have evolved submersed growth in adaption to special light and CO_2 concentrations [51] and could therefore have been selected to meet the demands of customers using this liverwort for greening of aquarium bottoms. We generated a novel axenic *R. fluitans* BoGa ecotype culture with known origin and compared it with the *R. fluitans* 001TC ecotype. While *R. fluitans* 001TC WF lacks air chambers which likely mediate a submersed growth, BoGa ecotype WF thalli develop air chambers and float below the water surface. These minor morphological differences could be adaptations to individual ecotype habitats with varying light availability, as floating forms might inhabit deeper, turbid lakes whereas submersed growing forms inhabit shallow, clear waters. It has been reported that *R. fluitans* can sink to the bottom of ponds to survive colder temperatures during winter time [24,28], which further strengthens an adaptability by likely adjusting air chamber formation and thus thallus impetus to altered environmental conditions.

3.4. Outlook

The amphibious life style of *R. fluitans* liverwort enables the investigation of dynamic transitions between water and land plant forms as well as adaptations to different stress factors in a single basal land plant. Two ecotypes, *R. fluitans* 001TC and *R. fluitans* BoGa, are available as axenic cultures and the establishment of transformation and sexual induction protocols make *R. fluitans* accessible for future molecular analyses. Drought, UV light, and strongly fluctuating temperatures have been central abiotic stressors for plants during the conquest of land over 470 MYA until today, where climate change represents a challenge. Investigating the impact of these diverse stresses on *R. fluitans* growth and unravelling the adaptive molecular mechanisms will provide insights in plant terrestrialization processes and support climate-change focused plant breeding.

4. Materials and Methods

4.1. Plants and Growth Conditions

An axenic in vitro culture of *R. fluitans* was purchased from Tropica Aquarium Plants (Egå, Denmark) (https://tropica.com/en/plants/plantdetails/Ricciafluitans(001TC)/4386) and accordingly named *R. fluitans* 001TC. For LF growth *R. fluitans* 001TC was cultivated on damp autoclaved aquarium scaper's soil (Dennerle GmbH, Münchweiler an der Rodalb, Germany) and WF thalli were grown in containers with a bottom layer of scaper's soil filled with ddH_2O. For vitro cultivation, LF was grown on solid half strength Gamborg's B5 medium including vitamins ($\frac{1}{2}$ GB5) (Duchefa, Haarlem, The Netherlands) supplemented with 14 g/L agar-agar Kobe I (Roth, Karlsruhe, Germany). WF thalli were grown on solid $\frac{1}{2}$ GB5 medium overlaid with liquid $\frac{1}{2}$ GB5 medium.

Slanted growth containers were prepared by fixing cylindrical growth containers at a 45° angle while the agar solidified, which was afterwards overlaid with liquid $\frac{1}{2}$ GB5 medium. Induction of reproductive tissues by flooding was carried out axenically on agar or on aquarium scaper's soil. Antibiotics for selection of transgenic lines were added at indicated concentrations. Plants were grown in a climate chamber at 22 °C under long day conditions (16 h light:8 h dark) and cool white (840) fluorescent tubes emitting 90 $\mu mol\ m^{-2}\ s^{-1}$ photons.

R. fluitans thalli were also collected from the main pond of the Botanical Garden of the University of Osnabrück, referred to as *R. fluitans* BoGa. Tips from WF thalli were isolated and rinsed under running water for 15 min to remove debris. Tissues were then surface sterilized using 50 µL NaOCl (11–15% available chlorine) in 1 mL ddH_20 for 4 min and afterwards washed with ddH_2O. Sterilized thalli were transferred to solid $\frac{1}{2}$ GB5 plates containing 100 µg/mL cefotaxime (Duchefa, Haarlem, The Netherlands) for removal of remaining bacteria.

Voucher specimens are deposited at the OSBU herbarium under accession numbers OSBU 28206 (*R. fluitans* BoGa) and OSBU 28207 (*R. fluitans* 001TC).

4.2. Microscopy and Sectioning

Habitus pictures were taken with a Canon EOS 750D DSLR camera or Leica M165 FC stereo microscope equipped with a Leica DFC490 camera. Pictures of cross sections, archegonia, and chromosome analyses were taken with a Leica DM5000 B microscope and Leica DFC490 camera.

80 to 130 µm thallus cross sections were prepared using a Leica VT1000S vibratome. Thalli were immersed in 1% liquid agarose and vacuum was applied for 30 s. Degassed thalli were transferred to 3% liquid agarose and hardened by cooling to room temperature.

Tissues for scanning electron microscopy were fixed in FAE fixative (2% formaldehyde, 5% acetic acid, 50% ethanol) at 4 °C for 2 days and then washed and dehydrated using an ethanol series. Critical point drying was conducted in the Leica EM CPD300 system. The chamber was cooled to 10 °C, 36 CO_2 changes were applied to the tissue, and the chamber was afterwards heated to 40 °C. Dried samples were gold sputtered in the Leica EM ACE600 sputter coater. Sputtering was performed twice at 0° and 15° angles resulting in a gold coat of 10 nm. Samples were analyzed with a Jeol JSM-IT200 scanning microscope.

Fluorescence pictures of transgenic thalli were taken with a Zeiss LSM 880 in airy scan mode. For mCherry detection, thalli were excited with laser DPSS 561-10 at a wavelength of 561 nm and maximum emission was detected at 615 nm. Chlorophyll a autofluorescence was exited at 633 nm using the laser HeNe633 and maximum emission was detected at 687 nm. Raw pictures were processed and assembled using the Fiji image processing program (version 2.0.0-rc-69/1.52i) [52].

4.3. Chromosome Counting

Chromosome counting was carried out according to Schwarzacher [53]. Briefly, thallus tips from freshly started *R. fluitans* 001TC and *R. fluitans* BoGa cultures ensuring high cell division rates, were collected and transferred to ice water for 24 h and afterwards incubated in saturated 8-hydroxyquinoline solution for 3 h. Thallus pieces were then fixed overnight in Carnoy's solution. The next day, thalli were washed five times with 10 mM citric acid buffer (pH 4.6) to remove the fixative. Thalli were then digested with 2.5% pectolyase (Duchefa, Haarlem, The Netherlands), 2.5% pectinase (Sigma-Aldrich, Darmstadt, Germany), and 2.5% cellulase (Duchefa, Haarlem, The Netherlands) for 30 min and washed with ddH_2O. DAPI (10 µg/mL) staining was performed on object slides, tissue was then squashed with a cover slip, and staining results were immediately observed using a Leica DM5000 B microscope.

4.4. Generation of *R. fluitans* Callus Tissue and Transformation Procedure

First, 250 mg of axenic grown LF thallus was chopped into 1 mm pieces with a flame-sterilized razor blade and spread with a Drigalski glass spatula on one $\frac{1}{2}$ GB5 plate supplemented with 200 µM NAA (Duchefa, Haarlem, The Netherlands). Plates were gown for 3–4 weeks until calli and curled thalli were formed. These tissues were then chopped again into 1 mm pieces and material from one plate was split up into three 100 mL Erlenmeyer flasks each containing 25 mL $\frac{1}{2}$ GB5 medium supplemented with 0.03% L-glutamine (Roth, Karlsruhe, Germany), 0.1% casamino acids (Roth, Karlsruhe, Germany), and 2% sucrose (Roth, Karlsruhe, Germany).

Transformation experiments were carried out with the pGWB2 vector derivate pMpGWB-mCherry comprising the *hpt* hygromycin resistance gene and mediating mCherry expression [31] under the control of the *M. polymorpha EF1α* promoter, known to confer a strong and almost ubiquitous expression in *Marchantia* [32]. Co-cultivation with *Agrobacterium tumefaciens* strain C58C1 (pGV2260) carrying the pMpGWB-mCherry plasmid was performed as described by Ishizaki et al. [18] with slight modifications according to Althoff et al. [32]. Tissues from the three flasks were transferred after 2 days of co-cultivation to three 50 mL falcon tubes, passively sedimented and the supernatant, containing *Agrobacterium* cells, was replaced by three washing steps with ddH_2O. Next, plant tissues were divided on nine (three per falcon) square petri dishes (10 × 10 cm) containing $\frac{1}{2}$ GB5 medium supplemented with hygromycin (Roth, Karlsruhe, Germany) (2.5, 5, or 10 µg/mL) and 100 µg/mL cefotaxime (Duchefa, Haarlem, The Netherlands) to remove *Agrobacterium* cells and sealed with micropore tape (3M).

Selection plates were cultured under long day conditions at 22 °C for 3 weeks. Whole young regenerated transformants were transferred to fresh plates with 10 µg/mL hygromycin and 100 µg/mL cefotaxime selection and mCherry expression was analyzed after 3 weeks of selective growth conditions.

Supplementary Materials: Supplementary materials can be found at http://www.mdpi.com/1422-0067/21/15/5410/s1.

Author Contributions: F.A. and S.Z. designed the study and experiments. F.A. performed the experiments and generated and documented the data. Both authors analyzed the data and wrote the manuscript. All authors have read and agreed to the published version of the manuscript.

Funding: This research was supported by the Deutsche Forschungsgemeinschaft (SFB944 to SZ).

Acknowledgments: We thank Lucia Maß from our lab and Katherina Psathaki (iBiOs, Osnabrück) for microscopy support and Nikolai Friesen (Botanical Garden, Osnabrück) for help with chromosome analyses. We are grateful to Claudia Gieshoidt for *Riccia* tissue culture support. SZ thanks Henrik Buschmann (Botany, Osnabrück) for stimulating discussions on plant terrestrialization.

Conflicts of Interest: The authors declare no conflict of interest

References

1. Bar-On, Y.M.; Phillips, R.; Milo, R. The biomass distribution on Earth. *Proc. Natl. Acad. Sci. USA* **2018**, *115*, 6506–6511. [CrossRef] [PubMed]
2. Pires, N.D.; Dolan, L. Morphological evolution in land plants: New designs with old genes. *Philos. Trans. R. Soc. Lond. B Biol. Sci.* **2012**, *367*, 508–518. [CrossRef] [PubMed]
3. Wickett, N.J.; Mirarab, S.; Nguyen, N.; Warnow, T.; Carpenter, E.; Matasci, N.; Ayyampalayam, S.; Barker, M.S.; Burleigh, J.G.; Gitzendanner, M.A.; et al. Phylotranscriptomic analysis of the origin and early diversification of land plants. *Proc. Natl. Acad. Sci. USA* **2014**, *111*, E4859–E4868. [CrossRef] [PubMed]
4. One thousand plant transcriptomes and the phylogenomics of green plants. *Nature* **2019**, *574*, 679–685. [CrossRef]
5. Harris, B.J.; Harrison, C.J.; Hetherington, A.M.; Williams, T.A. Phylogenomic Evidence for the Monophyly of Bryophytes and the Reductive Evolution of Stomata. *Curr. Biol.* **2020**. [CrossRef]
6. Becker, B.; Feng, X.; Yin, Y.; Holzinger, A. Desiccation tolerance in streptophyte algae and the algae to land plant transition: Evolution of LEA and MIP protein families within the Viridiplantae. *J. Exp. Bot.* **2020**, *71*, 3270–3278. [CrossRef]
7. Hori, K.; Maruyama, F.; Fujisawa, T.; Togashi, T.; Yamamoto, N.; Seo, M.; Sato, S.; Yamada, T.; Mori, H.; Tajima, N.; et al. *Klebsormidium flaccidum* genome reveals primary factors for plant terrestrial adaptation. *Nat. Commun.* **2014**, *5*, 3978. [CrossRef]
8. De Vries, S.; de Vries, J.; von Dahlen, J.K.; Gould, S.B.; Archibald, J.M.; Rose, L.E.; Slamovits, C.H. On plant defense signaling networks and early land plant evolution. *Commun. Integr. Biol.* **2018**, *11*, 1–14. [CrossRef]
9. Holzinger, A.; Pichrtová, M. Abiotic Stress Tolerance of Charophyte Green Algae: New Challenges for Omics Techniques. *Front. Plant. Sci.* **2016**, *7*, 678. [CrossRef]
10. Hetherington, A.J.; Dolan, L. Bilaterally symmetric axes with rhizoids composed the rooting structure of the common ancestor of vascular plants. *Philos. Trans. R. Soc. Lond. Ser. B Biol. Sci.* **2018**, *373*, 20170042. [CrossRef]
11. Ishizaki, K.; Mizutani, M.; Shimamura, M.; Masuda, A.; Nishihama, R.; Kohchi, T. Essential role of the E3 ubiquitin ligase nopperabo1 in schizogenous intercellular space formation in the liverwort *Marchantia polymorpha*. *Plant. Cell* **2013**, *25*, 4075–4084. [CrossRef] [PubMed]
12. Honkanen, S.; Jones, V.A.S.; Morieri, G.; Champion, C.; Hetherington, A.J.; Kelly, S.; Proust, H.; Saint-Marcoux, D.; Prescott, H.; Dolan, L. The Mechanism Forming the Cell Surface of Tip-Growing Rooting Cells Is Conserved among Land Plants. *Curr. Biol.* **2016**, *26*, 3238–3244. [CrossRef] [PubMed]
13. Jones, V.A.S.; Dolan, L. MpWIP regulates air pore complex development in the liverwort *Marchantia polymorpha*. *Development* **2017**, *144*, 1472–1476. [CrossRef] [PubMed]
14. Shimamura, M. *Marchantia polymorpha*: Taxonomy, Phylogeny and Morphology of a Model System. *Plant. Cell Physiol.* **2016**, *57*, 230–256. [CrossRef]
15. Ishizaki, K.; Johzuka-Hisatomi, Y.; Ishida, S.; Iida, S.; Kohchi, T. Homologous recombination-mediated gene targeting in the liverwort *Marchantia polymorpha* L. *Sci. Rep.* **2013**, *3*, 1532. [CrossRef]
16. Kopischke, S.; Schussler, E.; Althoff, F.; Zachgo, S. TALEN-mediated genome-editing approaches in the liverwort *Marchantia polymorpha* yield high efficiencies for targeted mutagenesis. *Plant. Methods* **2017**, *13*, 20. [CrossRef]

17. Sugano, S.S.; Nishihama, R.; Shirakawa, M.; Takagi, J.; Matsuda, Y.; Ishida, S.; Shimada, T.; Hara-Nishimura, I.; Osakabe, K.; Kohchi, T. Efficient CRISPR/Cas9-based genome editing and its application to conditional genetic analysis in *Marchantia polymorpha*. *PLoS ONE* **2018**, *13*, e0205117. [CrossRef]
18. Ishizaki, K.; Chiyoda, S.; Yamato, K.T.; Kohchi, T. Agrobacterium-mediated transformation of the haploid liverwort *Marchantia polymorpha* L., an emerging model for plant biology. *Plant. Cell Physiol.* **2008**, *49*, 1084–1091. [CrossRef]
19. Tsuboyama-Tanaka, S.; Kodama, Y. AgarTrap-mediated genetic transformation using intact gemmae/gemmalings of the liverwort *Marchantia polymorpha* L. *J. Plant. Res.* **2015**, *128*, 337–344. [CrossRef]
20. Kubota, A.; Ishizaki, K.; Hosaka, M.; Kohchi, T. Efficient Agrobacterium-mediated transformation of the liverwort *Marchantia polymorpha* using regenerating thalli. *Biosci. Biotechnol. Biochem.* **2013**, *77*, 167–172. [CrossRef]
21. Gratani, L. Plant Phenotypic Plasticity in Response to Environmental Factors. *Adv. Bot.* **2014**, 208747. [CrossRef]
22. Müller, K. *Beiträge zur Systematik der Lebermoose*; Hedwigia: Dresden, Germany, 1941.
23. Wyatt, R.; Davison, P.G. Notes on the Distribution and Ecology of *Riccia rhenana* in Georgia. *Evansia* **2013**, *30*, 130–134. [CrossRef]
24. Donaghy, F. The morphology of *Riccia fluitans* L. *Proc. Indiana Acad. Sci.* **1915**, *25*, 131–134.
25. Klingmüller, W.; Denffer, D.v. Zur Entwicklungsphysiologie der Ricciaceen. *Flora Oder Allg. Bot. Ztg.* **1959**, *147*, 76–122.
26. Hellwege, E.M.; Volk, O.H.; Hartung, W. A Physiological Role of Abscisic Acid in the Liverwort *Riccia fluitans* L. *J. Plant. Physiol.* **1992**, *140*, 553–556. [CrossRef]
27. Kwon, W.; Min, J.; Xi, H.; Park, J. The complete chloroplast genome of *Riccia fluitans* L. (Ricciaceae, Marchantiophyta). *Mitochondrial DNA* **2019**, *4*, 1895–1896. [CrossRef]
28. Paton, J.A. *Riccia fluitans* L. with sporophytes. *J. Bryol.* **1973**, *7*, 253–259. [CrossRef]
29. Crum, H.A. *Structural Diversity of Bryophytes*; University of Michigan Herbarium: Ann Arbor, MI, USA, 2001.
30. Selkirk, P.M. Effect of Nutritional Conditions on Sexual Reproduction in Riccia. *Bryologist* **1979**, *82*, 37–46. [CrossRef]
31. Gutsche, N.; Holtmannspötter, M.; Maß, L.; O'Donoghue, M.; Busch, A.; Lauri, A.; Schubert, V.; Zachgo, S. Conserved redox-dependent DNA binding of ROXY glutaredoxins with TGA transcription factors. *Plant. Direct* **2017**, *1*, e00030. [CrossRef]
32. Althoff, F.; Kopischke, S.; Zobell, O.; Ide, K.; Ishizaki, K.; Kohchi, T.; Zachgo, S. Comparison of the MpEF1alpha and CaMV35 promoters for application in *Marchantia polymorpha* overexpression studies. *Transgenic Res.* **2014**, *23*, 235–244. [CrossRef]
33. Tsuboyama, S.; Nonaka, S.; Ezura, H.; Kodama, Y. Improved G-AgarTrap: A highly efficient transformation method for intact gemmalings of the liverwort *Marchantia polymorpha*. *Sci. Rep.* **2018**, *8*, 10800. [CrossRef] [PubMed]
34. Zhang, Y.; Li, J.; Gao, C. Generation of Stable Transgenic Rice (*Oryza sativa* L.) by Agrobacterium-Mediated Transformation. *Curr. Protoc. Plant. Biol.* **2016**, *1*, 235–246. [CrossRef] [PubMed]
35. Yadava, P.; Abhishek, A.; Singh, R.; Singh, I.; Kaul, T.; Pattanayak, A.; Agrawal, P.K. Advances in Maize Transformation Technologies and Development of Transgenic Maize. *Front. Plant. Sci.* **2016**, *7*, 1949. [CrossRef] [PubMed]
36. Xu, C.; Cao, H.; Zhang, Q.; Wang, H.; Xin, W.; Xu, E.; Zhang, S.; Yu, R.; Yu, D.; Hu, Y. Control of auxin-induced callus formation by bZIP59-LBD complex in Arabidopsis regeneration. *Nat. Plants* **2018**, *4*, 108–115. [CrossRef] [PubMed]
37. Eklund, D.M.; Ishizaki, K.; Flores-Sandoval, E.; Kikuchi, S.; Takebayashi, Y.; Tsukamoto, S.; Hirakawa, Y.; Nonomura, M.; Kato, H.; Kouno, M.; et al. Auxin Produced by the Indole-3-Pyruvic Acid Pathway Regulates Development and Gemmae Dormancy in the Liverwort *Marchantia polymorpha*. *Plant. Cell* **2015**, *27*, 1650–1669. [CrossRef] [PubMed]
38. Carter, A.M. *Riccia fluitans* L.-A Composite Species. *Bull. Torrey Bot. Club* **1935**, *62*, 33–42. [CrossRef]
39. Borovichev, E.A.; Bakalin, V.A. Survey of the Russian Far East Marchantiales IV: A revision of Ricciaceae (Hepaticae). *Bot. Pac.* **2016**, *5*, 3–29. [CrossRef]
40. Ishizaki, K.; Nishihama, R.; Yamato, K.T.; Kohchi, T. Molecular Genetic Tools and Techniques for *Marchantia polymorpha* Research. *Plant. Cell Physiol.* **2016**, *57*, 262–270. [CrossRef]

41. Bowman, J.L.; Kohchi, T.; Yamato, K.T.; Jenkins, J.; Shu, S.; Ishizaki, K.; Yamaoka, S.; Nishihama, R.; Nakamura, Y.; Berger, F.; et al. Insights into Land Plant Evolution Garnered from the *Marchantia polymorpha* Genome. *Cell* **2017**, *171*, 287–304. [CrossRef]
42. Du, D.; Jin, R.; Guo, J.; Zhang, F. Infection of Embryonic Callus with Agrobacterium Enables High-Speed Transformation of Maize. *Int. J. Mol. Sci.* **2019**, *20*, 279. [CrossRef]
43. Ikeuchi, M.; Sugimoto, K.; Iwase, A. Plant callus: Mechanisms of induction and repression. *Plant. Cell* **2013**, *25*, 3159–3173. [CrossRef] [PubMed]
44. Beck, C.F.; Acker, A. Gametic Differentiation of *Chlamydomonas reinhardtii*: Control by Nitrogen and Light. *Plant. Physiol.* **1992**, *98*, 822–826. [CrossRef] [PubMed]
45. Takeno, K. Stress-induced flowering: The third category of flowering response. *J. Exp. Bot.* **2016**, *67*, 4925–4934. [CrossRef] [PubMed]
46. Nishiyama, T.; Sakayama, H.; de Vries, J.; Buschmann, H.; Saint-Marcoux, D.; Ullrich, K.K.; Haas, F.B.; Vanderstraeten, L.; Becker, D.; Lang, D.; et al. The Chara Genome: Secondary Complexity and Implications for Plant Terrestrialization. *Cell* **2018**, *174*, 448–464. [CrossRef]
47. McGregor, R.L. Vegetative propagation of *Riccia rhenana*. *Bryologist* **1961**, *64*, 75–76.
48. Villarreal, A.J.; Crandall-Stotler, B.J.; Hart, M.L.; Long, D.G.; Forrest, L.L. Divergence times and the evolution of morphological complexity in an early land plant lineage (Marchantiopsida) with a slow molecular rate. *New Phytol.* **2016**, *209*, 1734–1746. [CrossRef]
49. Boisselier-Dubayle, M.C.; Lambourdiere, J.; Bischler, H. Molecular phylogenies support multiple morphological reductions in the liverwort subclass Marchantiidae (Bryophyta). *Mol. Phylogenet Evol.* **2002**, *24*, 66–77. [CrossRef]
50. Söderström, L.; Hagborg, A.; von Konrat, M.; Bartholomew-Began, S.; Bell, D.; Briscoe, L.; Brown, E.; Cargill, D.C.; Costa, D.P.; Crandall-Stotler, B.J.; et al. World checklist of hornworts and liverworts. *PhytoKeys* **2016**, *59*, 1–828. [CrossRef]
51. Andersen, T.; Pedersen, O. Interactions between light and CO2 enhance the growth of *Riccia fluitans*. *Hydrobiologia* **2002**, *477*, 163–170. [CrossRef]
52. Schindelin, J.; Arganda-Carreras, I.; Frise, E.; Kaynig, V.; Longair, M.; Pietzsch, T.; Preibisch, S.; Rueden, C.; Saalfeld, S.; Schmid, B.; et al. Preparation and Fluorescent Analysis of Plant Metaphase Chromosomes. *Methods Mol. Biol.* **2016**, *1370*, 87–103.
53. Klingmüller, W. Zur Systematik der Ricciaceen des fluitans-Formenkreises: *Riccia media* n. sp. *Flora Oder Allg. Bot. Ztg.* **1958**, *146*, 616–624. [CrossRef]

International Journal of
Molecular Sciences

MDPI

Article

Metabolomics Analysis Reveals Tissue-Specific Metabolite Compositions in Leaf Blade and Traps of Carnivorous *Nepenthes* Plants

Alberto Dávila-Lara [1,2,†], Carlos E. Rodríguez-López [3,†], Sarah E. O'Connor [3] and Axel Mithöfer [1,*]

1. Research Group Plant Defense Physiology, Max Planck Institute for Chemical Ecology, 07745 Jena, Germany; adavila-lara@ice.mpg.de
2. Departamento de Biología, Universidad Nacional Autónoma de Nicaragua-León (UNAN), 21000 León, Nicaragua
3. Department of Natural Product Biosynthesis, Max Planck Institute for Chemical Ecology, 07745 Jena, Germany; clopez@ice.mpg.de (C.E.R.-L.); oconnor@ice.mpg.de (S.E.O.)

* Correspondence: amithoefer@ice.mpg.de

† These authors contributed equally to this work.

Received: 18 May 2020; Accepted: 17 June 2020; Published: 19 June 2020

check for updates

Abstract: *Nepenthes* is a genus of carnivorous plants that evolved a pitfall trap, the pitcher, to catch and digest insect prey to obtain additional nutrients. Each pitcher is part of the whole leaf, together with a leaf blade. These two completely different parts of the same organ were studied separately in a non-targeted metabolomics approach in *Nepenthes x ventrata*, a robust natural hybrid. The first aim was the analysis and profiling of small (50–1000 *m*/*z*) polar and non-polar molecules to find a characteristic metabolite pattern for the particular tissues. Second, the impact of insect feeding on the metabolome of the pitcher and leaf blade was studied. Using UPLC-ESI-qTOF and cheminformatics, about 2000 features (MS/MS events) were detected in the two tissues. They showed a huge chemical diversity, harboring classes of chemical substances that significantly discriminate these tissues. Among the common constituents of *N. x ventrata* are phenolics, flavonoids and naphthoquinones, namely plumbagin, a characteristic compound for carnivorous Nepenthales, and many yet-unknown compounds. Upon insect feeding, only in pitchers in the polar compounds fraction, small but significant differences could be detected. By further integrating information with cheminformatics approaches, we provide and discuss evidence that the metabolite composition of the tissues can point to their function.

Keywords: *Nepenthes*; carnivorous plants; UPLC-qToF-MS; metabolomics; tissue specificity; cheminformatics

1. Introduction

Metamorphosis of plant organs is a common feature in higher plants and often an adaptation to the particular environment. Metamorphosis covers genetically fixed changes in both morphology and anatomy leading to new structural or functional modifications. In higher plants, leaves are mainly involved in photosynthesis and transpiration, but many leaf metamorphoses are also known for exhibiting new functions. Examples are spines as protection against herbivores (cacti), needles to reduce water loss (conifers), bulbs for storage of water and nutrients (onion), and tendrils for climbing (pea). Striking structures of leaf metamorphosis are found in many carnivorous plants that live on nutrient-poor soil and catch animal prey to get additional nutrients, such as nitrogen and phosphate [1,2]. Here, the leaves are employed in catching prey, mainly insects. For instance, in Venus flytrap (*Dionaea muscipula*), rapidly closing snap traps are found, in sundew (*Drosera*) species sticky flypaper traps, and in bladderwort (*Utricularia)* species sucking bladder traps [1,2]. Another type of trap is realized in so-called pitcher traps that can be found in the genus *Nepenthes* (Figure 1), occurring in Southeast Asia.

Figure 1. *Nepenthes x ventrata*. Natural hybrid of *N. ventricosa* and *N. alata*.

These passive traps attract prey to the pitcher opening, the peristome, which is extremely slippery for insects causing them to fall into the pitcher. The lower part of the pitcher is filled with a fluid where the prey drowns. Subsequently, plant-derived hydrolytic enzymes inside the fluid digest the prey and generate absorbable forms of nutrients, which are taken up and delivered further to the plant body through bi-functional glands [2,3]. In *Nepenthes* species, the whole leaf underwent an extensive metamorphosis: the typical leaf lamina (synonym: leaf blade) turned into a pitcher for catching prey, the petiole into a tendril to climb, and the leaf base into a basal leaf-derived leaf blade (from now on: leaf blade) substituting the lamina to ensure photosynthesis (Figure 2) [4,5].

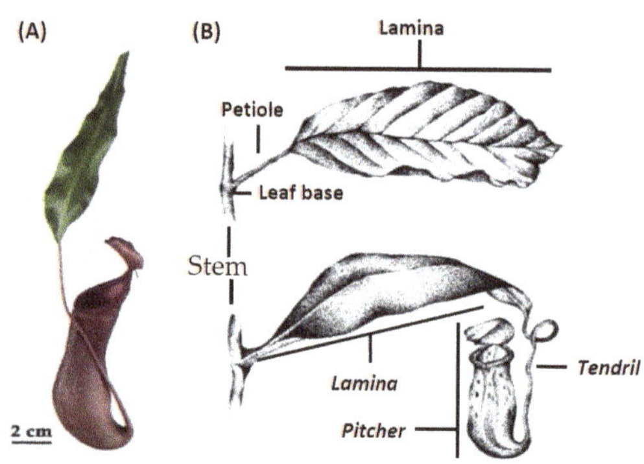

Figure 2. Comparison of leaf morphology. (**A**) *Nepenthes x ventrata* leaf. (**B**) Typical foliage leaves (upper), *Nepenthes* leaf (below). In italics, the leaf parts developed in *Nepenthes* as result of metamorphosis of the typical leaf parts. For further explanation, see the text. Copyright © of drawing (**B**) held by Sarah Zunk.

For many years, scientists studied the different trapping mechanisms in order to understand their function and biomechanics. However, changes and adaptations in leaf morphology and anatomy also come along with changes in the physiology, biochemistry, and molecular biology of carnivorous plants. Thus, in recent years, many studies in carnivorous plants focused more and more on molecular aspects and "omics" approaches, except metabolomics. Those studies have produced more and deeper insights in the molecular events accompanying the various steps necessary for successful prey hunting and digestion, suggesting, for example, that plant carnivory originates from defense mechanisms [6–12]; however, most studies are still related to the particular traps.

In *Nepenthes*, the pitcher fluid was investigated in detail, including its proteome [13–15] and the composition of organic and inorganic low-molecular-weight compounds [16]. Based on such studies, we learned that the pitcher fluids consist of enzymes necessary for digestion and also defensive proteins belonging to the group of pathogenesis-related proteins [17]. Moreover, the pitcher fluid is poor in inorganic nutrients and contains secondary metabolites with antimicrobial properties, i.e., naphthoquinones; droserone and 5-*O*-methyl droserone are described for *N. khasiana* [18] and plumbagin and 7-methyl-juglon for *N. ventricosa* [16]. These compounds are not widespread in plants but very often occur in carnivorous plants of the order Nepenthales [19], a *sensu stricto* sister group to Caryophyllales [5]. For *Nepenthes*, some of these naphthoquinones were described as inducible by chitin and prey [18,20], suggesting a functional role after prey catch. Naphthoquinones are highly bioactive compounds with defense-related properties [21]. Therefore, it has for a long time been suggested that these compounds are involved in protection against various microbes and pest attack and preserving prey during digestion [16–19]. Plumbagin and some other naphthoquinone derivatives have also been found in various tissues of *Nepenthes* species including the pitchers [16,20,22,23]. In addition, in the literature, the presence of carotenoids, flavonoids, sterols and triterpenes was mentioned for *Nepenthes* leaves [2,24,25].

As many carnivorous plants, including *Nepenthes*, harbor a huge chemical diversity, many secondary metabolites from carnivorous plants are currently isolated for pharmaceutical, biotechnological and pseudo-medical use [2,26,27]. This approach *per se* has led to pharmacologically valuable molecules, and, notably in times of an ongoing pandemic, its value is obvious. However, metabolomics studies to better understand the role of metabolites concerning their ecological function in a carnivorous plant are not available but nevertheless important. As suggested by Hatcher and colleagues [19], the metabolite diversity may represent a mechanism supporting the evolution of carnivory and the ability to cope with new and harsh environments. In addition, regarding the metabolome, carnivorous plants' responses to the assimilation of animal-derived nutrients remain largely unknown. Thus, the examinations of metabolite changes in pitcher and leaf blade tissues before and after prey digestion may also provide insight into dynamic processes in plant metabolism.

In order to address these questions, we used a non-targeted approach to analyze and compare, in *Nepenthes x ventrata*, the ionizable metabolites of specialized tissues; i.e., pitcher traps that are involved in prey catch and (basal) leaf blades involved in photosynthesis. In addition, we analyzed changes in the metabolite composition upon insect prey digestion. Besides these ecological aspects, the unique metamorphosis of a typical leaf organ into highly specialized tissues adds a fascinating developmental aspect.

2. Results

2.1. Metabolomics Reveals a Loss in Metabolite Load and Diversity in the Specialized Pitcher Organ

Drosophila melanogaster-fed and non-fed pitchers and related leaf blades of *N. x ventrata* were subject to independent polar and non-polar extractions. Extracts were analyzed by UPLC-ESI-qTOF in positive mode, with data-dependent fragmentation. Data was acquired in positive mode due to higher sensitivity and the higher quality of fingerprint predictions of SIRIUS+CIS-FingerID in positive as compared to negative mode. Since, in polar extractions, the chromatograms were dominated by a few peaks, to increase the coverage the samples were injected twice; as concentrated extracts and as ten-fold dilution. Using MetaboScape®, in the non-polar extraction 1396 peaks were detected and adducts grouped into 1226 features, 984 of which had at least one MS/MS event. In the polar extracts, 1398 and

560 peaks were detected, grouped in 1250 and 509 features, with 1012 and 383 fragmentation events in concentrated and diluted samples, respectively; both matrices of polar features were concatenated.

To gain an overview of the metabolomics changes, non-supervised analysis was performed separately on both polar and non-polar extracts. For both extractions, a Principal Component Analysis (PCA) showed that the main source of variation is the tissue, separated by the first component, explaining 35% and 25% of the variance in polar and non-polar metabolites, respectively (Figure 3). Interestingly, only the polar features of fed and non-fed pitchers were separated in the PCA (by the third component), explaining around 5% of the variance (Figure 3a). None of the other combinations of PCs, cumulatively explaining up to 95% of the variance, managed to separate samples by feeding status. Remarkably, a consistent trend can be seen in the score plots (Figure S1), where leaf-specific features have a higher *m/z* than pitcher-specific peaks in both polar and non-polar extracts.

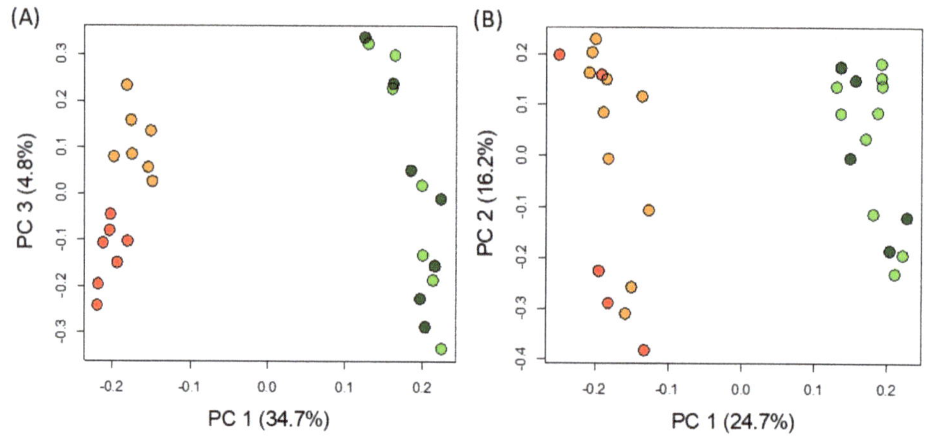

Figure 3. Unsupervised analysis of all detected features. PCA analysis of features detected in polar (**A**) and non-polar (**B**) extracts. Tissue and feeding status are indicated by the colors dark green and light green, showing fed and not-fed leaves, and red and orange, showing fed and not-fed pitchers, respectively.

To complement the non-supervised analysis and to estimate the effect of tissue type and feeding status, two-way ANOVA tests were run on the features. Ratifying the previous observation, only tissue had features that were significantly different (FDR < 0.01). After removing duplicated signals, in the polar fraction 797 differentially accumulated features (DAFs) were found, with the vast majority (634) being highly accumulated in leaf compared to pitcher (163 features; Figure S2). Correspondingly, the non-polar fraction had 449 DAFs that were more balanced, with 272 and 177 over-accumulated in leaf and pitcher, respectively (Figure S3). The DAFs are shown in the cloud plot of Figure 4, where the trend hinted at by the PCA score plots is confirmed: in both polar and non-polar extracts, features over-accumulated in leaf are of higher *m/z* than those over-accumulated in pitcher, with a difference of medians of 122 Th and 121 Th, respectively (Figure S4).

Moreover, besides the finding that leaves show more significantly accumulated features, the fold-change of those features is also remarkably higher (size of the circles in Figure 4) than the features over-accumulated in pitchers (Figure S5).

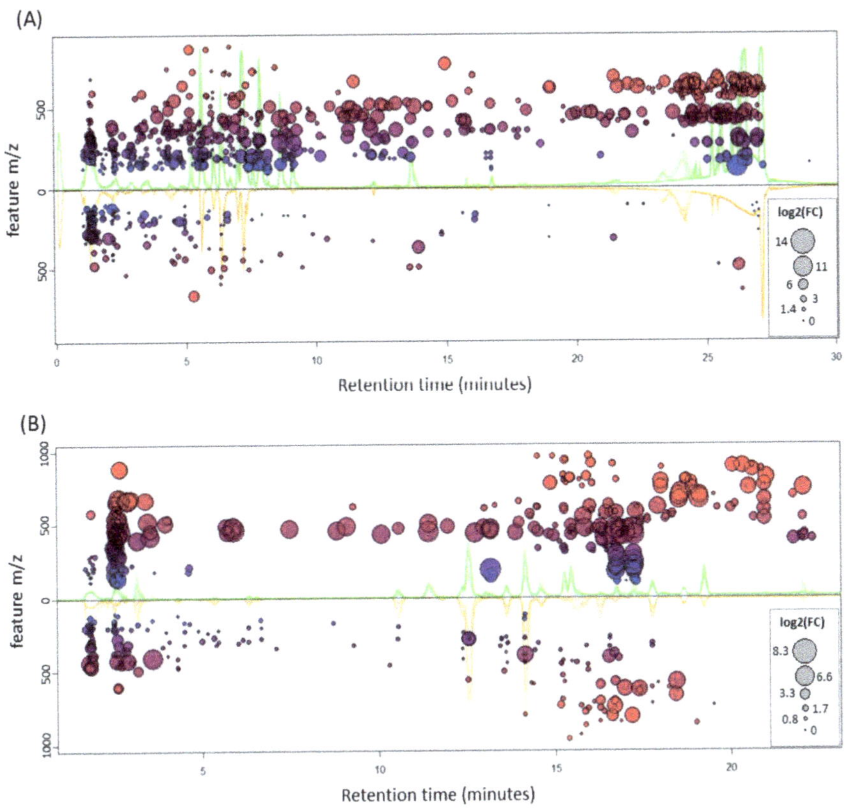

Figure 4. Mirror plots of differentially accumulated features (DAFs). DAFs (FDR < 0.01) in polar (**A**) and non-polar (**B**) extracts are shown for leaf (top) and pitcher (bottom). Circle size depicts the absolute value of the $\log_2$ of the average fold change, on the top if it is over-accumulated in leaf, and on the bottom otherwise. Color and y-axis value depict the *m*/*z* value of the feature, with blue being low- (100) and red high- (1000) *m*/*z* features; the further away from the origin, the higher the *m*/*z*, as indicated by the y-axis. The superimposed, raw base-peak chromatograms (BPC) of all runs are shown in the background, colored accordingly: green, all leaf BPCs; orange, all pitcher BPCs.

2.2. Database-Independent Spectral Analysis Identifies Key Substructures in DAF

Assignation of feature identity is a complicated endeavor, which in MS-based metabolomics relies heavily on compound databases. Unequivocal identification of a compound requires isolation and analysis by NMR, and putative identification by fragmentation patterns requires manual curation of candidate lists, generated by algorithms that automate comparisons to databases. Given that *Nepenthes* is an understudied genus, we expect few of the detected compounds to be present in chemical databases; however, some structural information can be directly extracted from the MS/MS spectra.

With that purpose, for every adduct of all DAFs, we collected fragmentation spectra and analyzed it using SIRIUS [28–30] and CSI-FingerID [31], from which the best-predicted fingerprint vectors for each DAF were selected for analysis. In total, 580 DAFs (72%) from the polar and 212 DAFs (47%) from the non-polar fractions were each assigned a vector of chemical fingerprints. For reference, only 11 DAFs (2%) of the non-polar fraction had a hit using the extended database LipidBlast [32]. CSI-FingerID vectors contain 2937 chemical fingerprints [31] to which we assigned one of three values (present, absent, and uncertain) based on their posterior probabilities. We then calculated enrichment probabilities of the presence and absence of each fingerprint in each tissue, separately for polar and non-polar; the significantly enriched ones (FDR < 0.05) are shown in Tables S1 and S2.

Strikingly, pitcher DAFs have an increased presence of phosphate groups (Figure 5). They also mostly lack tertiary and quaternary carbons and rings, which would point at acyl lipids and phospholipids as those lipids in pitchers that best differentiate them compared with leaves. Accordingly, leaf DAFs have a distinctive annulated ring structure, along with fingerprints of at least two six-carbon rings, ternary carbons and branched fatty acyl chains, all typical fingerprints of sterol lipids. Indeed, analyzing the heatmap of the selected vectors (Figure 5) it can be seen that the right-most clusters, with most of the leaf DAFs, show typical sterol fingerprints. In contrast, the left-most clusters, with the majority of the pitcher DAFs, have at most one ring. In addition, this cluster harbors the prominent PO_2-containing cluster, consisting almost entirely of pitcher DAFs.

Concerning the fingerprints of the polar extracts, there are many more DAFs in leaves than in pitchers. Because structural variability is strikingly higher in polar compounds, interpretation is less straightforward. However, pitcher DAFs are seemingly enriched in compounds with heteroatoms, such as nitrogen or phosphate, and pentose fingerprints. Some diimines are found naturally in purines and ureides—both soluble molecule families that have a high nitrogen load. Given that there are five times more DAFs with fingerprints in leaf than in pitcher, not many characteristic fingerprints can be robustly assigned to be leaf-specific. Nevertheless, one of the main DAFs found in leaf blades, which appears to be 32 times higher in leaf blades than in pitchers, has been tentatively identified as the naphthoquinone plumbagin. In sum, in the corresponding fingerprint heatmap (Figure 6) the enrichment is not as clear cut as in the lipids, given the low abundance of pitcher DAFs. However, it is still noticeable that the right-most cluster concentrates almost exclusively pitcher DAFs: of the 11 DAFs simultaneously having four of these five fingerprints, only one is from leaf. Only one of these compounds had a biologically relevant database hit, resembling a uridine bisphosphate. In addition, interestingly, only five out of the 16 DAFs with a pentose fingerprint are accumulated in leaf.

2.3. Differences in Pitcher Due to Feeding Status

As the PCA suggested that only the polar extract of pitchers had a difference depending on feeding status, and to avoid interference with external variance, a one-way ANOVA was performed specifically in the polar extract of fed and non-fed pitchers. Thus, we found 73 DAFs due to the feeding status, with 27 features accumulating in fed pitchers, and 46 accumulating in non-fed pitchers (Figure S6). Unlike the above-mentioned examples, fold changes appear to be balanced, although the features accumulating in fed pitchers appear to have a higher m/z than those in non-fed pitchers (Figure S7). Notably, almost all of the DAFs (69 out of 73) are present only in the concentrated extract, and even there with low intensity.

Since most compounds do not have fragmentation due to low intensity, the full pipeline of SIRIUS+CSI-FingerID was followed, and the candidate list was manually curated. The results are shown in Table S3, where it can be seen that only 11 DAFs had a fragmentation pattern that allowed structural interpretation. Although the largest DAF-containing group is the one of non-fed pitchers (46 DAFs), only four features have assignations. Interestingly, three are nitrogenated: a putative nitrogenated heptose ($C_7H_{15}NO_9$), an unidentified, densely nitrogenated compound ($C_{13}H_{17}N_9O_{12}$), and a third that appears to be a nucleotide phosphate with an either cyclic ($C_{10}H_{17}N_4O_7P$) or acyclic ($C_{10}H_{15}N_4O_6P$) attachment. As for the fed pitcher, seven DAFs were identified, four likely to be phenolic compounds and three nitrogenated compounds. The phenolics were likely three simple phenolics ($C_{10}H_{10}O_3$, $C_{17}H_{22}O_8$, and $C_{13}H_{14}O_{11}$, the latter two glycosylated) and a flavonoid ($C_{17}H_{14}O_7$). The nitrogenated compounds had no hits in biologically relevant databases, only in PubChem; of those, two were compounds with four nitrogen atoms ($C_{22}H_{24}N_4O_7$ and $C_{27}H_{18}N_4O_6$) with very similar fingerprints, with more than two aromatic rings and nitrogen atoms in heterocycles, and the remaining one ($C_{14}H_{16}NO_5$) had a single aromatic ring and a single nitrogen.

(A)

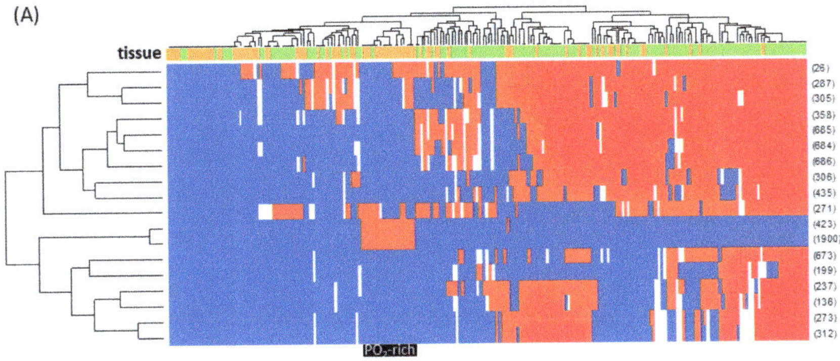

(B)

(26) – [R] (204 Ring)
(136) – (Annelated_rings)
(199) – [#6]~[#6](~[#6])(~[#6])~*
(237) – *@*(@*)@*
(271) – [CH3] >= 3
(273) – *1~*~*~*~*~*~1 >= 2
(287) – *1~*~*~*~*~*~1
(305) – >= 1 any ring size 6
(306) – >= 1 saturated or aromatic carbon-only ring size 6
(312) – >= 2 any ring size 6
(358) – [#6](~[#6])(~[#6])(~[#6])
(423) – [#15](~[#8])(~[#8])
(435) – [#6](-,:[#6])(-,:[#6])(=,:[#6])
(673) – [#6]-,:[#6]-,:[#6]-,:[#6]-,:[#6]-,:[#6](-,:[#6])-,:[#6]
(684) – [#6]-,:[#6](-,:[#6])-,:[#6]-,:[#6]
(685) – [#6]-,:[#6](-,:[#6])-,:[#6]-,:[#6]-,:[#6]
(686) – [#6]-,:[#6]-,:[#6](-,:[#6])-,:[#6]-,:[#6]
(1900) – ECFP6:-1610362824

(C)

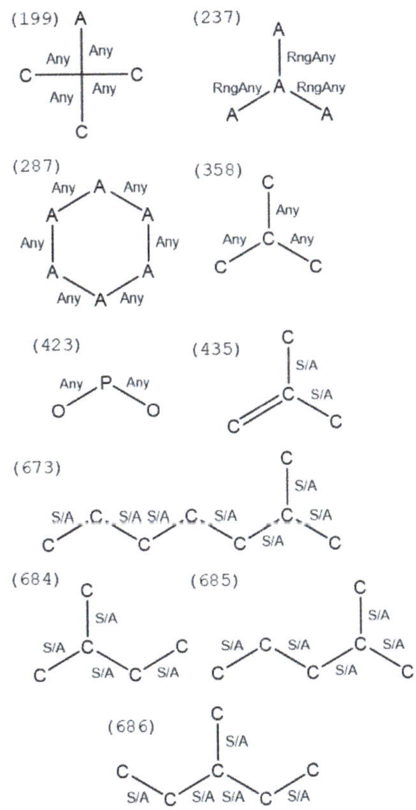

Figure 5. Fingerprint heatmaps of non-polar DAFs. A heatmap (**A**) is shown of the DAFs (columns) that had a fingerprint vector assigned, colored by tissue (green: leaf; orange: pitcher) on the top band. Only the enriched fingerprints (rows) are shown, named by CSI-FingerID relative index position (**A**). Based on posterior probabilities, the fingerprints were determined to be absent (blue), present (red), or uncertain (white). A cluster of DAFs almost exclusively accumulated in pitchers is highlighted in black, with the enriched fingerprints being described in (**B**) and, if graphical representation is possible, in (**C**). Any means it can be any kind of bond, RngAny means the bond is in a ring (of any kind), S/A means it is a single bond that can be anywhere (within a ring or not).

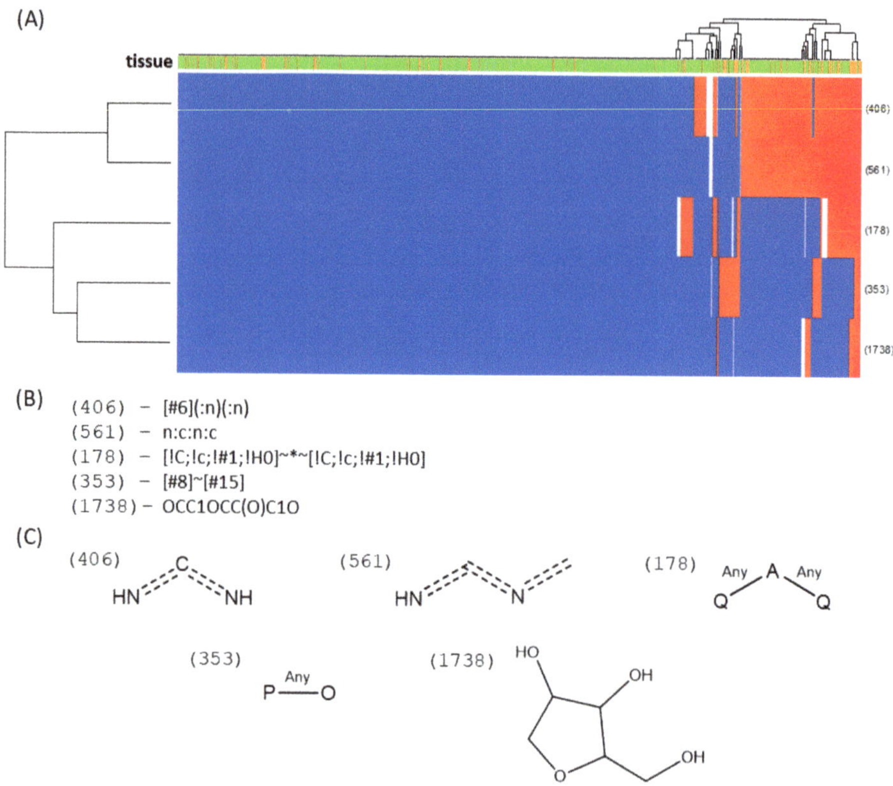

Figure 6. Fingerprint heatmaps of polar DAFs. A heatmap (**A**) is shown of the DAFs (columns) that had a fingerprint vector assigned, with blue cells being present, red being absent, and white being uncertain fingerprints. Given the nature of the sample, being mostly leaf DAFs, only the positive fingerprints enriched in pitcher and absent in leaf are shown. These fingerprints are described in (**B**) and the graphical approximation of their substructure in (**C**). It is important to note that the right-most cluster is unusually enriched in pitcher DAFs, with a high number of positive assignations of most of the selected fingerprints.

3. Discussion

Many low-molecular-weight compounds identified so far in carnivorous plants are volatile compounds suggested to be involved in prey attraction [19,33]. For instance, in *N. rafflesiana*, more than 50 volatiles have been found [34]. Less information is available for non-volatile compounds. Thus, we performed an untargeted metabolomics approach to determine which compounds might be related to carnivory in the metabolism of *Nepenthes x ventrata*, used here as a model plant. Two different questions have been addressed; first, we wanted to see whether or not the leaf blade and the pitcher contain different tissue- and function-specific metabolite patterns; second, we looked for differences in the tissues before and after insect feeding. This is the first study where a metabolomic profiling of the carnivory process in the genus *Nepenthes* is performed. Due to the technical design of this untargeted metabolomics work, the vast majority of primary metabolites fall inside the exclusion range for fragmentation (50–150 *m*/*z*); therefore, no meaningful assignation of identity or fingerprints could be performed on primary metabolites.

3.1. Metabolite Differences in Nepenthes Tissues: Leaf Blade vs. Pitcher

Overall, the number of features observed in leaves was much higher compared with pitcher tissue. In particular, there is a clear trend for the presence of polar compounds with $m/z > 300$ and of non-polar compounds with $m/z > 400$ in leaves. In addition, more over-accumulated features were found in leaves, with higher fold changes compared to pitcher. This means that both metabolite levels and diversity are lower in pitchers.

In the non-polar phase, the DAFs that best discriminate between pitcher and leaf are very likely acyl lipids and phospholipids, which are preferentially found in pitchers, and sterol derivatives, which are preferentially accumulated in leaves. The different membrane composition of these two tissues may be reflective of the differing functions. Sterols affect membrane fluidity and permeability, making the membranes more rigid, and are considered membrane reinforcers [35]. In addition, sterols are critical for the formation of lipid "rafts", which regulate biological processes such as signaling and transport across the membrane [36]. In *Nepenthes*, first, nutrient uptake from the pitcher fluid is performed by the bi-functional glands localized inside the pitcher. Besides carriers, clathrin-mediated endocytosis is involved in this process [37]. Specific for the vesicles of the clathrin-mediated pathway are phospholipids, favoring vesicle formation in contrast to sterols [38]. This might be another point that explains the different distribution of lipophilic metabolites in pitchers and leaves. In addition, a unique feature of *Nepenthes* pitchers is the waxy coating of the inner part of the pitcher, making it slippy for any prey trying to escape. This might also explain the difference in lipophilic metabolites in the pitcher compared with the leaf.

Interestingly, there is a family of polar compounds that simultaneously have a methylene-interrupted heteroatom, diimine-like structure (*~N=C=N~* and *~N=C=N=C-*), and phosphate and pentose fingerprints, and are exclusive to pitchers (10 out of the 11 DAFs with at least four of the five fingerprints). This finding was surprising as the carnivorous plants actually are limited in nitrogen and phosphate, and none of these DAFs are changing significantly due to feeding status. Nevertheless, since pitchers need to be ready for catching and digesting prey, they might be active in transport of both phosphate- and nitrogen-containing compounds. The presence of nucleotide phosphates supports the view at the pitcher as an active tissue ready to start de-novo synthesis of all necessary biosynthetic pathways. As long as no prey or not enough prey has been caught, even the pitcher must be seen as a sink tissue, and transport can occur in any direction. The putative nitrogen- and phosphate-containing glycosylated compounds are not present in biological databases and may hold valuable information on nitrogen and phosphate transport. The nature of these compounds, which might be mobile within the plant, is still an open question. Nitrogenous bases, like ureides, are well known to undergo long-distance transport in rhizobia–legume symbioses [39] as well as in non-nodulated plants [40]. Interestingly, the final enzymatic step to release ammonia from ureides is catalyzed by a urease. Its presence and activity were recently demonstrated for *Nepenthes* and other carnivorous plants [41]. Whether or not this scenario mirrors the nitrogen translocation and distribution that occurs in *Nepenthes* remains to be elucidated.

3.2. Insect Feeding Causes Changes in Polar Metabolite Pattern in Pitchers

In order to better understand the dynamics of the metabolic processes of carnivory in *Nepenthes* plants, immediately after opening, the pitchers were fed with fruit flies or not fed for 72 h. Results of the MS-based untargeted metabolomics analysis determined small but significant changes only in the pitcher tissue and, moreover, only in the fraction containing the polar metabolites. No significant changes in the leaf blade and no changes in the pitchers' non-polar metabolites were found as a result of feeding. Nevertheless, there was a trend showing that fed pitchers accumulated more compounds with higher molecular weight compared with non-fed pitchers, indicating a modulated, increased metabolic activity. Without knowing the exact structures of the compounds, the ecological relevance of changes in metabolite composition remains speculative. It might be due to higher physiological activities, in the sense that mobile compounds are built which can more easily be distributed within the plant or that the pitcher tissue needs to be more defended against detrimental organisms showing up together with caught prey. This would explain an increase in, for example, some phenolic compounds. For example, in our experiment, the fed pitchers were found having an around four times higher concentration of a flavonoid-related feature (c_331.0809-12.16; $C_{17}H_{14}O_7$; Table S3) compared with non-fed pitchers. It is also suggested that *Nepenthes* is a slowly digesting plant [42]. For example, prey-initiated induction of digestive enzymes such as the protease nepenthesin can take days [43]. Thus, it is conceivable that the selected 72 h of prey digestion were not sufficient to detect more induced metabolites, qualitatively or quantitatively. Following this thread, it may also explain why no effect of feeding was found in the leaf blades. Experiments with *N. hemsleyana*, a coprophagous *Nepenthes* species that does not catch prey any more but feeds on bat feces [44], showed that upon ^{15}N-enriched urea application into pitchers, after only four days, ^{15}N was significantly detectable in protein fractions of leaf blades [41].

These data suggest the lipid composition of pitcher appears to favor vesicle formation, while leaf blade lipids promote rafts and membrane rigidity; pitcher-specific DAFs contain nitrogen and phosphorus, with typical fingerprints of molecules known to undergo long-distance transport; and changes in leaf and pitcher features are weak due to feeding status. We may further speculate that prey-derived nutrients are taken up via vesicles in the pitcher, further degraded, fixed in organic N- and P-rich compounds, and eventually systemically distributed, thereby passing the proximal leaf blades. This is supported by research showing that developing leaves incorporated a higher level of ^{15}N, being preferentially supplied compared with a leaf that carries a fully developed pitcher [45]. Additional future experiments with different time points of harvesting may provide more insight into the dynamics of prey-induced changes in the *Nepenthes* metabolome in different tissues. However, as carnivorous plants mainly hunt for nitrogen and phosphate, it was not surprising to find prey-induced metabolite changes in the fraction containing polar, water-soluble compounds.

LC-MS-based metabolomics is a powerful tool for assessing chemical diversity in an un-biased manner, and is particularly useful for characterizing non-model plants, for which available data is scarce. However, the very nature of understudied plants complicates interpretation of the results, as most methods of putative identification rely heavily on databases, suffering greatly from popularity bias, and require manual curation, hindering analysis of systemic changes, such as those in pools of metabolites. Cheminformatics has long been used to extract information from large databases in an automated manner, but usually requires the existence of a chemical structure. We used a cheminformatics-aided metabolomics approach for characterizing the carnivorous plant *N. x ventrata*, using CSI-FingerID [31] fingerprint vectors directly, entirely bypassing structure assignation, the weakest link in the metabolomics pipeline. This minimizes false positives, and produces a robust, evidence-based approach for exploring systemic changes in metabolites.

In order to elucidate the real structures of the numerous compounds, further analyses are necessary, such as NMR. However, the compounds we found occur at low abundance, and this makes it extremely difficult to isolate enough material for analysis. However, the methods employed in the present study highlight general tissue-specific metabolites and their changes upon prey digestion.

Nevertheless, the fact that many features could not be identified in biologically relevant databases highlights the need to characterize non-model plant species to increase our knowledge of chemical diversity and find still-unknown compounds, which might be biologically or pharmaceutically relevant. In particular, *Nepenthes* species are well known in traditional medicine. Various reports are available describing curative effects of extracts from different *Nepenthes* species and tissues on diseases, for example, on cough, fever, hypertension, urinary system infections [46], malaria [47,48], asthma, pain [48]; *Staphylococcus* infection [49], celiac disease [50], and recently on different kinds of oral cancer cells [51]. Thus, further work on the isolation and structure elucidation of *Nepenthes* metabolites as well as the analysis of their putative pharmaceutical uses seems promising in order to find new structures and therapeutics.

In conclusion, the studied *Nepenthes x ventrata* plant contains a huge variety of different metabolites. We focused on MS-based and data mining approaches to visualize the metabolic differences between leaf and pitcher tissues, and between fed and un-fed plants. Leaf metamorphosis into pitchers and leaf blades generated new tissues that are different in function, which is also clearly represented in their respective DAFs. Surprisingly, insect prey feeding has a much smaller impact on the measured metabolites. Cheminformatics approaches suggest the presence of many structurally unknown compounds which might be of therapeutic interest, bearing in mind that *Nepenthes* species have been long used in traditional medicine. Further research should be carried out addressing the remaining questions of metabolite identification, biosynthetic pathways and the ecological relevance of *Nepenthes* metabolites.

4. Materials and Methods

4.1. Plant Material, Treatment, and Sampling

We used the natural hybrid *Nepenthes x ventrata* (*N. alata x N. ventricosa*) as a model organism. *N. x ventrata* plants were grown in the greenhouse of the MPI for Chemical Ecology at 21–23 °C, 50–60% relative humidity and a 16/8 h light/dark photoperiod. To avoid contamination, still-closed pitchers were covered with a mesh. Once the pitchers opened, they were left untreated for controls or prey degradation was induced by adding 30 wild-type *Drosophila melanogaster*, representing ca. 31 mg fresh weight. Individual pitchers represent independent biological replicates from different plants. After 72 h, pitchers were emptied, i.e., the digestive fluid with or without the remains of fruit flies was discarded, and subsequently rinsed 3 times with sterile distilled water. Next, both the tissue from the glandular zone (lower third part of the pitcher) and the related leaf blade were dissected and sampled in 50-mL Falcon tubes and immediately frozen in liquid nitrogen. The plant material was finely ground in liquid nitrogen using a mortar and pestle. Then, ground material was stored in screw-cap Eppendorf tubes and stored at −80 °C until further processing.

4.2. Metabolomic Extraction

Altogether, 28 individual samples were examined—7 *D. melanogaster*-treated and 7 untreated pitchers—and their corresponding leaf blades harvested after 72 h. Samples were extracted following a procedure derived from [52,53] with some modifications. In short, double extractions of 100 mg FW tissue powder were performed in 2-mL Eppendorf tubes at room temperature, using 500 µL MeOH:ammonium acetate buffer (pH 4.8). Therefore, after 5 min shaking, a 15 min sonication in water bath followed (3× for 5 min and 3 min resting in between). Extracts were centrifuged at 20,000× *g* for 10 min. Clear supernatants were combined and filtrated using a PTFE syringe filter (hydrophilic 0.22 µm pores, 13 mm diameter, Fisherbrand, Cat.# 15161499, Fisher Scientific, Schwerte, Germany). This extract was diluted 1:10 with 75% MeOH and further analyzed.

4.3. Lipidomics Extraction

Here, altogether 30 individual samples were examined: 5 non-treated control pitchers and leaf blades were taken directly after pitcher opening at 0 h; 5 *D. melanogaster*-treated and 5 untreated pitchers and their corresponding leaf blades taken after 72 h. Each sample represents an independent biological replicate. Extractions were done following a procedure derived from Matyash et al. (2008) [54] and Chen et al. (2013) [55] with some modifications. All steps were performed in glass test tubes and kept at room temperature. In short, an adjusted volume of methanol was added to 100 mg FW of tissue powder, based on a ratio of 150:1 *v/w* DW. Milli-Q water was added to a final ratio of 3:1 MeOH:H_2O, taking the water content (87%) of the tissues into consideration, which was determined before. Next, samples were vortexed followed by 5 min sonication in a water bath (5× for 1 min and 1 min resting in between). Thereafter, methyl-*tert*-butyl ether (MTBE) was added to achieve a ratio of 10:3:1 (MTBE:MeOH:H_2O). Samples were vortexed again, sonicated as described and shaken at 100 rpm for 1 h. Afterwards, milli-Q water was added to reach a total ratio of 20:6:7 (MTBE:MeOH:H_2O). Samples were vortexed, sonicated as previously described, and shaken for 10 min. To separate them into two phases, samples were centrifuged at 100× *g* for 20 min. The organic phase was recovered, while the aqueous phase was extracted again in 2 mL, keeping the ratio of MTBE:MeOH (20:6:7). Both organic phases were combined and evaporated under vacuum at 45 °C. The dry aqueous and organic samples were resuspended in acetonitrile:isopropanol (50:50) to a concentration equivalent to 1 g/L DW and filtrated using a PTFE syringe filter. This extract was diluted 1:10 with acetonitrile:isopropanol (50:50) and further analyzed.

4.4. Metabolic Profiling Using HPLC-qToF-MS

Samples were analyzed using an Elute LC system (Bruker Daltonik, Bremen, Germany) coupled via ESI to a Maxis II q-TOF (Bruker Daltonik, Bremen, Germany). Polar compounds were separated using a Kinetex® XB-C18 column (100 × 2.1 mm, 2.6 µm, 100 Å; Phenomenex, Aschaffenburg, Germany) at 40 °C with a gradient from water to acetonitrile, both modified with 0.1% formic acid, according to [52] with minor modifications. Namely, there was a flow of 0.2 mL/min, a linear gradient from 5% to 75% acetonitrile over 20 min, increased linearly to 95% acetonitrile over 5 min, followed by a 5-min equilibration at the initial conditions. Non-polar compounds were separated using a Luna® Omega PS C18 column (150 × 2.1 mm, 3 µm, 100 Å; Phenomenex, Aschaffenburg, Germany) at 50 °C. Mobile phase A was a mixture of water and acetonitrile (4:1 *v*/*v*) and mobile phase B was an isopropanol:acetonitrile mixture (9:1 *v*/*v*); both phases were modified to a final concentration of 10 mM ammonium acetate and 0.1% formic acid. The gradient was as previously published [56] with minor modifications: at a flow of 0.2 mL/min, a linear increase from 40% B to 45% B in 2 min, then to 55% B in 8 min, followed by an immediate step increase to 70% B, then a linear increase to 99% B in 10 min, holding at 99% B for 5 min, and finally returning to the initial conditions for 5 min. For analysis of the extracts, 5 µL of a 10-fold dilution was injected, and, for the polar extracts, a second batch of 5 µL of concentrated extract was injected. Injections in each of these three batches were randomized, with 5 evenly interleaved quality control injections of pooled samples, preceded by 4 "dummy" injections of pooled quality control samples to passivate the column, which was extensively washed after each batch. Analyses of the quality control samples are shown in Figures S8–S10.

Acquisition of MS data was done using the same conditions for both polar and non-polar compounds. Ionization was performed via pneumatic-assisted electrospray ionization in positive mode (ESI+) with a capillary voltage of 4.5 kV and an end plate offset of 500 V; a nebulizer pressure of 3 bar was used, with nitrogen at 350 °C and a flow of 12 L/min as the drying gas. Acquisition was done at 12 Hz following a mass range from 50 to 1000 *m*/*z*, with data-dependent MS/MS and an active exclusion window of 0.2 min, a reconsideration threshold of 1.8-fold change, and an exclusion range of 50–150 *m*/*z*. Fragmentation was triggered on an absolute threshold of 400 and acquired on the most intense peaks using a target intensity of 20,000 counts, with MS/MS spectra acquisition between 12 and 20 Hz, and limited to a total cycle time range of 0.5 s. Collision energy was determined automatically by the software depending on *m*/*z* value. At the beginning of each run, an injection of 20 µL of a sodium formate–isopropanol solution was performed in the dead volume of the injection, and the *m*/*z* values were re-calibrated using the expected cluster ion *m*/*z* values.

4.5. Feature Detection

Peak detection was done using Metaboscape software (Bruker Daltonik, Bremen, Germany) with the T-Rex 3D algorithm for qTOF data. For the non-polar runs, parameters for detection were an intensity threshold of 500 with a minimum of 7 spectra, and features were kept if they were detected in at least 3 replicates of the same treatment, tissue and time (60% of n). Adducts of $[M+H]^+$, $[M+Na]^+$, $[M+K]^+$, and $[M+NH_4]^+$ were grouped as a single feature if they had an EIC correlation of 0.8. For the polar runs, the intensity threshold was set to 1000, the features were kept if detected in at least 5 replicates of the same treatment and tissue (70% of n), and adducts were grouped in the same manner, only excluding the ammonium adduct, which was not expected in the polar runs.

4.6. Spectral Analysis

Proprietary MS Bruker files were re-calibrated with cluster ions of sodium formate in the dead-volume injection time and converted to mzXML [57–59] using Bruker DataAnalyst software (Bruker Daltronik, Bremen, Germany). Access to the raw data in mzXML files was done in R with the aid of the *mzR* library [60]. MS/MS data was extracted for selected features using an in-house built code that searched in all samples for fragmentation events triggered in a window of 0.5 min within the feature retention time (RT). To avoid misassignation of closely eluting isobaric compounds within the RT window, the maximum of intensity in the MS1 extracted ion chromatogram (XIC) of the feature *m*/*z* (with 5 ppm error) that was closest to the feature RT was searched. Only contiguous peaks decreasing in intensity from the previous point in the MS1 XIC and with intensity higher than 10% of the maximum were kept. The new RT window was determined by the time in the first and last events. Within this new RT window, all fragmentation events whose parent ions matched the feature *m*/*z* within a 5 ppm error were stored. The fragmentation events of the most abundant 5 (non-polar) and 7 (polar) peaks for each feature adduct were merged using previously published in-house binning algorithm [61], and saved as MASCOT generic format (MGF) files.

Candidate structures and database-independent fingerprint vectors were obtained by loading the above-mentioned MGF files into the SIRIUS [28–30] and CSI-FingerID [31] pipeline. Candidate structures for the DAFs of fed and non-fed pitchers were obtained by searching the top hit of CSI-FingerID in all databases and manually curating the results; for all the other analyses, fingerprint vectors of the top 10 candidates of all predicted formulas were exported and loaded in R. When more than one adduct was present in a feature, only the formulas that matched the formulas of the adducts were kept. Then, only fingerprints that explained more than 3 peaks and more than one third of the intensity were kept. The final selection of the fingerprint vectors was made by collapsing all the adducts per feature, only keeping the fingerprint vectors corresponding to the top-scoring candidate and those that were less than 30% different. Fingerprints were assigned as present if the highest posterior probability of fingerprint vectors and adducts was greater than 0.75, as absent if the lowest posterior probability was less than 0.25, and as uncertain otherwise. Enrichment for presence and absence were calculated via a hypergeometric test, with uncertain assignations not being considered in the probability calculations as either hits or fails. The p-values of the hypergeometric tests were corrected for multiple testing.

4.7. Statistical Analysis

All statistical analyses were performed using the R 3.6.1 *base* package [62] and graphics using a combination of the *ggplot2* [63] and *gplots* [64] libraries, unless otherwise specified. Analysis of polar and non-polar fractions was done separately, given the nature of the experiments. Since the maximum signal-to-noise ratio was assumed to be 1/3, the zeroes in the matrices were replaced by their respective minimum measured area, divided by three, and then $\log_{10}$-transformed. The resulting matrices, estimated as Normal by Q-Q plots, were used for ANOVAs. For principal component analysis, these $\log_{10}$-transformed matrices were z-scaled by subtracting the mean and dividing by the standard deviation in a feature-wise manner. For the non-polar analysis, a two-way ANOVA was done on samples after 72 h, taking tissue and treatment as factors, and blocking by extraction batch. Since no difference was found by treatment, the 0 h control was added to analysis discriminating tissue, blocking by all other variables. For the polar analysis, a two-way ANOVA was done on the concatenated matrix of concentrated and diluted injections, taking tissue and treatment as factors. The features were de-duplicated only after statistical testing and false discovery rate correction, and this deduplication was only performed on significantly different peaks. Features were considered duplicated if they shared the same *m*/*z* (within 10 ppm or an absolute 0.0025 difference) and retention time (within 0.15 min) and were not detected as different features in either the concentrated or diluted injections. That is, if 3 (significantly different) features were detected in the concentrated batch within that window (10 ppm, 0.15 min), and 2 (significantly different) features were detected in the diluted sample, the deduplication would keep all 3 (significantly different) features in the concentrated sample because, even when they share *m/z* and RT, they were detected as different features by MetaboScape. This is a conservative approach for calculating both FDR and fold change. All statistical testing was controlled for multiple testing by Benjamini and Hochberg's (1995) [65] false discovery rate correction.

Supplementary Materials: The following are available online at http://www.mdpi.com/1422-0067/21/12/4376/s1, Figure S1: PCA Scores vs. *m/z*; Figure S2: Heatmap of polar DAFs; Figure S3: Heatmap of non-polar DAFs; Figure S4: The *m/z* plots; Figure S5: Fold-change density plot; Figure S6: Heatmap of polar DAFs in pitchers due to feeding status; Figure S7: Feeding fold-change density plots; Figure S8: Quality control injections in the lipidomics experiment; Figure S9: Quality control injections in the polar experiment, injecting the raw extracts; Figure S10: Quality control injections in the polar experiment, injecting the ten-fold diluted extracts; Table S1: Polar fingerprints; Table S2: Non-polar fingerprints; Table S3: Features. Raw data was deposited in Metabolights Study MTBLS1783, as well as in the EDMOND database (DOI: https://dx.doi.org/10.17617/3.42).

Author Contributions: A.D.-L., C.E.R.-L. and A.M. conceived the study and experiments. A.D.-L. and C.E.R.-L. performed the experiments and analyzed data. A.D.-L., C.E.R.-L., S.E.O. and A.M. discussed the data and wrote the manuscript. All authors read and agreed to the present version of the manuscript.

Funding: This work was supported by a PhD fellowship from the DAAD (German Academic Exchange Service) to A.D.-L.

Acknowledgments: We thank Birgit Arnold and the whole greenhouse team of the MPI for cultivating the *Nepenthes* plants.

Conflicts of Interest: The authors declare no conflict of interest.

References

1. Darwin, C. *Insectivorous Plants*; John Murray Press: London, UK, 1875.
2. Juniper, B.E.; Robins, R.J.; Joel, D.M. The Carnivorous Plants. *Plant Sci.* **1989**, *63*, 116–117.
3. Ellison, A.M.; Adamec, L. *Carnivorous Plants: Physiology, Ecology, and Evolution*; Oxford University Press: New York, NY, USA, 2018.
4. Owen, T.P.; Lennon, K.A. Structure and development of the pitchers from the carnivorous plant *Nepenthes alata* (Nepenthaceae). *Am. J. Bot.* **1999**, *86*, 1382–1390. [CrossRef]
5. Fleischmann, A.; Schlauer, J.; Smith, S.A.; Givnish, T.J. *Evolution of Carnivory in Angiosperms*; Oxford University Press (OUP): New York, NY, USA, 2018; pp. 22–42.
6. Ibarra-Laclette, E.; Albert, V.A.; Pérez-Torres, C.-A.; Zamudio-Hernández, F.; Ortega-Estrada, M.D.J.; Herrera-Estrella, A.; Herrera-Estrella, L.R. Transcriptomics and molecular evolutionary rate analysis of the bladderwort (*Utricularia*), a carnivorous plant with a minimal genome. *BMC Plant Biol.* **2011**, *11*, 101. [CrossRef] [PubMed]
7. Ibarra-Laclette, E.; Lyons, E.; Hernández-Guzmán, G.; Pérez-Torres, C.A.; Carretero-Paulet, L.; Chang, T.-H.; Lan, T.; Welch, A.J.; Juárez, M.J.A.; Simpson, J.; et al. Architecture and evolution of a minute plant genome. *Nature* **2013**, *498*, 94–98. [CrossRef] [PubMed]
8. Leushkin, E.V.; Sutormin, R.A.; Nabieva, E.; Penin, A.A.; Kondrashov, A.S.; Logacheva, M. The miniature genome of a carnivorous plant *Genlisea aurea* contains a low number of genes and short non-coding sequences. *BMC Genom.* **2013**, *14*, 476. [CrossRef]
9. Schulze, W.X.; Sanggaard, K.W.; Kreuzer, I.; Knudsen, A.D.; Bemm, F.; Thøgersen, I.B.; Bräutigam, A.; Thomsen, L.R.; Schliesky, S.; Dyrlund, T.F.; et al. The Protein Composition of the Digestive Fluid from the Venus Flytrap Sheds Light on Prey Digestion Mechanisms. *Mol. Cell. Proteom.* **2012**, *11*, 1306–1319. [CrossRef] [PubMed]
10. Bemm, F.; Becker, D.; Larisch, C.; Kreuzer, I.; Escalante-Perez, M.; Schulze, W.X.; Ankenbrand, M.J.; Van De Weyer, A.-L.; Krol, E.; Al-Rasheid, K.A.; et al. Venus flytrap carnivorous lifestyle builds on herbivore defense strategies. *Genome Res.* **2016**, *26*, 812–825. [CrossRef]
11. Böhm, J.; Scherzer, S.; Krol, E.; Kreuzer, I.; Von Meyer, K.; Lorey, C.; Mueller, T.D.; Shabala, L.; Monte, I.; Solano, R.; et al. The Venus Flytrap *Dionaea muscipula* Counts Prey-Induced Action Potentials to Induce Sodium Uptake. *Curr. Biol.* **2016**, *26*, 286–295. [CrossRef]
12. Fukushima, K.; Fang, X.; Alvarez-Ponce, D.; Cai, H.; Carretero-Paulet, L.; Chen, C.; Chang, T.-H.; Farr, K.M.; Fujita, T.; Hiwatashi, Y.; et al. Genome of the pitcher plant *Cephalotus* reveals genetic changes associated with carnivory. *Nat. Ecol. Evol.* **2017**, *1*, 59. [CrossRef]

13. Hatano, N.; Hamada, T. Proteome Analysis of Pitcher Fluid of the Carnivorous Plant *Nepenthes alata*. *J. Proteome Res.* **2008**, *7*, 809–816. [CrossRef]
14. Hatano, N.; Hamada, T. Proteomic analysis of secreted protein induced by a component of prey in pitcher fluid of the carnivorous plant *Nepenthes alata*. *J. Proteom.* **2012**, *75*, 4844–4852. [CrossRef] [PubMed]
15. Rottloff, S.; Miguel, S.; Biteau, F.; Nisse, E.; Hammann, P.; Kuhn, L.; Chicher, J.; Bazile, V.; Gaume, L.; Mignard, B.; et al. Proteome analysis of digestive fluids in *Nepenthes* pitchers. *Ann. Bot.* **2016**, *117*, 479–495. [CrossRef] [PubMed]
16. Buch, F.; Rott, M.; Rottloff, S.; Paetz, C.; Hilke, I.; Raessler, M.; Mithöfer, A. Secreted pitfall-trap fluid of carnivorous *Nepenthes* plants is unsuitable for microbial growth. *Ann. Bot.* **2012**, *111*, 375–383. [CrossRef] [PubMed]
17. Mithöfer, A. Carnivorous pitcher plants: Insights in an old topic. *Phytochemistry* **2011**, *72*, 1678–1682. [CrossRef]
18. Eilenberg, H.; Pnini-Cohen, S.; Rahamim, Y.; Sionov, E.; Segal, E.; Carmeli, S.; Zilberstein, A. Induced production of antifungal naphthoquinones in the pitchers of the carnivorous plant *Nepenthes khasiana*. *J. Exp. Bot.* **2009**, *61*, 911–922. [CrossRef]
19. Hatcher, C.R.; Ryves, D.B.; Millett, J. The function of secondary metabolites in plant carnivory. *Ann. Bot.* **2019**, *125*, 399–411. [CrossRef]
20. Raj, G.; Kurup, R.; Hussain, A.A.; Baby, S. Distribution of naphthoquinones, plumbagin, droserone, and 5-O-methyl droserone in chitin-induced and uninduced *Nepenthes khasiana*: Molecular events in prey capture. *J. Exp. Bot.* **2011**, *62*, 5429–5436. [CrossRef]
21. Devi, S.P.; Kumaria, S.; Rao, S.R.; Tandon, P. Carnivorous Plants as a Source of Potent Bioactive Compound: Naphthoquinones. *Trop. Plant Boil.* **2016**, *9*, 267–279. [CrossRef]
22. Rischer, H.; Hamm, A.; Bringmann, G. *Nepenthes insignis* uses a C2-portion of the carbon skeleton of l-alanine acquired via its carnivorous organs, to build up the allelochemical plumbagin. *Phytochemistry* **2002**, *59*, 603–609. [CrossRef]
23. Schlauer, J.; Nerz, J.; Rischer, H. Carnivorous plant chemistry. *Acta Bot. Gallica* **2005**, *152*, 187–195. [CrossRef]
24. Aung, H.H.; Chia, L.S.; Goh, N.K.; Chia, T.F.; Ahmed, A.A.; Pare, P.W.; Mabry, T.J. Phenolic constituents from the leaves of the carnivorous plant *Nepenthes gracilis*. *Fitoterapia* **2002**, *73*, 445–447. [CrossRef]
25. Wan, A.; Aexel, R.; Ramsey, R.; Nicholas, H. Sterols and triterpenes of the pitcher plant. *Phytochemistry* **1972**, *11*, 456–461. [CrossRef]
26. Miguel, S.; Hehn, A.; Bourgaud, F. *Nepenthes*: State of the art of an inspiring plant for biotechnologists. *J. Biotechnol.* **2018**, *265*, 109–115. [CrossRef]
27. Legendre, L.; Darnowski, D.W. *Biotechnology with Carnivorous Plants*; Oxford University Press (OUP): New York, NY, USA, 2018; pp. 270–282.
28. Böcker, S.; Letzel, M.C.; Lipták, Z.; Pervukhin, A. SIRIUS: Decomposing isotope patterns for metabolite identification. *Bioinformatics* **2008**, *25*, 218–224. [CrossRef] [PubMed]
29. Böcker, S.; Dührkop, K. Fragmentation trees reloaded. *J. Cheminform.* **2016**, *8*, 5. [CrossRef] [PubMed]
30. Dührkop, K.; Fleischauer, M.; Ludwig, M.; Aksenov, A.A.; Melnik, A.V.; Meusel, M.; Dorrestein, P.C.; Rousu, J.; Böcker, S. Sirius 4: Turning tandem mass spectra into metabolite structure information. *Nat. Methods* **2019**, *16*, 299–302. [CrossRef]
31. Dührkop, K.; Shen, H.; Meusel, M.; Rousu, J.; Böcker, S. Searching molecular structure databases with tandem mass spectra using CSI:FingerID. *Proc. Natl. Acad. Sci. USA* **2015**, *112*, 12580–12585. [CrossRef]
32. Kind, T.; Liu, K.-H.; Lee, Y.; DeFelice, B.; Meissen, J.K.; Fiehn, O. Lipid Blast in silico tandem mass spectrometry database for lipid identification. *Nat. Methods* **2013**, *10*, 755–758. [CrossRef]
33. Jürgens, A.; El-Sayed, A.M.; Suckling, D.M. Do carnivorous plants use volatiles for attracting prey insects? *Funct. Ecol.* **2009**, *23*, 875–887. [CrossRef]
34. Di Giusto, B.; Bessière, J.; Guéroult, M.; Lim, L.B.L.; Marshall, D.; Hossaert-Mckey, M.; Gaume, L. Flower-scent mimicry masks a deadly trap in the carnivorous plant *Nepenthes rafflesiana*. *J. Ecol.* **2010**, *98*, 845–856. [CrossRef]
35. Dufourc, E.J. Sterols and membrane dynamics. *J. Chem. Boil.* **2008**, *1*, 63–77. [CrossRef] [PubMed]
36. Mongrand, S.; Stanislas, T.; Bayer, E.M.F.; Lherminier, J.; Simon-Plas, F. Membrane rafts in plant cells. *Trends Plant Sci.* **2010**, *15*, 656–663. [CrossRef] [PubMed]
37. Adlassnig, W.; Bauer, S.; Koshkin, E.; Lendl, T.; Lichtscheidl, I.K.; Koller-Peroutka, M. Endocytotic uptake of nutrients in carnivorous plants. *Plant J.* **2012**, *71*, 303–313. [CrossRef] [PubMed]

38. Fan, L.; Li, R.; Pan, J.; Ding, Z.; Lin, J. Endocytosis and its regulation in plants. *Trends Plant Sci.* **2015**, *20*, 388–397. [CrossRef]
39. Todd, C.D.; Tipton, P.A.; Blevins, D.G.; Piedras, P.; Pineda, M.; Polacco, J.C. Update on ureide degradation in legumes. *J. Exp. Bot.* **2005**, *57*, 5–12. [CrossRef]
40. Thu, S.W.; Lu, M.-Z.; Carter, A.M.; Collier, R.; Gandin, A.; Sitton, C.C.; Tegeder, M. Role of ureides in source-to-sink transport of photoassimilates in non-fixing soybean. *J. Exp. Bot.* **2020**. [CrossRef]
41. Yilamujiang, A.; Zhu, A.; Ligabue-Braun, R.; Bartram, S.; Witte, C.-P.; Hedrich, R.; Hasabe, M.; Schöner, C.R.; Schöner, M.G.; Kerth, G.; et al. Coprophagous features in carnivorous *Nepenthes* plants: A task for ureases. *Sci. Rep.* **2017**, *7*, 11647. [CrossRef]
42. Yilamujiang, A.; Reichelt, M.; Mithöfer, A. Slow food: Insect prey and chitin induce phytohormone accumulation and gene expression in carnivorous *Nepenthes* plants. *Ann. Bot.* **2016**, *118*, 369–375. [CrossRef]
43. Buch, F.; Kaman, W.E.; Bikker, F.; Yilamujiang, A.; Mithöfer, A. Nepenthesin Protease Activity Indicates Digestive Fluid Dynamics in Carnivorous *Nepenthes* Plants. *PLoS ONE* **2015**, *10*, e0118853. [CrossRef]
44. Grafe, U.; Schöner, C.R.; Kerth, G.; Junaidi, A.; Schöner, M.G. A novel resource–service mutualism between bats and pitcher plants. *Biol. Lett.* **2011**, *7*, 436–439. [CrossRef]
45. Schulze, W.; Schulze, E.D.; Pate, J.S.; Gillison, A.N. The nitrogen supply from soils and insects during growth of the pitcher plants *Nepenthes mirabilis*, *Cephalotus follicularis* and *Darlingtonia californica*. *Oecologia* **1997**, *112*, 464–471. [CrossRef] [PubMed]
46. Chi, V.V. *Dictionary of Vietnamese Medicinal Plants*; Publishing House Medicine: Hanoi, Vietnam, 2012; Volume 2.
47. Likhitwitayawuid, K.; Kaewamatawong, R.; Ruangrungsi, N.; Krungkrai, J. Antimalarial Naphthoquinones from *Nepenthes thorelii*. *Planta Med.* **1998**, *64*, 237–241. [CrossRef] [PubMed]
48. D'Amoto, P. *The Savage Garden*; Ten Speed Press: Berkeley, CA, USA, 1998.
49. Wiard, C.; Morgana, S.; Khalifah, S.; Mahan, M.; Ismael, S.; Buckle, M.; Narayana, A.K.; Sulaiman, M. Antimicrobial screening of plants used for traditional medicine in the state of Perak, Peninsula Malaysia. *Fitoterapia* **2004**, *75*, 68–73.
50. Rey, M.; Yang, M.; Lee, L.; Zhang, Y.; Sheff, J.G.; Sensen, C.W.; Mrazek, H.; Halada, P.; Man, P.; McCarville, J.; et al. Addressing proteolytic efficiency in enzymatic degradation therapy for celiac disease. *Sci. Rep.* **2016**, *6*, 30980. [CrossRef]
51. Tang, J.-Y.; Peng, S.-Y.; Cheng, Y.-B.; Wang, C.-L.; Farooqi, A.A.; Yu, T.-J.; Hou, M.-F.; Wang, S.-C.; Yem, C.-H.; Chan, L.-P.; et al. Ethyl acetat extract of *Nepenthes adrianii x clipeata* induces antiproliferation, apoptosis, and DNA damage against oral cancer cells through oxidative stress. *Environ. Toxicol.* **2019**, *34*, 891–901. [CrossRef]
52. De Vos, R.; Schipper, B.; Hall, R.D. *Plant Metabolomics: Methods and Protocols*; Springer Science and Business Media LLC: New York, NY, USA, 2012; pp. 111–128.
53. Calf, O.W.; Huber, H.; Peters, J.L.; Weinhold, A.; Van Dam, N.M. Glycoalkaloid composition explains variation in slug resistance in *Solanum dulcamara*. *Oecologia* **2018**, *187*, 495–506. [CrossRef]
54. Matyash, V.; Liebisch, G.; Kurzchalia, T.V.; Shevchenko, A.; Schwudke, D. Lipid extraction by methyl-*tert*-butyl ether for high-throughput lipidomics. *J. Lipid Res.* **2008**, *49*, 1137–1146. [CrossRef]
55. Chen, S.; Hoene, M.; Li, J.; Li, Y.; Zhao, X.; Häring, H.-U.; Schleicher, E.D.; Weigert, C.; Xua, G.; Lehmann, R. Simultaneous extraction of metabolome and lipidome with methyl tert-butyl ether from a single small tissue sample for ultra-high performance liquid chromatography/mass spectrometry. *J. Chromatogr. A* **2013**, *1298*, 9–16. [CrossRef]
56. Rodríguez-López, C.; Hernández-Brenes, C.; Treviño, V.; De La Garza, R.I.D. Avocado fruit maturation and ripening: Dynamics of aliphatic acetogenins and lipidomic profiles from mesocarp, idioblasts and seed. *BMC Plant Boil.* **2017**, *17*, 159. [CrossRef]
57. Martens, L.; Chambers, M.; Sturm, M.; Kessner, D.; Levander, F.; Shofstahl, J.; Tang, W.H.; Römpp, A.; Neumann, S.; Pizarro, A.; et al. mzML—a community standard for mass spectrometry data. *Mol. Cell. Proteom.* **2010**, *10*, 1–7. [CrossRef]
58. Keller, A.; Eng, J.; Zhang, N.; Li, X.; Aebersold, R. A uniform proteomics MS/MS analysis platform utilizing open XML file formats. *Mol. Syst. Biol.* **2005**, *1*. [CrossRef] [PubMed]
59. Pedrioli, P.G.A.; Eng, J.K.; Hubley, R.; Vogelzang, M.; Deutsch, E.W.; Raught, B.; Pratt, B.; Nilsson, E.; Angeletti, R.H.; Apweiler, R.; et al. A common open representation of mass spectrometry data and its application to proteomics research. *Nat. Biotechnol.* **2004**, *22*, 1459–1466. [CrossRef] [PubMed]

60. Chambers, M.C.; MacLean, B.; Burke, R.; Amodei, D.; Ruderman, D.L.; Neumann, S.; Gatto, L.; Fischer, B.; Pratt, B.; Egertson, J.; et al. A cross-platform toolkit for mass spectrometry and proteomics. *Nat. Biotechnol.* **2012**, *30*, 918–920. [CrossRef] [PubMed]
61. Yamamoto, K.; Takahashi, K.; Caputi, L.; Mizuno, H.; Rodriguez-Lopez, C.E.; Iwasaki, T.; Ishizaki, K.; Fukaki, H.; Ohnishi, M.; Yamazaki, M.; et al. The complexity of intercellular localisation of alkaloids revealed by single-cell metabolomics. *New Phytol.* **2019**, *224*, 848–859. [CrossRef]
62. Anonymous. The R Project for Statistical Computing. Available online: http://www.r-project.org/ (accessed on 13 February 2012).
63. Wickham, H. *ggplot2: Elegant Graphics for Data Analysis*; Springer: New York, NY, USA, 2016.
64. Warnes, G.R.; Bolker, B.; Bonebakker, L.; Gentleman, R.; Huber, W.; Liaw, A.; Lumley, T.; Maechler, M.; Magnusson, A.; Moeller, S.; et al. gplots: Various R Programming Tools for Plotting Data. *R Package Version*. 2015. Available online: https://cran.r-project.org/package=gplots/index.html (accessed on 14 May 2019).
65. Benjamini, Y.; Hochberg, Y. Controlling the False Discovery Rate: A Practical and Powerful Approach to Multiple Testing. *J. R. Stat. Soc. Ser. B Met.* **1995**, *57*, 289–300. [CrossRef]

International Journal of *Molecular Sciences*

MDPI

Review

Recent Progress Regarding the Molecular Aspects of Insect Gall Formation

Seiji Takeda [1,2,3,†], Tomoko Hirano [1,3,†], Issei Ohshima [1,3] and Masa H. Sato [1,3,*]

1 Graduate School of Life and Environmental Sciences, Kyoto Prefectural University, Shimogamo-Hangi-cho, Sakyo-ku, Kyoto 606-8522, Japan; seijitakeda@kpu.ac.jp (S.T.); thirano@kpu.ac.jp (T.H.); issei@kpu.ac.jp (I.O.)
2 Biotechnology Research Department, Kyoto Prefectural Agriculture Forestry and Fisheries Technology Center, Kitainayazuma Oji 74, Seika, Kyoto 619-0244, Japan
3 Center for Frontier Natural History, Kyoto Prefectural University, Shimogamo-Hangi-cho, Sakyo-ku, Kyoto 606-8522, Japan
* Correspondence: mhsato@kpu.ac.jp
† These authors contributed equally to this work.

Abstract: Galls are characteristic plant structures formed by cell size enlargement and/or cell proliferation induced by parasitic or pathogenic organisms. Insects are a major inducer of galls, and insect galls can occur on plant leaves, stems, floral buds, flowers, fruits, or roots. Many of these exhibit unique shapes, providing shelter and nutrients to insects. To form unique gall structures, gall-inducing insects are believed to secrete certain effector molecules and hijack host developmental programs. However, the molecular mechanisms of insect gall induction and development remain largely unknown due to the difficulties associated with the study of non-model plants in the wild. Recent advances in next-generation sequencing have allowed us to determine the biological processes in non-model organisms, including gall-inducing insects and their host plants. In this review, we first summarize the adaptive significance of galls for insects and plants. Thereafter, we summarize recent progress regarding the molecular aspects of insect gall formation.

Keywords: adaptive significance of insect galls; gall-inducing insects; gall formation mechanism; insect effectors

Citation: Takeda, S.; Hirano, T.; Ohshima, I.; Sato, M.H. Recent Progress Regarding the Molecular Aspects of Insect Gall Formation. *Int. J. Mol. Sci.* **2021**, *22*, 9424. https://doi.org/10.3390/ijms22179424

Academic Editor: Tomotsugu Koyama

Received: 12 June 2021
Accepted: 27 August 2021
Published: 30 August 2021

1. Introduction

Galls are induced on plants by viruses, mycoplasma, bacteria, fungi, nematodes, insects, mites, and other plants. They are defined by an abnormal plant organ development with ectopic cell proliferation and expansion, generating a wide range of gall morphologies [1,2]. Among them, insect-induced galls have attracted the attention of many researchers because of their unique shapes and wide range of variation. The estimated number of gall-inducing insects ranges from 21,000 to 211,000 [3,4]. Furthermore, host plant species span numerous phylogenetic lineages, suggesting that gall-inducing systems have evolved independently during the insects evolution [3–6].

Insect galls can be induced on plant leaves, stems, floral buds, flowers, fruits, or roots, and exhibit unique shapes (Figure 1).

Figure 1. Examples of insect galls generated on leaves. (**A**) *Glochidion obovatum* gall by the micromoth *Caloptilia cecidophora*. (**B**) *Eurya japonica* gall by the micromoth *Borboryctis euryae*. (**C**) *Distylium racemosum* gall by the aphid *Neothoracaphis yanonis*. (**D**) *Rhus javanica* gall by the aphid *Schlechtendalia chinensis*. (**E**) *Artemisia montana* gall by the gall midge *Rhopalomyia yomogicola*. (**F**) *Ulmus parvifolia* gall by *Tetraneura akinire*. Panels (**A**,**D**) are from a previous study [7].

Although galls often resemble fruits or floral buds, their morphologies are generally considerably different from those of plant structures. Therefore, gall-inducing insects appear to hijack the plant developmental system to generate a novel structure in plants. Gall-inducing insects produce stimuli that initiate the development and maintenance of gall tissue. Interestingly, phytohormones, such as indole acetic acid and cytokinins, are detected at higher concentrations in gall-inducing insects than in galls generated on host plants [8–12]. There is also evidence that certain amino acids and proteins are possible signals for gall induction [6]. These results suggest that substances transferred to plants via an insect ovipositor, secreted from larval saliva and/or excrement from larvae, induce the reprogramming of plant cells [5,6,13].

The unique shapes and wide range of variation in galls have attracted the attention of researchers in entomology, botany, and ecology, as well as amateur nature enthusiasts. Each specific insect species generates more or less a fixed gall shape on their respective host plants, indicating that gall-inducing insects activate specific developmental pathways of each host plant and tightly control these pathways. However, the molecular and cellular mechanisms of gall development remain poorly understood, and further studies are needed that focus on a variety of insect and plant taxa.

In this review, we assess insect galls from the perspective of plant molecular biology, with the hope of promoting future molecular plant studies on insect galls. First, we review the evolutionary aspects of the gall-inducing life histories of insects and plants and their adaptive significance. We then briefly review the recent advances made in the molecular aspects of insect gall formation.

2. Significance of Morphology and Function of Galls for Insects

The shape and cell status of galls have raised three major hypotheses for their adaptive significances of gall-inducing insects: nutrient hypothesis, microenvironment hypothesis, and enemy hypothesis [3,6]. These are profits of galls for insects, rather than host plants. The nutrient hypothesis explains that galls play a role as a nutrient source for insects, and in turn, insects somehow keep regulating plant cell differentiation to provide nutrients, since gall-inducing insects can live within galls for several months. For instance, aphid–induced galls on leaves accumulate a much higher amount of amino acids than intact leaves, suggesting that the galls provide a nitrogen source to gall-inducing insects [14,15]. Photoassimilates accumulate in some aphid-induced galls, suggesting that galls act as a sink organ [16,17]. Structurally, leaf galls of *Glochidion obovatum*, induced by micromoth *Caloptilia cecidophora*, generate several cell layers with different characters surrounding the central chamber, one of which is speculated to be nutritive tissue (Figure 2) because the tissue is almost completely consumed by the pupation of inside larvae [18]. Another type of leaf gall on *Eurya japonica* induced by micromoth *Borboryctis euryae* are much thinner compared to the *G. obovatum* galls and are called mine-galls, but larvae actively induce callus proliferation within the leaves, and feeding on the induced tissues are essential for the pupation [19]. These results strongly support the hypothesis of galls as nutrient source for gall-inducing insects. Interestingly, both *C. cecidophora* and *B. euryae* show mixed feeding habits, where early instars are normal leaf miners, but larvae suddenly begin inducing galls after molting into species specific instars [18,19]. Recent phylogenetic studies clearly show that the two gall-inducing leaf-miner species evolved independently from normal leaf miners [20,21]. The fact that both *C. cecidophora* and *B. euryae* cannot complete their larval period without feeding on gall tissues, coupled with their phylogenetic positions, indicates that these gall-inducing insects lost their ability to survive by feeding only on normal leaf tissues during the course of the evolution of their gall-inducing ability, and that some nutrients obtained from gall tissues are not supplemental but vital for their survival.

The enemy hypothesis explains that the galls protect the residential insects from their enemies and pathogens. Many galls have enclosed environments surrounded by the lignified cell layer for protecting the insects inside of the gall against attack by nonspecialist predators and pathogens [22]. However, the gall structures never provide complete enemy-free space: there are many specialist enemies including fungi, wasps, beetles, moths, and flies, which confer higher mortality [3,6,23]. Nevertheless, the enemy hypothesis is supported for the case of gall-inducing sawflies which are attacked by fewer parasitoid species with lower mortalities than free-living forms [24].

The microenvironment hypothesis states that gall tissues would protect gall-inducing insects from unfavorable abiotic stress conditions such as temperature and humidity [3,25,26]. For instance, when a hole is made on the closed gall of *Distylium racemosum* generated by social aphid *Nipponaphis monzeni*, many soldier nymphs inside the gall gather to the hall and discharge their body fluid to plug the hall, as well as activating plant cell proliferation around the hall [27]. This strongly suggests that the gall-inducing insects actively maintain the microenvironment within the gall. Humidity is suggested to be important for insects, since desiccation can be lethal for their lives [3,6]. There are ambiguous supports for temperature and protection: the temperature does not differ significantly between the inside and outside of galls, and there are many species of both parasitoids and inquilines targeting the gall-inducing insects [3,6,28]. Although the lignified cell layer of galls suggests the function as a shelter from predators, more investigation would be required to confirm this hypothesis.

It is interesting that the outer surface of some galls turns red, due to accumulation of secondary metabolites such as anthocyanins (Figure 1). Red is the contrast color of green and must be remarkable to birds or mammalian animals, a candidate predator of insects. Three hypotheses have been suggested for significance of gall coloring for distinct reasons: (i) representing aposematic or warning coloration to potential enemies of gall-inducing insects, (ii) causing early senescence of the plant tissue for nutrient translocation, and (iii) an adjunct during gall induction caused by cytokinin and sugar [23,29,30]. Another possibility is that, if galls affect plant growth negatively, turning the surface color with accumulation of anthocyanin could be a plant response to remove the galls by predators such birds, but evidence supporting this hypothesis has not been reported thus far. Alternatively, since galls carry molecular features such as flowers and fruits (shown by transcriptome analyses: see below), coloration of galls are one of the characteristics of these reproductive organs. There are several other hypotheses, therefore the coloration of galls would be an interesting phenomenon worth investigating its adaptive significance.

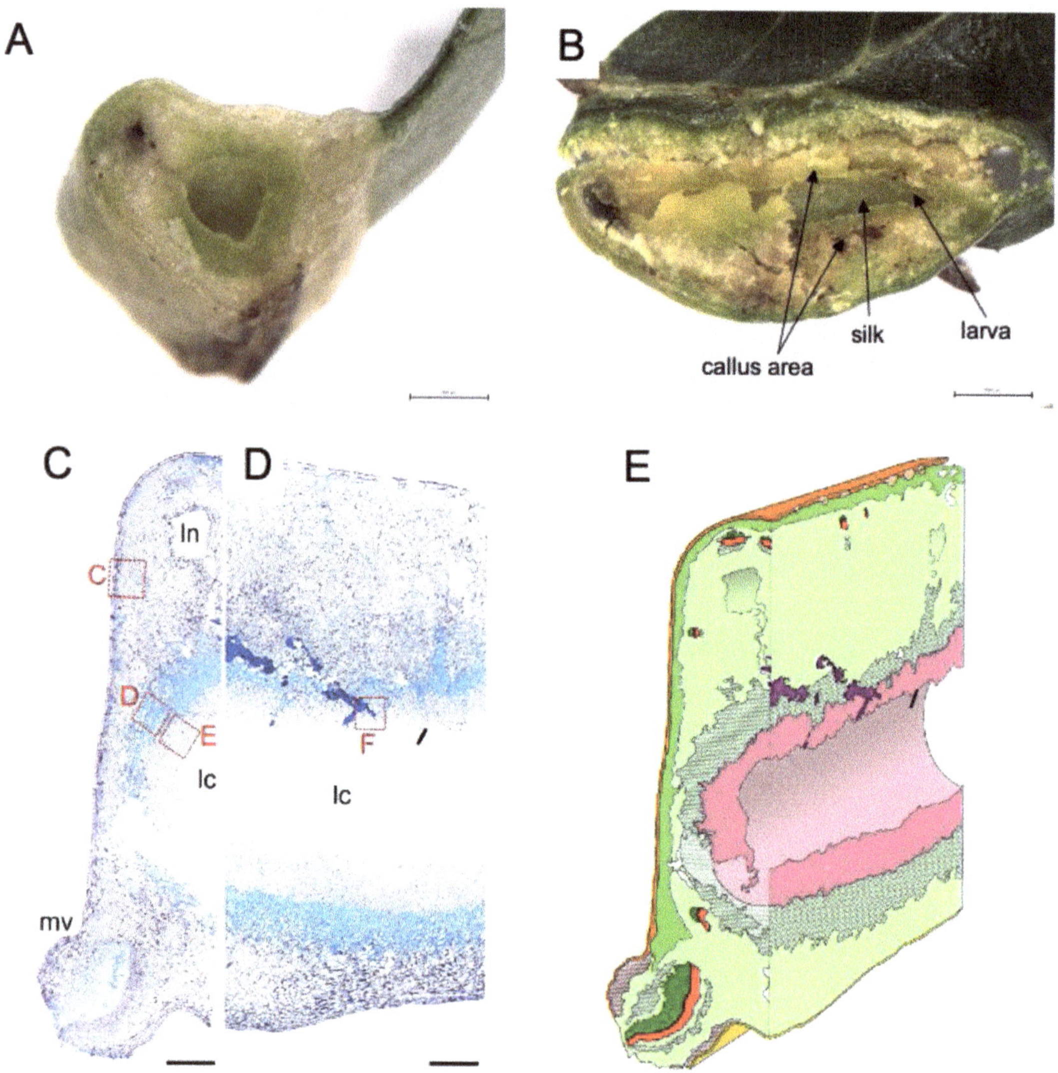

Figure 2. Inside of leaf gall on *Glochidion obovatum*. (**A**) Transverse section. (**B**) Sagittal section showing callus-like cells, silk covering the chamber wall, and larva inside the chamber. (**C**,**D**) Histological transverse (**C**) and sagittal (**D**) sections. Ln: lacuna, lc: larva chamber, mv: midvein. (**E**) 3D representation of gall inside based on histology sections. Dark green: xylem, red: phloem, pink: nutritive tissue, green cross: sclerenchyma, red cross: collenchyma, purple: frass. Bars: A, B, 1 mm; C, D, 0.5 mm. Panels are cited from [7,18].

3. Benefits of Gall Formation for Plants

Galls do not appear to carry adaptive benefits for plants; on the contrary, they either negatively affect plant growth or have no effect, as observed in many cases. Plants may generate galls to localize parasitic gall-inducing insects in a limited space, allowing plants to inhibit the spread of damage, although evidence supporting this hypothesis is poor [1,3]. Another hypothesis is that plants use excrement from gall-inducing insects by absorbing them from galls. For instance, in closed galls of *Distylium racemosum* and *Syrax japonicus*, honeydew from gall-inducing aphids is absorbed by the galls through the plant vascular system, and this mechanism helps aphids avoid drowning in their secretions within the gall. This suggests the possibility that plants use secretions from gall-inducing insects; however, the sugar concentration of honeydew from these aphids is low and devoid of sucrose [31]. In the *S. chinensis-Rhus chinensis* galling system, the gall and the aphids inside are composed of a metabolic and nutrient exchange hub that benefits not only the aphid but also its host plant. Consequently, host plants provide both shelter and nutrients, and in turn, aphid-derived metabolites compensate host metabolism to a certain extent [32]. A gall-inducing weevil, *Smicronyx madaranus* forms spherical galls on the shoots of an obligate parasitic plant, *Cuscuta campestris*. Parasitic *C. campestris* usually shows low chlorophyll content and has less photosynthetic activity because it usually depends on host plants for its nutrients. However, *S. mandaranes*-induced galls have significantly increased chlorophyll content and photosynthetic activity, suggesting that the gall-inducing weevil enhances the photosynthetic activity to modify the plant developmental programs for producing insect nutrients [33].

The galled shoot of *Silphium integrifolium* formed by the cynipid wasp, *Antistrophus silphii* increases photosynthesis activity, stomatal conductance, and xylem water potential in drought-stress conditions compared with the ungalled shoot, suggesting that the cynipid wasp enhances photosynthesis and water potential not only in gall tissues but also in whole shoots attached to galls [34]. As another example, *Eucalyptus* plants with galls exhibited indirect effects, resulting in tolerance to cold injury, compared to those without galls [35]. Although these examples show the benefits of galls for plants, direct or indirect evidence is required to elucidate the benefits derived from galls in plants.

4. Molecular Biology in Host Plants in Insect Gall Formation

4.1. Changes in Plant Hormonal Regulation during Gall Development

The common characteristics of complex insect gall structures include the lignified outside cell layer, which has a protective function against natural enemies and the outside environment, and the nutritive tissues of callus-like cells, which contain insect nutrients and vascular tissues to transport nutrients to these tissues [1,6,18,19]. To construct gall structures, many gall-inducing insects are believed to have the ability to secrete certain effectors, including plant hormones, into plant tissues using their mouthparts or ovipositors to induce gall formation in the host plants [5,8,11,36,37]. To induce complex gall structures in plant tissues, gall-inducing insects must control the cellular machinery of their hosts, including development, metabolism, chemistry, and physiology to hijack host gene expression programs in favor of those of the insect.

During the growing process of the gall tissues, a combination of cell division and cell growth occurs by the action of gall-inducing insects. In this process, phytohormones produced by gall-inducing insects have long been hypothesized to play a key role in inducing gall structures to exogenously control phytohormone regulatory pathways in host plants [9,12,38]. However, little was known regarding the gene expression profiles in developing galls because of the difficulty of studying non-model plants in the wild. Recent progress in next-generation sequencing (NGS) has allowed us to determine the structure of biological processes in non-model organisms, including gall-inducing insects and their host plants. Recently, several transcriptome analyses indicated that a significant number of phytohormone-regulatory genes are upregulated in various developing galls.

For instance, a transcriptome analysis of psyllid galls on the Hawaiian *Metrosideros polymorpha* showed that auxin response genes were upregulated in galls [39]. In galls formed on *R. chinensis* and *R. javanica*, the genes involved in the metabolic and signal transduction pathways of auxin and cytokinins were found to be substantially activated [40,41].

Comparison of transcripts in four different galls, suggested that 38 genes are commonly up regulated. GO analysis showed that peptide biosynthesis and metabolism are commonly involved in the four different galls [7]. Among these, AtMYB77 is involved in lateral root formation via auxin signaling [42]. Since Arabidopsis cell regeneration mediates the process of lateral root development [43], the callus generation within the gall may be mediated by AtMYB77 and auxin signaling.

4.2. Attempts to Identification of Effector Molecules Involved in the Gall Formation

In addition to the effects of insect-derived phytohormones for inducing gall structures, other effector molecules are believed to be needed for constructing complex insect gall structures [5]. A gall midge, the hessian fly (*Mayetiola destructor*), produces nutritive tissue which acts as a sink tissue with its host plant [44]. The hessian fly was the first gall-inducing insect for which the genome and salivary gland transcriptome and proteome were published [45–47]. The genome provided evidence for hundreds of transcripts encoding candidate effectors [47]. Map-based cloning identified four candidate avirulence effectors (vH6, vH9, vH13, and vH14) [47–49]. RNA-interference-knockdown of the vH13 gene confirmed that knockdown of that the gene allows larvae to survive on H13-resisitant plants, suggesting that the vH13-encoded protein is an avirulence effector, but not a gall-inducing effector [48].

Gall-inducing effector candidates have been identified from the transcriptome analysis of ovaries and venom glands of two cynipid gall wasps, *Biorhiza pallida* and *Diplolepis rosae*, inducing galls on oak and rose, respectively [50], or the analysis of salivary gland proteome of root-galling grape phylloxera, *Daktulosphaira vitifoliae* [51]. However, there is no direct evidence showing that these effector candidates have gall-inducing activity in their host plants.

Recently, a novel insect secretory protein, designated as BICYCLE, was identified from the gall-inducing aphid *Hormaphis cornu*, which produces distinctive cone galls on the leaves of *Hamamelis virginiana* [52]. *Bicycle* genes were most strongly expressed in the salivary glands of gall-inducing generations. These results suggest that BICYCLE proteins regulate the gall development process in *H. virginiana*. A insect-derived plant regulators, Bruchins identified from both cowpea weevil (*Callosobruchus maculatus* F.) and pea weevil (*Bruchus pisorum* L.) were revealed to show neoplasm formation activity on pods of all of the pea line tested when few amounts of these compounds were applied to intact plants [53]. Thus, several effector candidates have been identified. However, because of the lack of model systems for analyzing gall-inducing insect and host plant interactions, it is not known whether these effector candidates indeed have a gall-inducing function. Future studies are needed to establish a model system for assessing gall induction and development processes at the molecular level.

4.3. Changes in the Expression Patterns of Genes Involved in the Biosynthesis of the Metabolic Process during Gall Formation

Tannins are crucial for the protection of host plants and gall-inducing insects from herbivores. Aphid galls on *R. chinensis* accumulate gallotannin, and genes involved in gallotannin biosynthesis [54], gallic acid synthesis [41], and lignin biosynthesis [40] have been identified. In the developing gall of the chestnut gall wasp, *Dryocosmus kuriphilus*, on the Chinese chestnut, *Castanea mollissima*, the expression of genes related to metabolic processes, such as phenylpropanoid biosynthesis, secondary metabolism, and plant-pathogen interactions, was altered compared to that of non-infested leaves [55]. Galls induced on elm leaves by a gall-inducing aphid, *Tetraneura akinire*, were shown to express the genes encoding lignocellulose synthase, suggesting the reinforcement of cell walls to improve resistance to damage by aphids [56].

4.4. Regulation of Transcriptional Factors for Reproductive Organ Development

Darwin proposed that the shapes of complex insect galls resemble those of flowers or fruits [57]. Indeed, many remarkable flower- and fruit-like structures have been observed in insect galls, particularly those induced by gall midges, aphids, and cynipids in various host plant species [58], suggesting that the formation of gall tissues is similar to the development of flowers or fruits [59,60]. It has recently been shown that genes involved in the development of flowers and fruits are activated in the leaf galls in *Vitis riparia* induced by the phylloxera, *D. vitifoliae* [60].

Activation of reproductive organ genes has also been reported during the early development of galls on *R. javanica* induced by a gall-inducing aphid, *S. chinensis*. Additionally, class-1 KNOX transcription factors were shown to be overexpressed ectopically in the early developmental gall tissue of *R. javanica*. Because the class-1 KNOX transcription factors are known to lead to the formation of de novo meristematic structures in leaves [61], the results support the hypothesis that gall-inducing insects convert source tissues into meristematic tissues by expressing the class-1 KNOX genes in leaf tissue cells.

This hypothesis was partly supported by comparative transcriptome analysis using four different insect galls generated on leaves: galls on *G. obovatum* induced by the micromoth *C. cecidophora*, on *E. japonica* by the micromoth *B. euryae*, *R. javanica* by the aphid *S. chinensis*, and on *Artemisia montana* by the gall midge *Rhopalomyia yomogicola*. Comparison of these different galls revealed that (i) photosynthetic genes are downregulated in galls, supporting the hypothesis that galls are converted to sink organs rather than source organs on leaves; (ii) developmental, cell cycle, and phytohormone genes are upregulated in galls; and (iii) approximately 40 genes are commonly upregulated in these four galls. In particular, several key regulators of stem cell generation (*CLE44*, *BAM3*, *WOXs)* [62,63] and maintenance floral organ development (*SEPALLATA*, *AGAMOUS*, and *APETALA1*) [64] were commonly upregulated in *A. montana*, *G. obovatum*, and *R. javanica* galls. Intriguingly, these floral organ regulatory transcription factors were not expressed in the *E. japonica* gall tissues. This gall has rather thin layers with a simple pouch-type gall structure compared with the other fruit-type galls, suggesting that these transcription factors do not induce the galls of *E. japonica*. We realize that the comparison of only four species is not sufficient to cover a wide range of variations in gall morphology. However, the accumulation of this type of comparable data will reveal the common mechanisms of gall development and those that generate diversity in galls.

Collectively, we propose following processes that occur during the early stage of gall formation: (i) unknown insect effectors alter the gene expression in host plant organs (e.g., leaves); (ii) source organs (leaves) change their cell fate to meristematic character; (iii) a different combination of floral homeotic genes generate various gall structures (Figure 3).

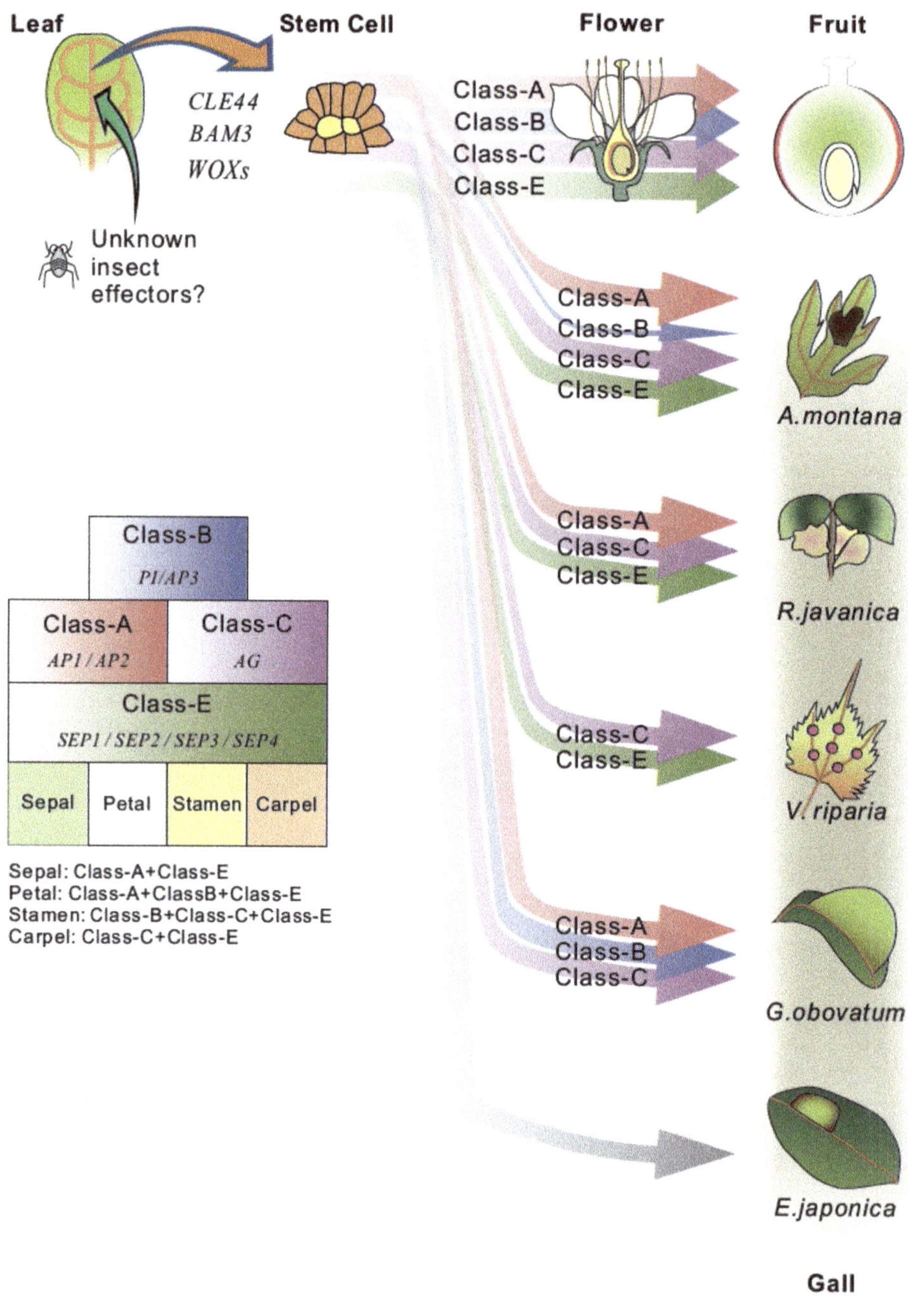

Figure 3. A hypothesis of gall formation processes in various host plants. Common candidate genes for stem cell generation and floral organ development were extracted from transcriptome data of developing galls of *A. montana*, *G. obovatum* [7], *R. javanica* [40] and *V. riparia* [60]. The thickness of arrow indicates the expression level of each gene. The gall formation hypothesis explains the gall formation mechanism that sepal and carpel-based galls are generated by the ectopic over expression of class-A, class-B, and class-E floral homeotic genes on leaves in *A. montana* and *R. javanica*, or carpel-based gall is generated by the expression of class-C and E genes in *V. riparia*. No floral homeotic genes need for generation of pouch-type thin gall of *E. japonica*.

5. Conclusions

Herein, we reviewed the molecular aspects of insect gall induction and development. The ancestral lifestyle of gall-inducing insects varies among taxa, from ectophagous herbivores (foliage feeders, sap-feeders, and pollen-feeders), plant-mining herbivores (leaf-miners and stem-borers), and fungus-feeders to parasitoids. Reports on the origin of leaf miners are scarce. Regarding the adaptive significance of galls for insects, the nutritive hypothesis has received the most experimental support through various direct approaches, including chemical analyses and bioassays. In contrast, enemy and microenvironment hypotheses are only supported by indirect evidence, such as comparisons of predation rates by enemies between taxa with and without a gall-inducing lifestyle and comparisons of the numbers of gall-inducing species between xeric and mesic habitats. Although gall morphologies have received attention in terms of protection from predators and harsh environments, few studies have been conducted on the coloration of galls. The adaptive significance of coloration is an interesting topic worthy of investigation. Although very few studies have been conducted on the plant benefits of galls, one study found that gall-induced plants have increased tolerance. Thus, more studies will be needed to elucidate the generality of this phenomenon. Finally, we note that an increasing number of studies challenge the molecular aspect of gall induction and development, and recent studies have revealed that the formation of gall tissues is similar to the development of flowers or fruits in terms of gene expression patterns. Therefore, studies using a wide range of gall-inducing insects and their host plant taxa would yield more detailed insights into molecular-level insect-plant interactions, which may be divided into subcategories. These studies should provide concise and precise descriptions of the experimental results, their interpretation, and the experimental conclusions drawn.

Author Contributions: M.H.S., S.T. and I.O. conceived and planned the manuscript. S.T., T.H., I.O. and M.H.S. wrote the manuscript. All authors have read and agreed to the published version of the manuscript.

Funding: This work was supported by JSPS KAKENHI Grant Numbers JP21K06234 to S.T., 19H00933 to M.H.S. and 17H06260 to I.O.

Conflicts of Interest: The authors declare no conflict of interest.

References

1. Mani, M.S. *Ecology of Plant Galls*; Junk, W., Ed.; Spring: The Hague, The Netherland, 1964.
2. Felt, E.P. Gall insects and their relations to plants. *Sci. Mon.* **1918**, *6*, 509–525.
3. Price, P.W.; Fernandes, G.W.; Waring, G.L. Adaptive nature of insect galls. *Environ. Entomol.* **1987**, *16*, 15–24. [CrossRef]
4. Espirito-Santo, M.M.; Fernandes, G.W. How many species of gall-inducing insects are there on earth, and where are they? *Ann. Entomol. Soc. Am.* **2007**, *100*, 95–99.
5. Giron, D.; Huguet, E.; Stone, G.N.; Body, M. Insect-induced effects on plants and possible effectors used by galling and leaf-mining insects to manipulate their host-plant. *J. Insect Physiol.* **2016**, *84*, 70–89. [CrossRef]
6. Stone, G.N.; Schönrogge, K. The adaptive significance of insect gall morphology. *Trends Ecol. Evol.* **2003**, *18*, 512–522. [CrossRef]
7. Takeda, S.; Yoza, M.; Amano, T.; Ohshima, I.; Hirano, T.; Sato, M.H.; Sakamoto, T.; Kimura, S. Comparative transcriptome analysis of galls from four different host plants suggests the molecular mechanism of gall development. *PLoS ONE* **2019**, *14*, e0223686.
8. Yamaguchi, H.; Tanaka, H.; Hasegawa, M.; Tokuda, M.; Asami, T.; Suzuki, Y. Phytohormones and willow gall induction by a gall-inducing sawfly. *New Phytol.* **2012**, *196*, 586–595. [CrossRef] [PubMed]
9. Tooker, J.F.; De Moraes, C.M. Feeding by a gall-inducing caterpillar species alters levels of indole-3-acetic and abscisic acid in *Solidago altissima* (Asteraceae) stems. *Arthropod-Plant Inte.* **2011**, *5*, 115–124. [CrossRef]
10. Andreas, P.; Kisiala, A.; Neil Emery, R.J.; De Clerck-Floate, R.; Tooker, J.F.; Price, P.W.; Miller, D.G.; Chen, M.S.; Connor, E.F. Cytokinins are abundant and widespread among insect species. *Plants* **2020**, *9*, 208. [CrossRef]
11. Ponce, G.E.; Fuse, M.; Chan, A.; Connor, E.F. The localization of phytohormones within the gall-inducing insect *Eurosta solidaginis* (Diptera: Tephritidae). *Arthropod Plant Interact.* **2021**, *15*, 375–385. [CrossRef]
12. Tanaka, Y.; Okada, K.; Asami, T.; Suzuki, Y. Phytohormones in japanese mugwort gall induction by a gall-inducing gall midge. *Biosci. Biotechnol. Biochem.* **2013**, *77*, 1942–1948. [CrossRef] [PubMed]
13. Barnewall, E.C.; De Clerck-Floate, R.A. A preliminary histological investigation of gall induction in an unconventional galling system. *Arthropod-Plant Interact.* **2012**, *6*, 449–459. [CrossRef]

14. Koyama, Y.; Yao, I.; Akimoto, S.I. Aphid galls accumulate high concentrations of amino acids: A support for the nutrition hypothesis for gall formation. *Entomol. Exp. Appl.* **2004**, *113*, 35–44. [CrossRef]
15. Suzuki, D.K.; Fukushi, Y.; Akimoto, S.I. Do aphid galls provide good nutrients for the aphids?: Comparisons of amino acid concentrations in galls among Tetraneura species (Aphididae: Eriosomatinae). *Arthropod. Plant. Interact.* **2009**, *3*, 241–247. [CrossRef]
16. Larson, K.C.; Whitham, T.G. Manipulation of food resources by a gall-forming aphid: The physiology of sink-source interactions. *Oecologia* **1991**, *88*, 15–21. [CrossRef] [PubMed]
17. Larson, K.C.; Whitham, T.G. Competition between gall aphids and natural plant sinks: Plant architecture affects resistance to galling. *Oecologia* **1997**, *109*, 575–582. [CrossRef]
18. Guiguet, A.; Ohshima, I.; Takeda, S.; Laurans, F.; Lopez-Vaamonde, C.; Giron, D. Origin of gall-inducing from leaf- mining in Caloptilia micromoths. *Sci. Rep.* **2019**, *9*, 6794. [CrossRef]
19. Guiguet, A.; Hamatani, A.; Amano, T.; Takeda, S.; Lopez-vaamonde, C.; Giron, D.; Ohshima, I. Inside the horn of plenty: Leaf-mining micromoth manipulates its host plant to obtain unending food provisioning. *PLoS ONE* **2018**, *13*, e0209485. [CrossRef]
20. Kawahara, A.Y.; Plotkin, D.; Ohshima, I.; Lopez-Vaamonde, C.; Houlihan, P.R.; Breinholt, J.W.; Kawakita, A.; Xiao, L.; Regier, J.C.; Davis, D.R.; et al. A molecular phylogeny and revised higher-level classification for the leaf-mining moth family Gracillariidae and its implications for larval host-use evolution. *Syst. Entomol.* **2017**, *42*, 60–81. [CrossRef]
21. Li, X.; Ryan, S.L.; Chandra, E.; Camiel, D.; van Erik, N.; Don, D.; Chris, J.; Atsushi, K.; Shigeki, K.; Andreas, Z.; et al. Phylogeny of gracillariid leaf-mining moths: Evolution of larval behaviour inferred from integrated phylogenomic and Sanger data. *Cladistics* **2021**, in press.
22. Hawkins, B.A.; Cornell, H.V.; Hochberg, M.E. Predators, parasitoids, and pathogens as mortality agents in phytophagous insect populations. *Ecology* **1997**, *78*, 2145–2152. [CrossRef]
23. Inbar, M.; Izhaki, I.; Koplovich, A.; Lupo, I.; Silanikove, N.; Glasser, T.; Gerchman, Y.; Perevolotsky, A.; Lev-Yadun, S. Why do many galls have conspicuous colors? A new hypothesis. *Arthropod. Plant. Interact.* **2010**, *4*, 1–6. [CrossRef]
24. Price, P.W.; Pschorn-Walcher, H. Are gall-inducing insects better protected against parasitoids than exposed feeders? A test using tenthredinid sawflies. *Ecol. Entomol.* **1988**, *13*, 195–205. [CrossRef]
25. Cornell, H.V. The secondary chemistry and complex morphology of galls formed by the Cynipiae (Hymenoptera): Why and How? *Am. Midl. Nat.* **1983**, *110*, 225–234. [CrossRef]
26. Blanche, K.R. Diversity of insect-induced galls along a temperature-rainfall gradient in the tropical savannah region of the Northern Territory, Australia. *Austral Ecol.* **2000**, *25*, 311–318. [CrossRef]
27. Kutsukake, M.; Shibao, H.; Uematsu, K.; Fukatsu, T. Scab formation and wound healing of plant tissue by soldier aphid. *Proc. R. Soc. B Biol. Sci.* **2009**, *276*, 1555–1563. [CrossRef]
28. Fujii, T.; Matsuo, K.; Abe, Y.; Yukawa, J.; Tokuda, M. An endoparasitoid avoids hyperparasitism by manipulating immobile host herbivore to modify host plant morphology. *PLoS ONE* **2014**, *9*, e102508. [CrossRef]
29. Connor, E.F.; Bartlett, L.; O'Toole, S.; Byrd, S.; Biskar, K.; Orozco, J. The mechanism of gall induction makes galls red. *Arthropod. Plant. Interact.* **2012**, *6*, 489–495. [CrossRef]
30. White, T.C.R. Why do many galls have conspicuous colours? An alternative hypothesis revisited. *Arthropod. Plant. Interact.* **2010**, *4*, 149–150. [CrossRef]
31. Kutsukake, M.; Meng, X.Y.; Katayama, N.; Nikoh, N.; Shibao, H.; Fukatsu, T. An insect-induced novel plant phenotype for sustaining social life in a closed system. *Nat. Commun.* **2012**, *3*, 1187. [CrossRef]
32. Chen, X.; Yang, Z.; Chen, H.; Qi, Q.; Liu, J.; Wang, C.; Shao, S.; Lu, Q.; Li, Y.; Wu, H.; et al. A Complex nutrient exchange between a gall-forming aphid and its plant host. *Front. Plant Sci.* **2020**, *11*, 811. [CrossRef]
33. Murakami, R.; Ushima, R.; Sugimoto, R.; Tamaoki, D.; Karahara, I.; Hanba, Y.; Wakasugi, T.; Tsuchida, T. A new galling insect model enhances photosynthetic activity in an obligate holoparasitic plant. *Sci. Rep.* **2021**, *11*, 13013. [CrossRef]
34. Fay, P.A.; Hartnett, D.C.; Knapp, A.K. Increased photosynthesis and water potentials in *Silphium integrifolium* galled by cynipid wasps. *Oecologia* **1993**, *93*, 114–120. [CrossRef]
35. Rocha, S.; Branco, M.; Boas, L.V.; Almeida, M.H.; Protasov, A.; Mendel, Z. Gall induction may benefit host plant: A case of a gall wasp and eucalyptus tree. *Tree Physiol.* **2013**, *33*, 388–397. [CrossRef]
36. Matsukura, K.; Matsumura, M.; Tokuda, M. Host manipulation by the orange leafhopper *Cicadulina bipunctata*: Gall induction on distant leaves by dose-dependent stimulation. *Naturwissenschaften* **2009**, *96*, 1059–1066. [CrossRef]
37. Stuart, J.J.; Chen, M.-S.; Shukle, R.; Harris, M.O. Gall midges (Hessian Flies) as plant pathogens. *Annu. Rev. Phytopathol.* **2012**, *50*, 339–357. [CrossRef]
38. Dorchin, N.; Hoffmann, J.H.; Stirk, W.A.; Novák, O.; Strnad, M.; Van Staden, J.; Ond, R. Sexually dimorphic gall structures correspond to differential phytohormone contents in male and. *Physiol. Entomol.* **2009**, *34*, 359–369. [CrossRef]
39. Bailey, S.; Percy, D.M.; Hefer, C.A.; Cronk, Q.C.B. The transcriptional landscape of insect galls: Psyllid (Hemiptera) gall formation in Hawaiian *Metrosideros polymorpha* (Myrtaceae). *BMC Genom.* **2015**, *16*, 943. [CrossRef]
40. Hirano, T.; Kimura, S.; Sakamoto, T.; Okamoto, A.; Nakayama, T.; Matsuura, T.; Ikeda, Y.; Takeda, S.; Suzuki, Y.; Ohshima, I.; et al. Reprogramming of the developmental program of *Rhus javanica* during initial stage of gall induction by *Schlechtendalia chinensis*. *Front. Plant Sci.* **2020**, *11*, 471. [CrossRef]

41. Wang, H.; Cui, K.; Shao, S.; Liu, J.; Chen, H.; Wang, C.; Yang, Z.; Lu, Q.; King-jones, K.; Chen, X.; et al. Molecular response of gall induction by aphid *Schlechtendalia chinensis* (Bell) attack on *Rhus chinensis* Mill. *J. Plant Interact.* **2017**, *12*, 465–479. [CrossRef]
42. Shin, R.; Burch, A.Y.; Huppert, K.A.; Tiwari, S.B.; Murphy, A.S.; Guilfoyle, T.J.; Schachtman, D.P. The Arabidopsis transcription factor MYB77 modulates auxin signal transduction. *Plant Cell* **2007**, *19*, 2440–2453. [CrossRef]
43. Sugimoto, K.; Jiao, Y.; Meyerowitz, E.M. Arabidopsis regeneration from multiple tissues occurs via a root development pathway. *Dev. Cell* **2010**, *18*, 463–471. [CrossRef]
44. Rani, D.D.; Pedada, S.; Singh, R. Hessian fly: New insights to develop a gall nutritive tissue. *J. Entomol. Zool. Stud.* **2018**, *6*, 720–728.
45. Chen, M.S.; Zhao, H.X.; Zhu, Y.C.; Scheffler, B.; Liu, X.; Liu, X.; Hulbert, S.; Stuart, J.J. Analysis of transcripts and proteins expressed in the salivary glands of Hessian fly (*Mayetiola destructor*) larvae. *J. Insect Physiol.* **2008**, *54*, 1–16. [CrossRef]
46. Chen, M.S.; Liu, X.; Yang, Z.; Zhao, H.; Shukle, R.H.; Stuart, J.J.; Hulbert, S. Unusual conservation among genes encoding small secreted salivary gland proteins from a gall midge. *BMC Evol. Biol.* **2010**, *10*, 296. [CrossRef]
47. Zhao, C.; Escalante, L.N.; Chen, H.; Benatti, T.R.; Qu, J.; Chellapilla, S.; Waterhouse, R.M.; Wheeler, D.; Andersson, M.N.; Bao, R.; et al. A massive expansion of effector genes underlies gall-formation in the wheat pest *Mayetiola destructor*. *Curr. Biol.* **2015**, *25*, 613–620. [CrossRef]
48. Aggarwal, R.; Subramanyam, S.; Zhao, C.; Chen, M.; Harris, M.O.; Stuart, J.J. Avirulence effector discovery in a plant galling and plant parasitic arthropod, the Hessian fly (*Mayetiola destructor*). *PLoS ONE* **2014**, *9*, e100958. [CrossRef]
49. Zhao, C.; Shukle, R.; Navarro-Escalante, L.; Chen, M.; Richards, S.; Stuart, J.J. Avirulence gene mapping in the Hessian fly (*Mayetiola destructor*) reveals a protein phosphatase 2C effector gene family. *J. Insect Physiol.* **2016**, *84*, 22–31. [CrossRef] [PubMed]
50. Cambier, S.; Ginis, O.; Moreau, S.J.M.; Gayral, P.; Hearn, J.; Stone, G.N.; Giron, D.; Huguet, E.; Drezen, J.M. Gall wasp transcriptomes unravel potential effectors involved in molecular dialogues with oak and rose. *Front. Physiol.* **2019**, *10*, 926. [CrossRef] [PubMed]
51. Eitle, M.W.; Carolan, J.C.; Griesser, M.; Forneck, A. The salivary gland proteome of root-galling grape phylloxera (*Daktulosphaira vitifoliae* Fitch) feeding on *Vitis* spp. *PLoS ONE* **2019**, *14*, e0225881. [CrossRef]
52. Korgaonkar, A.; Han, C.; Lemire, A.L.; Siwanowicz, I.; Bennouna, D.; Kopec, R.E.; Andolfatto, P.; Shigenobu, S.; Stern, D.L. A novel family of secreted insect proteins linked to plant gall development. *Curr. Biol.* **2021**, *31*, 1836–1849.e12. [CrossRef]
53. Doss, R.P.; Oliver, J.E.; Proebsting, W.M.; Potter, S.W.; Kuy, S.; Clement, S.L.; Williamson, R.T.; Carney, J.R.; DeVilbiss, E.D. Bruchins: Insect-derived plant regulators that stimulate neoplasm formation. *Proc. Natl. Acad. Sci. USA* **2000**, *97*, 6218–6223. [CrossRef] [PubMed]
54. Chen, H.; Liu, J.; Cui, K.; Lu, Q.; Wang, C.; Wu, H.; Yang, Z.; Ding, W.; Shao, S.; Wang, H.; et al. Molecular mechanisms of tannin accumulation in Rhus galls and genes involved in plant-insect interactions. *Sci. Rep.* **2018**, *8*, 9841. [CrossRef] [PubMed]
55. Zhu, C.; Shi, F.; Chen, Y.; Wang, M.; Zhao, Y.; Geng, G. Transcriptome analysis of Chinese chestnut (*Castanea mollissima* Blume) in response to *Dryocosmus kuriphilus* Yasumatsu infestation. *Int. J. Mol. Sci.* **2019**, *20*, 855. [CrossRef] [PubMed]
56. Lu, H.B.; Jin, L.P.; Wei, D.; Huang, Z.H. Study on the differential gene expression of elm leaves fed on by *Tetraneura akinire* Sasaki. *Genes Genom.* **2019**, *41*, 1505–1516. [CrossRef] [PubMed]
57. Darwin, C.R. *The Variation of Animals and Plants under Domestication Volumen I*; John Murray: London, UK, 1875.
58. Rohfritsch, O.; Shorthouse, J.D. Insect galls. In *Molecular Biology of Plant Tumors*; Kahl, G., Schell, J., Eds.; Academic Press: New York, NY, USA, 1982; pp. 121–152.
59. Kurosu, U.; Aoki, S. Formation of a "Cat's-paw" gall by the aphid *Ceratovacuna nekoashi* (Homoptera). *Jpn. J. Entomol.* **1990**, *58*, 155–166.
60. Schultz, J.C.; Edger, P.P.; Body, M.J.A.; Appel, H.M. A galling insect activates plant reproductive programs during gall development. *Sci. Rep.* **2019**, *9*, 1833. [CrossRef]
61. Lincoln, C.; Long, J.; Yamaguchi, J.; Serikawa, K.; Hakeaibi, S. A knotted1-like homeobox gene in *Arabidopsis* is expressed in the vegetative meristem and dramatically alters leaf morphology when overexpressed in transgenic plants. *Plant Cell* **1994**, *6*, 1859–1876.
62. Jun, J.H.; Fiume, E.; Fletcher, J.C. The CLE family of plant polypeptide signaling molecules. *Cell. Mol. Life Sci.* **2008**, *65*, 743–755. [CrossRef]
63. Yamaguchi, Y.L.; Ishida, T.; Sawa, S. CLE peptides and their signaling pathways in plant development. *J. Exp. Bot.* **2016**, *67*, 4813–4826. [CrossRef]
64. Ó'Maoiléidigh, D.S.; Graciet, E.; Wellmer, F. Gene networks controlling *Arabidopsis thaliana* flower development. *New Phytol.* **2014**, *201*, 16–30. [CrossRef] [PubMed]

MDPI
St. Alban-Anlage 66
4052 Basel
Switzerland
Tel. +41 61 683 77 34
Fax +41 61 302 89 18
www.mdpi.com

International Journal of Molecular Sciences Editorial Office
E-mail: ijms@mdpi.com
www.mdpi.com/journal/ijms

www.ingramcontent.com/pod-product-compliance
Lightning Source LLC
Chambersburg PA
CBHW061631210926
51850CB00009B/2509
9783036531632